Handbook of Innovation Ecosystems

Placemaking. Economics. Business. Governance

JOHN H HOWARD

October 2025

National Library of Australia - Cataloguing-in-Publication (CiP) Entry (Draft):
Howard, John H., 1949–

Handbook of innovation ecosystems: placemaking, economics, business, and governance
ISBN 978-1-7641407-1-3
Subjects:
Innovation districts—Australia.
Technological innovations—Economic aspects—Australia.
Knowledge management—Australia.
Urban renewal—Australia.
Regional planning—Australia.
Economic development—Australia.
Creative ability in business—Australia.
Research and development policy—Australia.
Public administration—Innovation policy—Australia.

Dewey Number: 338.0640994

Published by:
Acton Institute for Policy Research and Innovation
Canberra, ACT, Australia
www.actoninstitute.au

Published by the Acton Institute for Policy Research and Innovation, Australia
www.actoninstitute.au

APA Citation (7th edition):

Howard, J. H. (2025). Handbook of Innovation ecosystems: placemaking, economics, business, and governance. Acton Institute for Policy Research and Innovation.

Synopsis

Although *innovation* has become a central theme in public policy, the language of "innovation ecosystems" is often used loosely, risking loss of meaning and practical value. This *Handbook of Innovation Ecosystems* addresses this gap by providing conceptual clarity and a framework for policy and practice.

The *Handbook* suggests that successful ecosystems do not emerge from stand-alone programs or property developments; they evolve from the alignment of four interdependent domains: placemaking, economics, business, and governance.

Drawing on more than ninety international and Australian case studies, from Kendall Square in Boston and 22@Barcelona to Sydney's Tech Central and Melbourne's Parkville Biomedical Precinct, the *Handbook* demonstrates that ecosystems thrive where vision, leadership, and institutional capability are combined with long-term commitment and deliberate system integration.

The analysis challenges the assumption that funding alone delivers outcomes. Instead, it highlights the importance of translational infrastructure, absorptive capacity, and governance arrangements that can endure beyond electoral cycles.

By positioning innovation as both an economic and civic project, the *Handbook* emphasises inclusion, legitimacy, and sustainability alongside growth and competitiveness. It explores how ecosystems can generate value across health, advanced manufacturing, agrifood, creative industries, and clean energy, while also recognising the risks of exclusion, fragmentation, and short-termism.

This *Handbook* provides policymakers, university leaders, industry strategists, and community actors with a systemic vocabulary and evidence-based guide for designing and managing innovation ecosystems that deliver enduring national, urban, and regional value.

Contents

PREFACE

This *Handbook* is the product of a long engagement with the question of how innovation actually works in practice. Over three decades in public policy, research, and consulting, I have observed how innovation is often treated as a slogan, a project, or a set of structures, but rarely as a systemic process that integrates people, institutions, places, and ideas.

Too much of our national debate has focused on creating new entities or announcing funding programs, and too little on how to design the connective tissue that allows innovation to thrive.

The concept of the *innovation ecosystem* emerged as a way to capture this complexity. But the term is invoked so frequently and with such varied meanings that it risks losing practical value. The purpose of this *Handbook* is to bring greater conceptual clarity and practical relevance to the term.

I examine ecosystems through four interdependent domains, placemaking, economics, business, and governance, and argue that it is their alignment, not their separation, that ultimately determines success.

What the book adds

This work contributes a distinctive four-domain convergence framework that treats placemaking, economics, business, and governance as coequal components of ecosystem performance.

The framework offers a diagnostic lens for policymakers and practitioners to see gaps and imbalances, whether they be property-led developments without governance depth or research programs without translational infrastructure. It seeks to highlight patterns that explain why some ecosystems achieve scale and global recognition while others stall.

The Handbook challenges the prevailing input fallacy, *the assumption that innovation outcomes can be purchased through funding alone and reframes innovation policy as a task of system design, orchestration, and long-term stewardship.*

Finally, the *Handbook* positions innovation ecosystems as a civic and cultural projects, emphasising social inclusion, legitimacy, and public value. Frameworks such as the *Nine Forces of Ecosystem Convergence*, provided as an Addendum, provide practical tools for evaluation and strategy, making the *Handbook* relevant to policymakers, university leaders, and business strategists alike.

A Foundation for Practice

The *Handbook* builds on many years of innovation ecosystem research and hands-on experience, complemented by over 90 case studies of innovation districts, precincts, and hubs in Australia and around the world.

From Kendall Square in Boston to 22@Barcelona, from Toronto's MaRS to Sydney's Tech Central, the evidence shows that ecosystems cannot be engineered through property development alone, nor can they be sustained by research funding in isolation. They succeed through patient leadership, long-term commitment, and institutional expertise.

My aim is to provide readers with a framework for thinking systemically and acting coherently. Innovation is not merely about technology or markets; it is about social inclusion, cultural value, institutional integrity, and environmental responsibility. Ecosystems will not flourish if they exclude people, neglect context, or pursue growth without regard for resilience.

Each chapter is presented as a stand-alone narrative. Some chapters have appeared as *Innovation Insights* on the Acton Institute for Policy Research and Innovation website, but most have been written specifically for the *Handbook*.

My wish is that the *Handbook* helps move the national conversation beyond slogans, structures, and money towards systemic understanding and purposeful design.

Innovation ecosystems do not succeed by chance; they succeed through deliberate choices that connect people, places, and institutions in ways that generate enduring value for the economy and society.

John H. Howard
Director, Acton Institute for Policy Research and Innovation
Visiting Professor, University of Technology Sydney
Sydney, October 2025

FOREWORDS

Senator the Hon. Tim Ayres, Minister for Industry and Innovation, Minister for Science

Australia's prosperity has always been tied to our capacity to adapt, to build, and to innovate. From the earliest evidence of First Nations innovation and adaptation to the latest developments in advanced manufacturing, clean energy, and digital technologies, Australian living standards have improved when knowledge and creativity were turned into sustainable industries and secure jobs.

That is why the concept of the innovation ecosystem matters. It reminds us that innovation is not an isolated act of invention, nor a linear progression from laboratory to marketplace. It is the outcome of relationships—between people and institutions, between research and business, and between communities and governments. Strong ecosystems allow these connections to flourish, giving rise to industries that are globally competitive and nationally beneficial.

John Howard's book makes an important contribution to this task. It moves beyond slogans and structures to offer a clear framework for how innovation ecosystems work in practice. By examining innovation through the lenses of place and geography, economics, business, and governance, the book shows that success depends on alignment across these domains. When planning ignores economics, or when research is detached from commercial application, the result is fragmentation. When these perspectives converge, ecosystems become resilient, productive, and inclusive.

The case studies presented here, from Tech Central in Sydney to Kendall Square in Boston and 22@Barcelona, demonstrate both the opportunities and the risks. They show how vision, coordination, and long-term investment can transform regions. They also remind us of what happens when strategy is absent, governance is weak, or institutions fail to adapt. For policymakers like me, Howard's work is a not abstract analysis but a practical guide to future industrial strategy.

The Albanese Labor Government is committed to rebuilding the industrial capability, research and innovation system and the regional and outer urban communities where new manufacturing investments occur. Through the $15 billion National Reconstruction Fund, the $22.7 billion Future Made in Australia agenda, and investments in critical technologies and research infrastructure, this government is laying the foundations for industries that will deliver secure jobs and national resilience.

But as this book makes clear, funding alone is not enough. Alignment of effort is crucial to us delivering on our public purpose. Australia today needs to strengthen the connective tissue between research institutions, businesses, investors, and communities. Leaders in industry, government, the research sector and beyond must build ecosystems that are coordinated, inclusive, and adaptable to global change.

This book is timely, thoughtful, and relevant to modern Australia's industrial and innovation challenge. It provides policymakers, business leaders, and researchers with both the conceptual clarity and practical tools to approach innovation systemically. Most importantly, it highlights the role of institutional craft, long-term vision, and collaboration in shaping Australia's economic future.

I commend John Howard for his contribution and recommend this book to all Australians interested in the future of Australia's industrial, research and innovation system.

Senator the Hon. Tim Ayres
Minister for Industry and Innovation, Minister for Science
30 September 2025

Jeremy Gill, Policy Director, Committee for Sydney

Across Australia, the need for innovation has never been stronger. The Commonwealth, States and Territories are facing converging pressures: the need to create high-quality jobs in a competitive global economy, to build sovereign capability in key sectors, to strengthen health and education systems, and to transition industries to net zero.

Meeting these challenges demands more than ad hoc programs or stand-alone projects; it demands deliberate design and long-term stewardship and orchestration that connects research, business, government and communities.

This *Handbook* makes a timely and valuable contribution to Australia's innovation ecosystem discourse, drawing on years of research and dozens of global and local case studies. It calls out the Innovation

Imperative—where regions that fail to consistently generate new ideas and solutions will struggle to compete economically and address emerging challenges. Avoiding this requires both a high-level of intentionality and a systemic approach to embedding innovation in both what we do and how we do it.

Without a clear understanding of what fosters or inhibits innovation, and a framework for how to nurture it, at best we risk missing the opportunities that present themselves. At worst, we risk failing to adapt to the rapidly changing technological, social, economic, and environmental trends that will shape our future.

This book is both an exploration of how innovation ecosystems function and a how-to guide to ensure they flourish. It offers a systemic framework built around four domains—placemaking, economics, business, and governance—and shows how alignment between them is critical to sustained success. This alignment is important not only for how economies perform but how we plan for where they play out - in cities and regions. It highlights the significance of innovation districts as engines of R&D, industry emergence, productivity growth and job creation.

What stands out is the book's focus on innovation as both an economic and civic endeavour. It reminds us that thriving ecosystems are inclusive, resilient and globally connected. They nurture talent, encourage entrepreneurship, and deliver outcomes that matter to both society and the market.

Innovation, and its translation into economic, social and environmental outcomes requires intention, orchestration, investment and nurturing. It requires public, private, institutional, communal and individual engagement. For policymakers, business leaders, and researchers alike, this Handbook offers a clear, evidence-based foundation for action.

The Handbook challenges us to think in systems, act with purpose, and commit to the long-term work of building innovation ecosystems that can deliver enduring economic and social value for Australia.

Jeremy Gill, Policy Director
Committee for Sydney
October 2025

Professor Andrew Parfitt, Vice-Chancellor, University of Technology Sydney

For centuries, scholars have known and used the power of communities to shape new ideas and create knowledge. Universities are the archetypes for this process, where thoughts are developed and tested in an environment that thrives on challenging orthodoxy and pushing across boundaries.

In the scientific domain, the establishment of laboratories and the scientific processes of experimentation led to a framework that we now accept as the "scientific method", giving confidence to build on ideas and develop new fields of endeavour.

In the modern industrial age, we became used to the concept of processes such as manufacturing to expedite the development and commoditisation of products that became part of daily life and transformed our economics.

In the digital age, the knowledge economy has instilled the concept of the translation of ideas into value through a process of innovation, and that the process of innovation by its very nature requires an ecosystem of collaborators and co-creators, investors and business leaders, facilities and infrastructure. These ecosystems are now regarded as crucial to successful economic impact and growth.

But as John Howard reflects in this new book, the language and approach to innovation ecosystems is not well defined. Frequently there are assumptions made as how the system works or should work. Indeed, the very notion of system design as a critical enabler, whether through policy or the practice of knowledge transfer and translation, is often overlooked.

Tackling the system from the perspective of placemaking, economic factors, business enablement and, critically, the role of governance, this book provides a major resource for those charged with the responsibility of creating the innovation ecosystems of the future, and those wishing to maximise the benefit of being part of them.

Over the past several decades, successions of governments, businesses and academic institutions have sought to bring their communities together to harness the power of the ecosystem. Technology Parks have morphed into innovation precincts. Incubators and Accelerators have evolved to support start-up and scale-up communities on their journey to successful businesses, often providing valuable shared infrastructure and core capabilities.

Some of these endeavours have thrived as the participants of the ecosystem engage with one another and find mutual value. Others have withered, defaulting to convenient co-location and shared services. The ingredients that lead to one outcome over the other are often only learned in hindsight, if at all.

This book strives to demystify the success factors and provide a foundation for the visionaries who will imagine the next generation of innovation ecosystems primed to leverage the asset that communities can bring to making change and driving our future economic and social ambitions.

Andrew Parfitt, Vice-Chancellor and President
University of Technology Sydney,
October 2025

Professor Sharath Sriram, Chief Scientist for Western Australia

We are familiar with natural ecosystems – the lush wonder of the Amazon, the unique evolution at the Galapagos, or the size and complexity of the savannahs of Africa. Their beauty is always on a fine balance because of internal and external ecological, environmental, and human factors.

Innovation ecosystems are no different. To make them come together and work needs many elements to work in harmony. They have been considered the holy grail in unlocking long-term economic and social impact.

For such impact to ensure society can thrive, it relies on food security, health and wellbeing, environmental sustainability, energy supply, and defence. These grand challenges need scale, while building ecosystems around such missions will resonate with the public and give social licence for establishment.

Innovation ecosystems represent scale for impact. It brings together inventors, social scientists, policy makers, investors, manufacturers, businesses, and government to deliver outcomes.

This *Handbook* works to demystify the terminology behind innovation ecosystems, and that these need four pillars holding them up for success – location, money, markets, and strategy.

By examining how these domains interact and align, the author demonstrates that success does not come from structures or slogans, but from the coherence and integration of people, institutions, and places. With these combined with stable strategy.

What distinguishes this work is its breadth of evidence. Drawing on more than ninety case studies from Australia and across the world, it shows that ecosystems thrive where there is vision, leadership, and institutional dexterity. From Kendall Square in Boston to 22@Barcelona, from Toronto's MaRS to Melbourne's Parkville Precinct, the examples reveal consistent lessons. There are always lessons from failures and challenges, with this book highlighting these very effectively.

Innovation ecosystems must be challenging to implement – if easy, it is not ambitious or transformative. Unless there is a constant threat and risk of failure, ambitious plans will not take shape. Innovation ecosystems, as with most inventions and resulting innovations, rely on necessity and urgency.

As this book very effectively showcases, successful ecosystems become the nucleus of growth. It attracts talent, ideas, funding, and infrastructure. It ensures that the community and the country develop an innovation mindset – being bold and disruptive, to dramatically impact lives.

This *Handbook* is essential reading for policy makers, government leaders, industry, academics, and investors. There is a critical need to deliver successful precincts and innovation ecosystems in Australia. This will ensure we are creators and not just consumers.

Sharath Sriram, Chief Scientist of Western Australia
President, Science & Technology Australia
Professor, RMIT University,
October 2025

Emeritus Professor Roy Green AM, Vice-Chancellor's Innovation Adviser, UTS

In a world of 'influencers', Dr John Howard is distinguished by his use of evidence to enhance our understanding of the role of science, technology and innovation in advanced economies and to make the case for a supportive, future-focused policy framework. As well as being a recognised thought leader in the area, John is vitally interested in the practical measures needed to enable knowledge-driven productive performance at the enterprise level – for startups, scaling-up and established firms – in the context of more collaborative approaches by industry, government and research institutions.

In this *Handbook,* John turns his mind to place-based innovation ecosystems and the factors that may be identified in their success (or failure) both in Australia and globally. John's analysis couldn't come at a more relevant time as the Australian government grapples with a productivity slowdown and real wage stagnation along with the major challenges of climate change and energy transition. As Australia slides down the economic complexity rankings, due to its overweighted dependence on the export of unprocessed raw materials, the task of achieving net zero provides not just a challenge but the opportunity for transformation of its narrow trade and industrial structure.

Significantly, there is increasing recognition among researchers and policymakers that an effective, high performing research and innovation system will be essential to capitalising on this opportunity, and that this will require not only better policy coordination and funding, but also a different approach to research translation and commercialisation. John demonstrates the value of moving beyond the traditional linear 'lab to market' model to the development of a national network of more interactive and specialised innovation ecosystems that build on the many successful examples around the world. He notes that this does not necessarily mean replacing current institutional structures but rather modifying and adapting them to a new, more dynamic policy framework.

While the concept of innovation ecosystems is relatively straightforward, it becomes apparent that the implementation task is not. So we are very fortunate to have John Howard's *Handbook*, which provides an indispensable guide to both the concept and its implementation in collaborative research and technology precincts. As

Australia confronts a 'burning platform' of stalled productivity and climate change, John has helped us see a pathway to the jobs and industries of the future.

Professor Roy Green AM
Vice-Chancellor's Innovation Adviser,
University of Technology Sydney
October 2025

OVERVIEW

Innovation ecosystems are often described as if they are self-evident. But when examined closely, the frameworks that inform their study and practice are far from unified. Four domains—*placemaking, economics, business, and governance*—emerge from different intellectual traditions, each with its own assumptions, languages, and methods. This divergence helps explain why the policy debate around innovation ecosystems is often fragmented and sometimes incoherent

Domain	Disciplinary Foundations	Core Concepts	Key Thinkers, Works
Placemaking	Urban studies, human geography, architecture, cultural planning	Production of space; social ecology of cities; cultural economy; creative class; urban vibrancy	Jane Jacobs (1961); Henri Lefebvre (1974/1991); David Harvey (1989); Charles Landry (2000); Richard Florida (2002)
Economics	Innovation studies, evolutionary economics, industrial organisation, growth theory	Creative destruction; innovation systems; clusters; knowledge spillovers; endogenous growth	Joseph Schumpeter (1934, 1942); Christopher Freeman (1987); Bengt-Åke Lundvall (1992); Michael Porter (1990); Paul Romer (1990)
Business	Management studies, organisational theory, strategic management, entrepreneurship	Capabilities of firms; corporate strategy; business ecosystems; open innovation; entrepreneurial dynamics	Alfred Chandler (1962); Edith Penrose (1959); James F. Moore (1993, 1996); Henry Chesbrough (2003)
Governance	Political science, public administration, institutional economics, science policy	Collective action; Mode 2 knowledge; Triple Helix; network governance; institutional coordination	Elinor Ostrom (1990); Gibbons et al. (1994); Etzkowitz and Leydesdorff (1995, 2000); R. A. W. Rhodes (1997); Gerry Stoker (1998).

The problem of fragmentation

These traditions are not naturally aligned. Urbanists emphasise the qualities of place, economists model efficiency and growth, business scholars focus on firm capabilities, and governance researchers map institutional coordination. In isolation, each provides only a partial perspective. Policy debates often become fragmented, which risks people talking past one another.

Tradition	Contribution	Value	Risk
Placemaking	Frames innovation through space, design, and lived experiences.	Brings community and spatial conditions into the ecosystem	Urban policy may prioritise property and culture but overlook firm dynamics
Economics	Focuses on efficiency, competition, growth, and spillovers.	Explains productivity, incentives, and resource allocation	It may overlook cultural and spatial conditions
Business	Concentrates on firm capabilities, markets, and entrepreneurship.	Highlights scaling, strategy, and competitive dynamics	May neglect institutional coordination.
Governance	Examines institutions, authority, and legitimacy.	Provides coordination, legitimacy, and system integration	May underplay physical or economic factors

The Infrastructure Bedrock: The Capital Base

Innovation ecosystems depend on an underlying bedrock of capital that enables the four domains to operate. Physical, knowledge, financial, and social capital each provide essential resources: infrastructure and spaces, intellectual and data assets, investment and risk-bearing capacity, and networks of trust and reciprocity.

No single form is sufficient on its own. Their combined strength determines how effectively ecosystems function and why performance varies across different locations.

Capital Type	Description	Significance
Physical Capital	Tangible assets that support innovation activities, including laboratories, shared workspaces, incubators, test facilities, transport, energy, and	High-quality, accessible physical capital generally determines whether ecosystems can attract and

Capital Type	Description	Significance
	digital infrastructure, create the contexts and environments in which people interact, experiment, and scale ideas.	retain firms, talent, and investment.
Knowledge Capital	Intellectual resources generated by universities, research institutes, and collaborative centres, as well as tacit expertise, data infrastructures, and networks. It builds cumulatively through partnerships and experience.	It underpins discovery, problem-solving, and skill formation, making it essential for sustaining long-term innovation and equipping ecosystems with adaptive capabilities.
Financial Capital	Funding and investment flows that provide risk-bearing capacity for innovation include grants, venture financing, angel investments, corporate capital, and broader markets.	Effective financial capital supports long-term horizons, absorbs uncertainty, and enables scaling. It influences whether promising ideas can be translated into sustainable enterprises
Social Capital	Relationships, trust, and shared norms enable cooperation across sectors and institutions. Social capital is found in networks, partnerships, and communities of practice.	Social capital reduces transaction costs, creates reciprocity, and fosters legitimacy and collective problem-solving. Without it, other forms of capital are less effectively mobilised and coordinated

The capital foundation provides the tangible, intellectual, financial, and relational resources on which the domains draw. Differences in these forms of capital often explain why ecosystems vary so significantly between places.

Integration Through a Four-Domain Framework

By identifying and integrating traditions, the four-domain framework clarifies the complex interactions that enable ecosystems to function. Ecosystems succeed when placemaking, economic incentives, business strategies, and governance operate interdependently.

This rarely happens spontaneously: it depends on the presence of systems integrators—people and organisations with the capacity to connect across disciplinary, institutional, and organisational boundaries. They ensure that the underlying capital base (physical, knowledge, financial, and social) is mobilised in ways that reinforce one another rather than working at cross purposes.

Recognising and supporting system integrators is, therefore, essential if ecosystems are to deliver sustained and scalable outcomes.

International practice highlights this role. The Fraunhofer Institutes in Germany, Catapult Centres in the United Kingdom, and Manufacturing USA all demonstrate the importance of structured intermediaries. In Australia, similar initiatives have been attempted only sporadically, leaving ecosystems vulnerable to fragmentation.

From Fragmentation to Systemic Clarity

When the bedrock of capital supports the four domains, and when the domains are brought into alignment through integration, innovation ecosystems transition from fragmentation towards coherence. They become systemic, adaptive, and capable of producing outcomes that are durable and scalable.

This perspective reframes the discussion. It shifts attention away from isolated projects or one-dimensional policy interventions towards a recognition that ecosystems thrive through convergence across traditions, supported by strong foundations of capital.

The discussion shows how these domains rest on a shared bedrock of capital and demonstrates the importance of systems integrators in aligning them. Case studies and comparative analyses highlight the opportunities created when they converge around systemic innovation and illustrate the risks of fragmentation when domains operate in isolation. This framing is illustrated below.

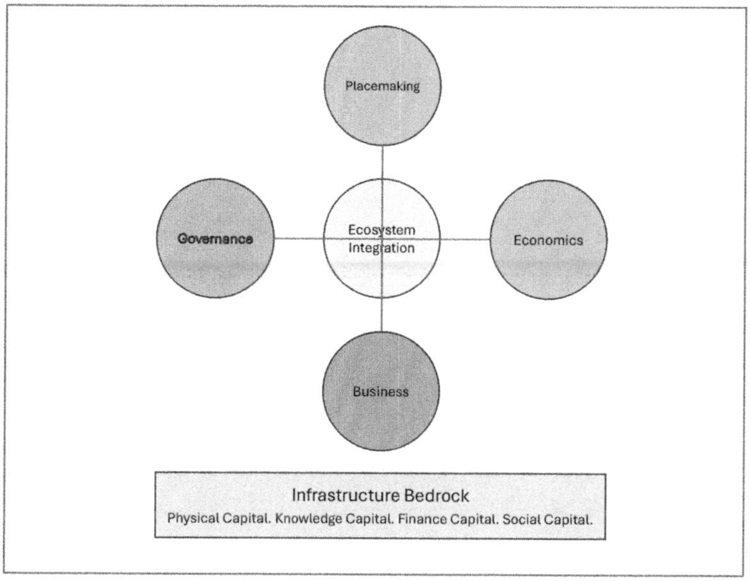

Each chapter which follows uses the four-domain framework as its organising principle, exploring how placemaking, economics, business, and governance shape the formation and performance of innovation ecosystems.

INTRODUCTION: POLICY IMPERATIVES FOR NATIONAL INNOVATION ECOSYSTEM DEVELOPMENT

Cities around the world are transforming underutilised urban areas into thriving hubs of research, entrepreneurship, and growth through innovation districts, precincts, and hubs that serve as powerful engines for economic transformation. Some districts grow organically, while government development authorities design others.

These ecosystems in places like MIT's Kendall Square, the Toronto MaRS District, and Singapore's One-North demonstrate how cities and regions can harness the power of proximity, knowledge sharing, and collaborative culture to drive innovation.

This research project, being funded by the Acton Institute for Policy Research and Innovation, is examining over 90 international innovation districts and precincts, revealing a complex picture of what works, what doesn't, and why some ecosystems thrive. In contrast, some people struggle to achieve their potential.

The analysis relies on four fundamental propositions that determine success: placemaking, economics, business development, and governance, and how they integrate in ecosystem practice. These propositions are summarised as follows.

- *Placemaking*: the quality of the built environment, public spaces, design thinking, and urban connectivity
- *Economics*: industry attraction, job creation, inward investment, and growth of high-value sectors
- *Business*: start-up activity, corporate engagement, and commercialisation of research
- *Governance*: coordination mechanisms, leadership, funding architecture, and institutional partnerships

The analysis aimed to identify patterns, differentiate emphasis, and benchmark performance to inform policy and practice in Australia. The analysis team also addressed practices related to integrating these elements.

The Architecture of Success

The research suggests that intentional integration of physical spaces, economic strategies, commercial outcomes, and institutional collaboration enables innovation ecosystems to thrive. Constructing impressive facilities and hoping that innovation emerges is more than just a goal. The research also suggests that policymakers need to understand at least seven key traits exhibited by the most successful districts.

1. *Thriving ecosystems are formed by anchor institutions* which provide stability and breadth and allow organic growth of intellectual capabilities across multiple partners. Translational research and commercial partnerships embraced by universities and hospitals generate the intellectual pull that draws talent and investment. MIT engages with industry, and the University of Cambridge has a long history of spin-out companies, showing how academic institutions can become economic drivers rather than ivory towers.

2. *The physical environment* is important in terms of visual appeal, but integrated urban design holds even greater importance. Successful districts combine laboratory and manufacturing infrastructure, which is often a focal point for deep tech growth. Digital infrastructure is becoming increasingly important. The physical environment also covers mixed-use development, walkable precincts, and spaces for potential for chance encounters among researchers, entrepreneurs, and investors.

3. *Governance arrangements* coordinate what markets alone cannot deliver. Intermediary organisations, such as in Toronto's MaRS or Houston's Ion, connect funding sources, manage programs, and facilitate partnerships while acting as system integrators. These bodies actively convene in ecosystems where they must coexist in competition and collaboration.

4. *A long-term public commitment* builds the stability that private investment needs. Phased investment strategies and policies that last beyond electoral cycles require decades to grow innovation ecosystems. This patience enables districts to grow and adapt through various stages.

5. *International connectivity* prevents ecosystems from becoming isolated clusters. Global research collaborations, talent attraction programs, and export pathways link local capabilities to worldwide markets and knowledge networks. The most successful districts act as nodes in global innovation networks rather than isolated bubbles.

6. *Social inclusion and civic intent*, particularly evident in Latin American and African models, integrate social mobility objectives with

economic development. This approach acknowledges that sustainable innovation ecosystems should benefit broad communities instead of creating enclaves of privilege.

7. *Flexible zoning and land use controls* allow communities to co-locate research facilities, businesses, and amenities, while enabling adaptation as needs evolve. Rigid planning frameworks that separate different activities can stifle the cross-pollination that drives innovation.

Models and Approaches

The research points to three broad approaches to ecosystem development, each with particular strengths and limitations.

Mission-oriented ecosystems align their development with national or regional priorities. Clear missions can direct investment and coordinate stakeholders. This approach works particularly well when national frameworks provide direction and allow for local adaptation.

Anchor-based clustering builds around existing institutional strengths. Proximity to elite universities, hospitals or global technology firms creates the initial conditions for ecosystem development. Kendall Square's relationship with MIT, or Medicon Valley's connection to leading pharmaceutical companies, illustrates how existing capabilities can be leveraged for broader development.

Phased urban integration recognises that ecosystems evolve over long timeframes of staged investment and policy adjustment. Cambridge Science Park's decades-long evolution demonstrates how patient capital and trust-based governance can support organic growth while maintaining strategic direction.

Learning from Leaders

Some important insights have emerged from the case examples. For example, several international case studies offer particularly valuable lessons for Australian policymakers.

Kendall Square is often represented as "the gold standard" for ecosystem integration. The district fuses real estate strategy with MIT's research capabilities, venture capital networks, and corporate research and development facilities. Success stems from MIT's institutional capacity, its proximity to Boston's urban infrastructure, a rapid transit system, dense venture networks, and the flexible zoning system that has evolved.

Toronto's MaRS Discovery District showcases the benefits of non-profit platform management. MaRS balances government support, corporate tenancies, and public engagement while embedding placemaking and tenant curation within a broader economic development strategy. The model illustrates how intermediary organisations can serve multiple stakeholders while maintaining a clear mission focus.

Porto Digital in Recife, Brazil, integrates creative placemaking with digital sector growth through strong public-private governance. The district succeeds by combining digital inclusion with economic development and urban revitalisation, demonstrating how innovation can serve both social and economic objectives.

Singapore's One-North delivers exemplary alignment across urban planning, digital infrastructure, research-business linkages, and public sector governance. The development demonstrates how coordinated government action can create world-class innovation environments when supported by long-term commitments and substantial resources.

What Doesn't Work

The research also illuminates common pitfalls that Australian policymakers should avoid.

Property-led models that prioritise building infrastructure over institutional capabilities consistently underperform. Success is driven by relationships, capabilities, and institutional dynamics rather than by impressive facilities. Districts that initiate property development and aim to attract innovation often struggle to achieve critical mass.

Insufficient investment in system integration leaves stakeholders operating in isolation. Without adequately resourced intermediary organisations to connect, coordinate, and translate across different actors, ecosystems fail to achieve their potential synergies.

Neglecting translational infrastructure creates gaps between research and commercialisation. Facilities that support prototype testing, clinical translation, or digital proof-of-concept are crucial for moving innovations from the laboratory to the market.

Rigid governance structures that cannot adapt to changing circumstances limit ecosystem evolution. Innovation districts must strike a balance between coordination and flexibility, allowing for organic development while maintaining a strategic direction.

Implications for Australian Policy

These international experiences offer several key insights for Australian innovation policy, including:

Stability of strategy and investment with decadal outlook. Districts and precincts may take 12-15 years to hit critical mass, with first five years being particularly painful—much like a start-up. Public sector support must therefore be very patient.

System integrators require substantial and sustained investment. Governance intermediaries must be adequately resourced to connect stakeholders, coordinate activities, and translate between different institutional cultures. This is a crucial but often underestimated policy requirement.

Translational infrastructure deserves the same attention as basic research facilities. Supporting the movement of innovations from research to market requires specialised facilities and capabilities that bridge institutional boundaries.

National frameworks achieve maximum impact when grounded in local ecosystems. Top-down approaches work best when they build on existing institutional strengths and local capabilities rather than imposing standardised solutions.

Performance measurement should assess ecosystem integration rather than focusing solely on individual outputs. Monitoring frameworks need to evaluate how well districts integrate across the dimensions of placemaking, economics, business development, and governance.

The Path Forward

This research by the Acton Institute for Policy Research and Innovation demonstrates that innovation ecosystems succeed through deliberate and evolving integration across multiple dimensions. While no single model dominates, consistent factors include institutional depth, mission clarity, and long-term system building.

Australian policymakers can draw from this international diversity to design support mechanisms that are place-responsive, system-aware, and globally connected. The challenge lies in adapting these lessons to

Australian conditions while avoiding the temptation to try and replicate overseas models.

Success requires patience, coordination, and sustained commitment. Innovation ecosystems develop over decades rather than electoral cycles, requiring policy approaches that maintain focus amid changing political circumstances while adapting to evolving economic and technological conditions.

The opportunity for Australia lies in learning from both the successes and failures of international experiences to create innovation ecosystems that serve Australian needs while connecting to global networks of knowledge and opportunity.

PART I: PROJECT FRAMING: THE CHALLENGES AND THE OPPORTUNITIES

1. Seeking Clarity: Defining Innovation

Innovation is sometimes described in terms of how it's done, as in "ideas successfully applied". In simple terms, this means that innovation occurs as a result of discoveries, inventions, knowledge gained from professional experience, and ideas that emerge from numerous sources and activities. Combining and coordinating these sources improves innovation outcomes.

Innovation is not just a theoretical concept, or something studied in schools. It is the main way that human societies change, improve, and move forward.

Innovation is fundamentally about change: an expectation of improvement for the better across the economy, industry, society, and the circumstances of people's lives. It also finds expression in the service sector of the economy.

Change and the Dimensions of Innovation

When innovation succeeds, it results in clear advances in economic productivity, industry skills, social wellbeing, and personal opportunities, as well as advantages in different specific fields. We observe this through various means. The features of these dimensions are outlined below.

Characteristics of Innovation Across Industries and Sectors

- **Economic.** Economic growth comes from new industries, better productivity, and increased competitiveness, leading to noticeable improvements in GDP, job quality, and living standards.
- **Industrial.** Innovation helps industries grow, adjust to new circumstances, and stay important in global markets by constantly improving technology, processes, and their overall strategy.
- **Business.** Innovation changes how businesses work by introducing new ways to operate, engage with customers, and organise themselves.

- **Social.** Innovation finds new ways to tackle social issues in healthcare, education, the environment, and communities, leading to noticeable improvements in people's lives and overall wellbeing.
- **Individual opportunity.** Innovation creates new possibilities for personal growth, career development, and life improvement through expanded educational opportunities, enhanced healthcare options, and improved quality of life.
- **Cultural-Creative.** Innovation in arts, culture, media, and creative industries generates new forms of expression, communication, and cultural value while preserving and revitalising cultural heritage.
- **Public Policy.** Innovation transforms government operations, service delivery, and policy effectiveness through new approaches to governance, citizen engagement, and public problem-solving that improve democratic processes and public outcomes.
- **Community-Civil Society**. Innovation within community organisations, non-profits, and civil society groups creates new approaches to local problem-solving, community organising, resource sharing, and citizen engagement.
- **Environmental-Ecological**. Innovation addresses environmental challenges and creates sustainable practices across all sectors, including clean technologies, circular economy approaches, conservation methods, and adaptation strategies.
- **Institutional-Government**. Innovation in institutions, governance structures, and regulatory frameworks improves how societies organise, make decisions, and manage collective challenges.

This detailed listing highlights several key points about how innovation affects society.

- *Innovation goes beyond technology or business; it influences all areas of human organisation and experience.* It brings about positive change in many areas, including personal career growth, institutional management, environmental protection, and cultural expression.
- *Every area of innovation has positive side effects that enhance progress in other fields.* Economic innovations provide resources for social programs, cultural innovations boost community identity, which helps businesses grow, and environmental innovations lead to industrial changes that improve people's lives.
- *Because innovation affects all these areas simultaneously, a successful innovation policy needs to work across traditional boundaries rather than just improving individual sectors.* These complicated relationships show why straightforward solutions, such as boosting research funding or establishing innovation agencies, do not work. They can't address

the many different types of coordination needed to achieve results in economic, social, cultural, and environmental areas.

- *Personal advantages, such as better job prospects, depend on larger changes in the economy and industry, which require supportive systems and a culture that embraces change.* This means that successful innovation often requires ongoing teamwork over a long period, which makes it difficult for innovation policies to function effectively when political terms and election pressures are short.

To fully realise the benefits of innovation, it is essential to coordinate systematically across all areas of change. This supports the idea that innovation results from complex systems working together, rather than something that governments can simply buy through specific spending.

This broad perspective of innovation requires policies that can manage complex coordination across different areas.

The Innovation Imperative

In fast-changing economies around the world, a lack of innovation can result in economic decline, outdated industries, social stagnation, and fewer opportunities for individuals. Countries and regions that do not consistently generate new ideas and solutions will struggle to compete economically and address emerging challenges.

This need has grown significantly in recent years due to faster technological development, increased global competition, and increasingly complex societal issues. Countries that once benefited from their natural resources, location, or historical industrial strengths are now quickly losing those advantages unless they continue to innovate to remain relevant.

Changes are happening so quickly that industries can be affected in just a few years instead of over many decades. This makes the ability to innovate crucial for remaining economically viable, rather than merely helpful for growth.

Countries that lack strong innovation systems tend to experience slower economic growth, lower productivity, and reduced competitiveness in global markets. Their businesses become weak against more innovative competitors, resulting in lost market share and an eventual decline in their industry.

In a societal context, a lack of new ideas restricts the ability to tackle important issues such as healthcare, education, environmental protection, and social equity. If new ways are not found to tackle these issues, they will worsen over time, leading to social unrest and a lower quality of life.

When someone experiences failure in innovation, it means they have fewer opportunities to advance in their career, earn more money, and grow personally. When economies are not innovative, job growth slows down, wages stagnate, and more young people move to more innovative areas in search of better opportunities. This loss of skilled people makes it harder for the local area to develop new ideas, leading to a cycle of worsening conditions.

The lack of innovation weakens our ability to handle unexpected challenges, technological changes, and global emergencies. Recent events, such as the COVID-19 pandemic, climate change, and global tensions, have demonstrated that a society's ability to innovate affects how quickly it can adapt and how well it responds to challenges.

Countries with low-quality innovation systems are increasingly relying on customers from other places, which decreases their independence and control over their own affairs.

This complex need for innovation highlights why policies that support innovation are now crucial for a country's competitiveness and social wellbeing in today's global economy.

Innovation's Dual Nature

Innovation is not universally beneficial. Technological and organisational changes can lead to job loss, push people out of their neighbourhoods, create new inequalities, and cause social unrest.

The use of machines in factories and service jobs has eliminated millions of positions and primarily benefited those who possess wealth and advanced technology. Digital platforms have transformed industries such as transportation, hospitality, and retail. However, they often weaken worker protections and job security, leading to new types of unstable jobs that offer little in the way of benefits or opportunities for advancement.

Innovation often leads to people relocating to different places. Innovative activities in certain cities significantly increase property prices and living expenses, pushing out longtime residents and weakening the social connections and cultural institutions that once made these places lively. This type of gentrification, driven by innovation, leads to an uneven distribution of resources. Being close to innovation centres influences access to job opportunities, quality services, and the ability to advance socially.

Additionally, digital advancements have led to new forms of exclusion and discrimination. The digital divide means that some people have high-speed internet, modern devices, and the skills to use them, while others do not. This leads to unequal opportunities in education, job participation, healthcare, and community involvement. Algorithm-based systems used in hiring, loan applications, criminal justice, and social services can reinforce or increase existing biases, leading to widespread discrimination that occurs quickly and on a large scale.

Innovation can also be "dark," enabling and extending harmful activities such as gambling technologies that exploit psychological vulnerabilities, surveillance systems that erode privacy and democratic freedoms, social media algorithms that spread misinformation and polarise communities, and financial technologies that facilitate fraud, money laundering, and predatory lending practices. These new ideas generate profit, but they also create serious social problems that the market cannot solve on its own.

An effective innovation policy must recognise innovation risks and include measures to ensure that the benefits of innovation are shared widely, while also mitigating negative effects through suitable social, economic, and regulatory interventions.

This means planning for workforce changes, creating inclusive strategies that consider the diverse impacts on people and communities, establishing robust support systems for those affected by these changes, and developing adaptable regulations that can address emerging issues without hindering beneficial innovations.

The Systemic Nature of Innovation: From Ecosystems to Global Networks

This comprehensive perspective on the results of innovation highlights that innovation is essentially a system-wide process. Breakthrough innovations usually do not come from individual efforts alone; instead, they result from the active collaboration of people, organisations, and skills working together in organised systems.

Contemporary understanding recognises innovation as occurring within "ecosystems"—complex webs of relationships among universities, research institutions, established companies, start-ups, investors, government agencies, and support organisations. These ecosystems work best when all types of innovative outcomes can develop simultaneously and support different participants.

Traditionally, people have viewed innovation ecosystems as places where innovation activities are concentrated in specific geographic areas, such as regions, districts, clusters, and hubs. Well-known places like Silicon Valley, Cambridge (UK), Kendall Square, and other innovation hubs demonstrate that being in the same physical space can facilitate informal connections, knowledge sharing, and collaborative work, which leads to better innovation outcomes.

Recent improvements in transportation, communication, and digital technology are changing where innovation occurs. Innovation ecosystems are increasingly functioning over larger areas and, even more importantly, through online networks that connect different continents. The COVID-19 pandemic accelerated this change, demonstrating that people can work together, share knowledge, and coordinate innovation effectively, even when they are not in the same location.

The most active innovation environments today bring together local resources and global connections. They take advantage of being close to each other for certain interactions but also use digital technology to access knowledge, skills, markets, and resources globally. This change means that innovation policy needs to consider the growth of local ecosystems and the connections to the global network that facilitate large scale innovation.

This approach to innovation, which is more distributed and interconnected, highlights the importance of policies that prioritise coordination and collaboration.

Innovation policy should not just aim to create local resources for innovation or to bring innovative activities to specific areas. It should support the complex systems that help local ecosystems function effectively while also linking them to global networks of innovation, where new and significant results arise from collaboration and the sharing of knowledge on an international scale.

2. People in Places and Spaces: The Evolution of the Geographical Ecosystem Concept

The innovation ecosystem concept has transformed our understanding of the relationships between people, places, spaces, and innovation processes. What began as separate streams of business strategy and institutional analysis has evolved into a comprehensive examination of individuals operating within specific institutional frameworks across distinct geographical contexts.

This evolution signifies a broader transformation in innovation thought, transitioning from the perception of innovation as an activity confined to individual organisations to recognising it as a phenomenon arising from intricate institutional relationships within specific locations and social contexts. The ecosystem metaphor effectively encapsulates both the institutional frameworks that support innovation and the environments in which these frameworks are cultivated and thrive through human relationships.

The Spatial Turn in Innovation Thinking

Innovation policy shifted in the late 20th and early 21st centuries from concentrating solely on individual companies and their products to recognising broader systems of innovation within specific locations. This change began with European researchers studying regional innovation systems (RIS), notably Cooke, Uranga, and Etxebarria (1997), who were the first to use the term "ecosystem" to describe the institutional frameworks that support innovation in particular areas.

This marked a significant departure from American traditions. In the US, the emphasis has been on companies and products, whereas in Europe, research underscores the importance of institutions, tacit knowledge, social networks, and consistent policies in supporting innovation. This perspective acknowledges that innovation emerges from interconnected relationships influenced by specific geographical and cultural contexts, rather than occurring in isolation.

The concept of place-based innovation became evident in the 2009 Barca Report to the European Commission. This significant document clearly opposed "space-neutral" policies and advocated for innovation strategies attuned to local areas and involved local communities. The report introduced new terminology related to innovation, such as embeddedness, institutional thickness, and relational proximity, drawing on cluster theory, agglomeration economics, and planning theory.

By the mid-2010s, ecosystem thinking and locally based methods began to converge. The creation of innovation districts, smart specialisation strategies, and precinct planning has made the ecosystem concept more tangible.

Innovation is now viewed as something that occurs in locations with their own unique histories, cultures, and community resources, rather than simply within companies or laboratories.

This change marked a major shift in how innovation policies are created, moving from simply investing in resources to managing entire systems, from focusing solely on technology to involving multiple participants, and from using general models to employing strategies tailored to specific places and situations.

From Corporate Strategy to Institutional Analysis

James Moore's original concept of the business ecosystem (1993) primarily examined the co-evolution of companies, customers, suppliers, and regulators within competitive environments. Although groundbreaking in its of interdependence, Moore's framework focused on market relationships rather than encompassing the broader institutional contexts in which innovation occurs.

Parallel developments in innovation studies examined these institutional dimensions more directly. The triple helix model, established by Etzkowitz and Leydesdorff (1995), offers a framework for analysing interactions among universities, industry, and government in innovation processes. In contrast to Moore's market-oriented approach, the triple helix framework explicitly acknowledges that innovation relies on specific types of institutional relationships that extend beyond purely market-driven logic.

The triple helix model was originally sectoral in nature, concentrating on interactions among three institutional spheres without highlighting the geographical context or relationship mechanisms.

Nevertheless, its institutional perspective was essential for comprehending variations in innovation patterns across different contexts, thereby establishing a foundation for subsequent spatial analysis.

Adner's work (2006) effectively integrated these approaches by refocusing attention from internal company processes to external dependencies and introducing the concept of ecosystem-wide value propositions. This change facilitated the examination of how innovation is influenced by institutional relationships that extend well beyond individual firms—relationships inherently shaped by the geographical contexts in which they are formed.

Placemaking as Economic Infrastructure

Building on this theoretical foundation, the practical implementation of place-based innovation strategies has focused heavily on placemaking; however, this domain remains widely misunderstood.

The most obvious feature of innovation districts is their physical layout, but creating successful spaces goes beyond standard city design rules.

Instead of viewing public spaces merely as amenities, effective placemaking treats them as crucial economic infrastructure, helping to attract skilled workers, facilitate knowledge sharing, and launch new businesses.

This approach addresses changes in what workers want, particularly those in knowledge-based jobs. These workers are placing greater importance on neighbourhood character, walkability, and cultural activities when deciding where to work. This trend has strengthened as new generations prefer urban areas that offer both job opportunities and social and cultural activities, rather than suburban office parks that require cars.

This preference makes sense when viewed through the lens of agglomeration theory. Human capital spillovers are a key driver of urban economic growth, as knowledge workers are more productive when gathered in diverse, interactive settings. These spillovers occur through both official channels, such as conferences, partnerships, and shared spaces, as well as through informal meetings in cafés, parks, and public areas.

Successful innovation districts deliberately bring together industries in thoughtfully planned public areas, which have been linked to noticeable increases in patent applications, new businesses, and cooperation between different sectors. These results connect to specific design elements: plazas that serve multiple purposes, active ground floors, and adaptable spaces that encourage both organised and casual interactions.

Although more people recognise the importance of placemaking, we still don't fully understand how investing in it affects economic outcomes in many innovation strategies.

Cities often invest in attractive public spaces but lack clear metrics to measure their economic benefits. This can lead to attractive places that do not generate income or support original goals, highlighting the ongoing challenge of translating local innovation ideas into tangible actions.

European Contributions and Regional Innovation Systems

European innovation scholars made significant contributions by illustrating how institutional relationships function across spatial scales. The national innovation systems approach established by Lundvall (1992) and Nelson (1993) demonstrated how countries' institutional arrangements influence their innovative capacity, though this research primarily focused on the national level.

The regional innovation systems (RIS) perspective advanced by Cooke, Uranga, and Etxebarria (1997) is important for understanding how institutional relationships function within specific territories. RIS scholars analysed the contributions of knowledge institutions, governance systems, and social connections to regional innovation capacity, offering spatial frameworks that enhance triple helix institutional analysis.

Unlike American approaches that emphasised market relationships, RIS highlighted the importance of tacit knowledge, social capital, and the political dimensions of innovation policy. This European tradition recognised that different regions innovate differently, reflecting their particular histories, institutional arrangements, and social structures.

By the early 2000s, triple helix scholarship began integrating spatial dimensions, as scholars and policymakers acknowledged that the dynamics of university-industry-government relationships varied across locations.

Research parks, technology transfer offices, and regional development agencies have emerged as significant intermediary institutions facilitating triple helix relationships within innovation districts and precincts.

Expanding Institutional Frameworks

The transition from triple helix to ecosystem thinking was expedited by the recognition that innovation necessitates more expansive institutional frameworks beyond traditional university-industry-government relationships. The "Quadruple Helix" model introduced by Carayannis and Campbell (2009) added civil society as a fourth institutional actor, acknowledging that innovation increasingly involves broader social constituencies and reflects wider social purposes.

This expansion reflected practical considerations: effective innovation increasingly required public support, community engagement, and a focus on social and environmental impacts. Innovation processes that overlooked these broader institutional requirements frequently encountered failure or produced negative externalities that compromised long-term sustainability.

The "Quintuple Helix" incorporates environmental sustainability as a fifth institutional dimension, acknowledging that innovation must address ecological challenges and operate within environmental constraints. This evolution demonstrated the expansion of institutional frameworks to integrate different forms of knowledge and criteria for assessing innovation outcomes.

Contemporary Synthesis and Future Directions

By the 2010s, innovation ecosystem thinking had evolved to incorporate insights from business strategy, institutional analysis, and spatial development. The concept now encompasses elements from management studies, triple helix institutional frameworks, regional development, urban planning, and socio-technical transitions.

This convergence reflects the recognition that innovation processes cannot be understood through a single analytical lens; they require frameworks that account for institutional complexity, spatial variation, and social embeddedness. Systems theory and evolutionary economics provide important foundations for understanding ecosystems as dynamic, adaptive systems that evolve through feedback, learning, and adaptation among multiple institutional actors.

Autio and Thomas (2014) proposed distinguishing between innovation ecosystems (focused on knowledge generation and technological development), business ecosystems (concentrated on value creation and market entry), and entrepreneurial ecosystems (concerned with institutional, financial, and social conditions supporting new business formation). This classification helps clarify different aspects of how institutional actors interact within places to generate innovation outcomes.

Current ecosystem approaches increasingly recognise the need to address questions of inclusion, power, and legitimacy in institutional relationships.

Unless addressed, innovation ecosystems can reinforce existing inequalities if they only include established institutional actors or define 'innovation' in ways that exclude different kinds of knowledge and capability.

A reflexive ecosystem practice acknowledges that institutional arrangements reflect power relationships and social structures that may need active addressing rather than simple acceptance. This includes examining how marginalised groups can be better included in institutional relationships and how benefits are distributed across communities.

The continued relevance of innovation ecosystems as policy frameworks depends on their capacity to work with institutional complexity while remaining sensitive to local social dynamics. This requires approaches that support institutional relationship-building without trying to control or direct outcomes mechanistically.

Innovation ecosystems offer practical value as analytical frameworks that help policymakers examine institutional arrangements, evaluate different intervention options, and identify ways to strengthen collaborative relationships. Their strength lies in recognising that innovation emerges from institutional interactions between people working within particular places, shaped by local contexts that cannot be easily replicated elsewhere but can be developed over time.

3. From Ecology to Economics: What Do We Mean by the "Eco" in Innovation Ecosystems?

The innovation ecosystem concept has become ubiquitous in public policy, academic research, and practitioner discourse. Underneath this apparent consensus lies a fundamental tension between two metaphorical frameworks: the ecological and the economic. The choice of metaphor shapes our understanding of system dynamics, intervention strategies, and success metrics in ways that have profound implications for innovation policy and practice.

The ecological metaphor suggests a complex adaptive system characterised by interdependence, co-evolution, and emergent properties that resist direct control. The economic metaphor, by contrast, implies more mechanistic relationships, rational actors, and systems that are amenable to optimisation through targeted interventions.

These different conceptual lenses describe the same phenomenon in different ways and can fundamentally alter what we see, what we measure, and what we attempt to manipulate.

The Genesis of Metaphorical Confusion

The innovation ecosystem concept emerged from multiple intellectual traditions, creating inherent tension from its inception. James Moore's seminal 1993 Harvard Business Review article "Predators and Prey: A New Ecology of Competition" explicitly drew from biological ecology, describing business ecosystems as complex adaptive systems in which companies co-evolve capabilities around shared innovations.

Moore stressed that "biological ecosystems are characterised by a large number of loosely interconnected participants who depend on each other for their mutual effectiveness and survival."

As the concept migrated into innovation policy and regional development contexts, it became increasingly absorbed in economic thinking. The ideas from cluster theory, industrial districts, and regional innovation systems, which are based on economic geography and institutional economics, began to combine economic concepts with ecological imagery. This developed a mixed framework that incorporates terminology from ecology but primarily follows economic principles.

The outcome is a body of work that uses the same term, "innovation ecosystem", to refer to everything from naturally occurring, self-organising networks of connected individuals (based on ecological principles) to intentionally built regional innovation systems aimed at maximising economic benefits (based on economic principles). This is not just about having different focuses; these are truly distinct viewpoints on what innovation systems are.

The Ecological Metaphor: Complexity and Emergence

When we genuinely consider the ecological metaphor, we identify several important principles that distinguish this viewpoint from economic approaches.

In nature, different species change and adapt in response to their interactions, forming complex relationships in which they rely on one another. In simple terms, this means that organisations, institutions, and individuals in a system develop their skills, plans, and identities by responding to one another. That is, the features of the ecosystem arise from the interactions among its parts, rather than being created or managed by one individual or group.

Ecological systems have characteristics that cannot be predicted by examining individual parts in isolation. In addition, innovation ecosystems based on natural principles should exhibit emergent features, such as group behaviours, patterns of innovation, and the ability to recover from complex interactions that cannot be directly controlled or created.

Natural ecosystems are considered resilient and adaptable, surviving disruptions by adjusting rather than merely striving for efficiency. A nature-based approach to innovation emphasises making systems resilient and flexible, rather than solely focusing on short-term results. This means that things that seem inefficient, such as extra resources or unsuccessful projects, could actually help improve the system's overall health in the long-run.

Ecological systems undergo predictable stages of growth and change. Young ecosystems grow quickly and are populated by initial species, while mature ecosystems are more complex, specialised, and stable. This suggests that, as innovation ecosystems tend to develop naturally, it makes them difficult to alter through government policies.

Ecological systems can be challenging to control and exhibit unpredictable behaviour. This implies that ecosystem development cannot be engineered in the same manner as economic systems, and that attempts at direct control may actually harm ecosystem health.

The Economic Interpretation: Optimisation and Intervention

The economic perspective presents a distinct set of beliefs and consequences.

- Economic theories typically assume that individuals involved in ecosystems make rational decisions to achieve clear objectives, such as profit, growth, and competitive advantage. This makes it easier to understand how systems work and to predict how they will respond to changes in policy.
- Economic principles aim to identify and eliminate waste, allocate resources efficiently, and maximise earnings. Ecosystem policies aim to lower transaction costs, enhance information sharing, and eliminate unnecessary steps—methods that may conflict with the ecological principles of diversity and redundancy.
- Economic methods typically rely on market forces to organise ecosystem activities, believing that price changes and competition will enhance the system's development for optimal results. This differs from ecological approaches, which focus on cooperation and adaptation among individuals without relying on market mechanisms.
- Economic systems are closely tied to clear measures of success, such as the number of patents, the number of start-ups, investments in venture capital, the value of initial public offerings (IPOs), and job creation. These measures enable the comparison of different ecosystems, and the evaluation of how well policies are working, but they may overlook important features of ecosystems that are difficult to express in economic terms.

Economic views indicate that ecosystems can be intentionally designed, improved, and optimised through specific policy actions.

This approach views ecosystems as systems that can be designed and managed, rather than as natural occurrences that develop on their own.

Manifestations of Metaphorical Confusion

This disagreement manifests in various troubling ways within the literature and practice of innovation ecosystems.

Policy Prescriptions: Ecological and economic metaphors can lead to dramatically different policy recommendations. Ecological approaches suggest that policy should focus on creating conditions for ecosystem health, maintaining diversity, supporting experimentation, and building adaptive capacity. In contrast, economic approaches emphasise targeted interventions to address market failures, improve efficiency, and optimise outcomes.

Success Metrics: Economic measures such as GDP growth, the number of patents, and the rise of unicorn start-ups can indicate short-term economic success but may undermine the ecosystem's ability to adapt and remain strong in the long-run. Ecological metrics might focus on diversity, the strength of connections, and the ability to adapt—factors that are difficult to measure but could serve as better indicators of long-term success in innovation.

Intervention Strategies: Ecological thinking promotes a modest approach to developing ecosystems by fostering conditions that allow for natural growth rather than attempting to control the outcomes. Economic thinking encourages taking more action to address market problems, create incentives, and enhance the system's functioning. Different approaches can lead to conflicting policies and wasted resources.

Economic strategies typically focus on achieving quick results and clear outcomes, whereas ecological strategies consider longer time frames that account for natural growth patterns and the time required for complex systems to develop.

This temporal mismatch can lead to the premature abandonment of ecosystem development efforts or inappropriate interventions that disrupt natural evolutionary processes.

Case example: Silicon Valley through two lenses

These different metaphorical frameworks would explain the growth of Silicon Valley in different ways.

The *ecological metaphor* suggests that Silicon Valley is a well-developed hub of innovation that has grown over many years through collaboration among universities, defence companies, chip manufacturers, investors, and business creators. Its success comes from strong connections between people, a culture that accepts failure, a high movement of talent and ideas, and systems that encourage experimentation and risk-taking.

The strength of the ecosystem comes from its variety; it has successfully shifted from defence electronics to semiconductors, then to personal computers, followed by internet technologies, mobile devices, and now artificial intelligence. Its features are hard to replicate because they stem from unique historical situations and intricate relationships developed over time.

In *economic terms*, Silicon Valley is successful because it has access to investment capital, is close to top research universities, features flexible job markets that match talent with opportunities, has strong protections for ideas and inventions, and has regulations that are favourable to businesses. These successes can be recognised, evaluated, and possibly duplicated in other areas with the right policy changes. The ecosystem is an effective framework for generating new ideas and products that can be enhanced through targeted investments and institutional changes.

Both interpretations carry weight, but they result in very different ideas about what makes Silicon Valley successful and whether its approach can work in other places.

Policy and Practice Implications

This divergence in thinking has several key effects on innovation policies and practices:

Policies/ Practices	Economic Metaphor	Ecological Metaphor	Implication
Resource Allocation	Support significant public spending on ecosystem development by	A more cautious and patient approach to investment,	The type of metaphor used can influence priorities between investing in

Policies/ Practices	Economic Metaphor	Ecological Metaphor	Implication
	highlighting clear benefits and advantages	allowing nature to develop naturally rather than attempting to control its progress	infrastructure (roads, rapid transit systems, and buildings) or on developing relationships and culture
Evaluation Frameworks	Assessment based on measurable results of innovation and financial gains	Distinct evaluation approaches that assess ecosystem health, resilience, and adaptive capacity.	Ecological characteristics may be more crucial for long-term innovation performance but are more challenging to measure and compare across regions
Timing and strength of interventions	Effective interventions can accelerate ecosystem growth and improve performance	Acting too quickly or too forcefully can upset natural evolution and harm the health of ecosystems	Affects when and how governments should step in to support new innovation groups.
Realistic Expectations	Economic comparisons can set unrealistic hopes about how quickly and predictably ecosystems develop	Metaphors offer more realistic timelines, but they might be hard to maintain in situations that require quick results.	Economic approaches can cause frustration with natural changes and lead to giving up on long-term development plans too soon

In global contexts, the way metaphors are used affects how regions strive to lead in innovation.

Economic metaphors can promote a mindset that views attracting businesses and talent from other areas as a win-lose situation. Ecological metaphors suggest that healthy ecosystems can coexist and that the diversity of ecosystems may be more significant than direct competition among them.

Towards Conceptual Clarity

Instead of ignoring this tension as merely a semantic issue, policy requires clarity about the kind of system being referenced in using the term 'ecosystem.' There are a few ways to help clear up this conceptual confusion:

- Researchers and practitioners should clearly communicate which metaphor is in play and apply it consistently. This does not require choosing a single "correct" metaphor, but it does require clarity regarding the assumptions and implications of the chosen framework.
- Some researchers create mixed methods that draw on ideas from both metaphors and recognise the conflicts between them. That is, understanding that innovation systems consist of both planned components (such as infrastructure and institutions) and embedded qualities (like culture and networks) that require different management approaches.
- Different situations may require different metaphors. Economic metaphors can help understand well-developed innovation areas, while ecological metaphors may be better suited for describing new and groundbreaking innovation environments.
- Recognising that innovation systems may operate according to different logics at various stages of development, with economic optimisation becoming more relevant as ecosystems mature and stabilise.

Implications for Practice

For innovation professionals and policymakers, several practical recommendations emerge:

- *Diagnose Before Prescribing*: Before developing ecosystem interventions, explicitly diagnose whether the system in question is better understood through ecological or economic lenses. This should influence both intervention strategies and success metrics.

- *Portfolio Approaches*: Understand that effective innovation systems may need both planned components (economic reasoning) and natural developments (ecological reasoning). Policy plans should combine focused actions with support for patients to facilitate natural changes.
- *Adaptive Management*: Use flexible management techniques that track how ecosystems respond to actions taken and modify strategies based on what is learned. This is especially important because complex adaptive systems are inherently uncertain.

Building an ecosystem takes time and dedication. Economic demands for quick results should be weighed against the understanding of how long natural processes take.

Conclusion: Embracing Productive Tension

The use of metaphors in innovation ecosystems is confusing but useful. Rather than seeking a single "correct" metaphor, it is more useful to draw on the insights that both ecological and economic perspectives offer. The ecological metaphor emphasises diversity, adaptation, and resilience; factors that economic analysis frequently overlooks.

Conversely, the economic metaphor provides tools for measurement, evaluation, and intervention that ecology does not offer. The challenge is to work with the tension between these perspectives to generate a deeper understanding and design more thoughtful interventions. This requires researchers, practitioners, and policymakers to engage with innovation systems as both natural and planned phenomena. Effective approaches must integrate:

- The natural dynamics of complex systems with deliberate design of innovation support
- Evaluation of economic outcomes with assessments of ecosystem health
- Respect for evolutionary change with targeted interventions that address market failures and institutional gaps

Humility is essential. Innovation ecosystems are complex adaptive systems, not easily controlled or predicted. The apparent "confusion" in metaphors is valuable because it mirrors the complexity of the systems themselves and reminds us that no single theory or model is sufficient.

Progress depends less on perfecting metaphors than on developing clear conceptual frameworks, using strong evidence, and applying practical knowledge. A pluralistic approach that combines ecological and economic reasoning provides a more robust basis for policy and strategy, provided it remains grounded in real-world challenges and opportunities.

Addendum: Key Authors and Works by Metaphorical Framework—A Summary

Metaphorical Framework	Key Authors	Key Works	Main Contribution	Policy Relevance
Ecological Metaphor – Complex Adaptive Systems	James F. Moore	*Predators and Prey: A New Ecology of Competition* (1993); *The Death of Competition* (1996)	Introduced business ecosystem thinking; emphasised co-evolution, interdependence, and resilience.	Encourages policies that build diversity, adaptability, and long-term resilience rather than purely short-term efficiency.
"Rainforest" Metaphor	Victor W. Hwang & Greg Horowitt	*The Rainforest* (2012); *The Rainforest Blueprint* (2016)	Uses biological analogies to highlight culture, trust, and diversity as drivers of innovation.	Suggests interventions that focus on building social capital, trust, and network density alongside infrastructure investment.
Economic Metaphor – Markets, Clusters, and Optimisation	Michael E. Porter	*The Competitive Advantage of Nations* (1990); *Clusters and the New Economics of Competition* (1998); *On Competition* (1998)	Developed cluster theory; provided tools for competitiveness and regional development analysis.	Offers structured models for cluster mapping, benchmarking, and targeted investment programs to strengthen competitive industries.
National Innovation Systems School – Institutional Economics	Christopher Freeman; Bengt-Åke Lundvall; Richard R. Nelson	Freeman: *Technology Policy and Economic Performance* (1987); Lundvall: *National Systems of Innovation*	Positioned innovation within national institutional and policy frameworks; emphasised interactive learning.	Supports systemic, whole-of-government policy design, with emphasis on education, research infrastructure, and knowledge

Metaphorical Framework	Key Authors	Key Works	Main Contribution	Policy Relevance
		(1992); Nelson: *National Innovation Systems* (1993)		diffusion mechanisms.
New Economic Geography & Regional Development	Paul Krugman; Alfred Marshall	Krugman: *Geography and Trade* (1991); *The Spatial Economy* (1999); Marshall: work on industrial districts	Highlighted spatial concentration, agglomeration effects, and cluster dynamics.	Justifies place-based policies, transport and housing planning, and spatially targeted investment to maximise agglomeration benefits.
Hybrid / Bridging Works	Philip Cooke; AnnaLee Saxenian; Walter Powell & Laurel Smith-Doerr	Cooke: *Regional Innovation Systems* (1997); Saxenian: *Regional Advantage* (1994); Powell & Smith-Doerr: network studies	Combine institutional analysis with network theory and cultural insights.	Encourage policies that integrate formal institutions with informal networks, enhancing collaboration and trust-building measures.
Institutional Development-alism	Mariana Mazzucato	*The Entrepreneurial State* (2013); *The Value of Everything* (2018); *Mission Economy* (2021)	Advocates state-led, mission-oriented innovation; positions government as a market-shaper.	Supports mission-oriented programs, public investment in high-risk R&D, and strong state leadership to address grand challenges (e.g. climate, health).

4. What Matters in Innovation Ecosystems: Placemaking, Economics, Business, and Governance

The term "innovation ecosystem" has become a ubiquitous feature of policy discourse across regions, cities, and sectors. Once a metaphor borrowed from ecology and management theory, it now frames how governments, firms, universities, and communities think about innovation, growth, and resilience. But its real significance lies in how it reflects a deeper convergence of once-distinct traditions in economics, business, placemaking, and governance.

This chapter explores how the innovation ecosystem concept has evolved into a multidimensional framework built around a synthesis of spatial awareness (placemaking), economic logic, business practices, and system governance. These perspectives, originally developed in separate intellectual and policy domains, are now increasingly interdependent. Their convergence reflects both an intellectual evolution and a strategic necessity.

From Innovation Theory to Ecosystem Practice: The Craft of Convergence

Beyond Metaphor: Innovation Ecosystems as a Practical Framework

The concept of the innovation ecosystem has evolved. It has shifted from being merely a biological metaphor used to describe economic links to a comprehensive way of thinking, a detailed governance approach, and an effective means of stimulating inclusive economic growth. This analysis demonstrates that the mature ecosystem framework covers four key areas: placemaking and infrastructure, economics, business, and governance. These areas support and influence one another, forming a complex system in which their interactions can lead to either success or failure within the ecosystem.

Understanding this convergence is crucial for the future planning and management of innovation ecosystems, whether they are cities, regions, neighbourhoods, or specialised centres. In Australia, where scattered resources, divided institutions, and entrenched policy divisions hinder innovation, the ecosystem framework offers a hopeful solution, which does not involve creating more bureaucratic systems or coordination efforts; it calls for consideration of how innovation genuinely develops and persists over time.

The key is in developing "reflexivity"—an ecosystem approach that remains consistently alert to local contexts, genuinely responds to changing conditions, and is deeply committed to creating shared value across all stakeholder groups.

This advanced and flexible method is the only way to turn the concept of "innovation ecosystem" into something measurable, shift from mere talk in policies to real capabilities within institutions, and move from strategic goals to actual results that benefit communities and economies.

Innovation Districts as Strategic Capability

Innovation districts are an excellent way for cities and regions to adopt ecosystem thinking, offering real opportunities to enhance their standing in the global knowledge economy. They recognise that innovation stems from the interaction of location, economy, and community, rather than from isolated actions or efforts focused on just one aspect.

The success of innovation districts depends on having strategies, implementation methods, and governance that are as advanced as the ecosystem ideas that support them. This is not a case where good intentions, sufficient funding, or standard project management practices are sufficient. Instead, success demands what can best be described as a "craft"—a practical framework that combines technical expertise with contextual sensitivity, strategic vision with operational agility, and ambitious goals with pragmatic implementation.

Cities that succeed in developing innovation districts are those that view it as a skilled process requiring long-term collaboration across different sectors and lasting many years, rather than merely focusing on short-term political cycles. They consistently work on building the necessary infrastructure, skills and resources to handle complex situations, navigate uncertainties, and stay focused on long-term goals despite the challenges and setbacks that accompany major urban changes.

A Framework for Future Practice

This analysis offers cities and regions a helpful starting point for addressing this complex task. It shows that innovation districts can succeed in the long-term if they effectively integrate four key areas:

establishing genuine spaces for innovation, developing the economy to gain a competitive advantage, ensuring commercial success to attract and retain private investment, and having strong management to oversee these efforts.

Merely hoping for coordination or attempting to collaborate gently isn't sufficient to achieve integration. It requires careful planning by people and organisations to ensure that everyone shares similar objectives, takes responsibility, and genuinely relies on each other across diverse groups and sectors.

This involves creating innovative ways for the public and private sectors to collaborate, establishing management systems that promote efficiency and inclusivity, and developing methods to assess success based on the overall health of the ecosystem rather than just focusing solely on specific areas.

The Challenge and Opportunity Ahead

Cities, regions, and their leaders now bear the responsibility of establishing the institutions, skills, and partnerships necessary to make this ecosystem framework work in specific locations. This is not a problem that can be fixed simply by copying methods from elsewhere or by using ready-made solutions developed in other places. The focus should be on building local institutions that address the community's needs while also learning from global practices and staying connected to broader networks of innovation.

The potential benefits for cities aiming to be more competitive, communities striving for shared success, and the overall economy needing growth driven by innovation clearly make this significant investment in advanced systems and strategies worthwhile. The challenge is not so much knowing what needs to be done; it is in developing and maintaining the skills and systems required to deliver on the ecosystem's promises.

Success will be judged by how effectively environments and systems are created that genuinely support innovation and ensure its benefits reach everyone, rather than by how impressive the plans or ideas may be. This is the biggest challenge and the greatest opportunity for cities aiming to grow through innovation within a more complex and interconnected global economy.

5. Placemaking: How Infrastructure Design Drives Innovation Performance

The conventional wisdom surrounding innovation districts centres on attracting the right mix of talent, capital, and institutions. There is mounting evidence that suggests that the physical design of these precincts may be equally important in determining their success. The most effective innovation districts are pioneering infrastructure approaches that actively engineer the conditions for breakthrough thinking and collaboration.

This shift is a fundamental departure from traditional science park models, where efficiency and cost minimisation drove design decisions. Instead, leading innovation precincts are embracing "productive inefficiency"—deliberately creating spatial arrangements that increase friction in positive ways, encouraging chance encounters and extended interactions that drive knowledge spillovers.

Designing for Unplanned Interactions

Facilitating serendipitous encounters has become a central aspect of planning innovation districts. Melbourne Connect exemplifies this approach through its "collaboration spine"—a continuous internal street connecting different buildings while providing natural gathering spaces. Rather than designing the shortest routes between destinations, the precinct creates longer pathways that maximise opportunities for cross-pollination between organisations and disciplines.

Copenhagen's Ørestad district takes this concept further by integrating public transport directly into building structures. Daily commuting becomes an opportunity for cross-sectoral interaction, with tram stops functioning as informal networking nodes. This integration challenges traditional boundaries between public infrastructure and private innovation spaces, creating shared territories where different communities naturally intersect.

The key insight is that innovation requires both planned collaboration and unexpected connections.

While formal meeting rooms and conference facilities serve important functions, breakthrough ideas often emerge from casual conversations in corridors, cafeterias, and transit areas. Smart infrastructure design maximises these opportunities without forcing artificial interactions.

Adaptive Systems for Dynamic Needs

Innovation districts must accommodate rapid organisational change as start-ups scale, partnerships form and research directions evolve. Sydney's Tech Central design incorporates modular building systems and flexible utility distribution, allowing spaces to be rapidly reconfigured without major construction work. This infrastructure's adaptability reduces the friction costs associated with growth and organisational change.

The concept extends beyond physical spaces to digital infrastructure. Many contemporary innovation precincts function as living laboratories for smart city technologies, featuring integrated sensor networks and dynamic building management systems. These data-rich environments enable continuous optimisation of environmental conditions while providing testbeds for emerging technologies.

Barcelona's 22@ district demonstrates how infrastructure adaptation can support mixed-use integration. The precinct combines student housing, commercial amenities, research facilities, and start-up spaces within single developments, creating 24-hour activity cycles that sustain informal knowledge networks beyond standard working hours.

Innovation districts require infrastructure solutions that can simultaneously serve research laboratories needing specialised environmental controls, residential units requiring different utility configurations, and retail spaces demanding high-traffic accessibility.

The Invisible Architecture of Cognitive Enhancement

The most transformative infrastructure innovations operate below the threshold of conscious awareness, shaping behaviour and cognition through environmental psychology rather than visual spectacle. These subtle interventions can significantly impact innovation productivity while requiring minimal additional investment compared to dramatic architectural gestures.

Lighting systems represent a prime example. Collaboration spaces often feature circadian-responsive LED systems that adjust colour temperature and intensity throughout the day to support different types of cognitive work. Morning light emphasises blue spectrums that enhance analytical thinking, while afternoon lighting shifts towards warmer tones that facilitate creative and collaborative activities.

Lighting systems can meaningfully impact cognitive performance, mood regulation, and social interaction patterns.

Acoustic design has become equally important for innovation precincts seeking to attract knowledge workers. Sound-masking systems can use precisely calibrated background noise to create zones of conversational privacy within open collaborative spaces. This allows sensitive commercial discussions to occur in semi-public environments without requiring formal meeting rooms, thereby supporting the informal relationship-building that is essential for innovation partnerships.

Biophilic Integration as Competitive Strategy

Integrating natural systems into the infrastructure of innovation districts creates measurable productivity advantages while addressing environmental sustainability objectives. Precincts may feature extensive green corridors, rooftop gardens, and internal courtyards that provide microclimate control while supporting stress reduction and cognitive restoration.

It is now well known that access to natural environments enhances creative problem-solving and reduces mental fatigue associated with complex analytical work. Innovation districts that integrate these systems into their core infrastructure can create tangible competitive advantages in attracting and retaining talent.

The most advanced precincts use infrastructure to create distinctive sensory signatures that function as territorial markers and cultural signals. This may even feature a signature scent distributed through the ventilation system, creating unconscious memory associations that enhance creative thinking and reinforce a collaborative identity across different buildings and spaces.

Economic Returns Through Design Innovation

Infrastructure investments in innovation districts increasingly generate returns through design innovation. The development of new construction techniques, building materials, and environmental systems for innovation precincts often becomes commercially valuable intellectual property. Brainport Eindhoven's cleanroom construction innovations have become industry standards throughout the semiconductor sector.

This suggests that infrastructure investment for innovation districts should be understood as R&D expenditure as well as capital investment.

The design challenges posed by innovation districts and precincts drive innovations in construction, materials science, and building systems that generate broader economic benefits.

Survey data from multiple innovation districts show that environmental quality ranks as highly as salary considerations for knowledge workers choosing between employment opportunities. The infrastructure becomes a form of non-monetary compensation that allows innovation districts to compete for talent against organisations offering higher direct remuneration. For start-ups and research institutions operating with constrained budgets, access to premium environmental conditions through shared infrastructure provides significant competitive advantages.

Cultural Programming Through Physical Design

Effective placemaking innovations involve infrastructure that actively promotes cultural change as well as accommodating existing behaviours. Common areas designed to cater for different cultural practices help diverse innovation actors develop shared working relationships while maintaining their distinct identities.

Food service infrastructure is important for building social capital that sustains collaborative innovation. Shared kitchen facilities, diverse dining options, and flexible event spaces that accommodate different cultural and dietary requirements help bridge the gaps between academic, corporate, and start-up environments.

Digital archive systems integrated into common areas showcase the history of innovations developed within precincts, designed to create inspiration while providing practical information about previous research directions and partnership opportunities. These systems help new arrivals understand the culture of innovation while providing established residents with a perspective on their contributions to broader innovation trajectories.

Policy Implications for Infrastructure Investment

The most valuable infrastructure investments in innovation districts may be those that enhance cognitive performance, social interaction, and cultural development over and above dramatic visual impact or technological complexity.

Economic returns emerge through sustained improvements in innovation productivity rather than through immediate marketing benefits.

For policymakers, this analysis suggests several priorities. Infrastructure funding should explicitly consider cognitive and social outcomes in addition to traditional metrics of capacity and efficiency. Planning processes should incorporate insights from environmental psychology and neuroscience research. Procurement approaches should encourage experimentation with new materials, systems, and design concepts that can generate exportable innovations.

Research infrastructure deserves particular attention. Shared research facilities are one of the most cost-effective mechanisms for enabling cross-sector collaboration while reducing duplication of expensive equipment and expertise. Policymakers should fund models that incentivise shared access to research infrastructure rather than creating isolated facilities within individual organisations.

At the same time, the governance of shared research infrastructure requires careful attention to intellectual property frameworks, data security protocols, and access pricing mechanisms. Successful models typically involve transparent governance structures that provide all stakeholders with confidence in fair access while protecting commercial interests. Government funding can play a catalytic role by supporting the initial capital investment and establishing sustainable operating models.

Digital research infrastructure presents specific opportunities for policy intervention. Cloud computing resources, research data repositories, and collaborative platforms can operate at the district or regional scale, creating significant economies of scale while enabling more ambitious research projects.

The technical complexity of digital research infrastructure often requires government coordination and funding that individual organisations cannot efficiently provide.

The infrastructure design of innovation districts is a critical yet undervalued component of innovation ecosystem development. As competition for talent and investment intensifies, the precincts that succeed will be those that understand infrastructure as a tool for enhancing human creativity and collaboration, rather than simply providing functional workspace.

6. Economics: The Economic Foundation of Collaborative Innovation

Innovation ecosystems represent a fundamental departure from traditional economic thinking about how value is created and distributed in modern economies. While classical economics focused on individual firms competing for market share, ecosystem economics recognises that the most significant innovations emerge from collaborative networks where knowledge, resources, and capabilities flow between participants. This shift requires an understanding of new economic principles that represent how innovation districts, precincts, and knowledge hubs actually function.

The economic logic of innovation ecosystems rests on several key principles that challenge conventional wisdom about competition and value creation:

- These environments generate increasing returns to scale, meaning that larger, more connected networks become disproportionately more valuable than smaller ones.
- They produce positive externalities when the activities of one organisation create benefits for others without direct compensation.
- They depend on shared infrastructure and public goods that individual firms cannot efficiently provide on their own.

Understanding these principles is essential for designing effective economic development strategies.

From Linear Models to System Economics

Traditional economic models viewed innovation as a linear production process, where research and development inputs yielded predictable outputs, such as patents, products, and revenue, with outcomes representing new or substantially improved products, processes, and business models.

This mechanistic view assumes that economic actors operate independently, with clear boundaries between firms and limited spillover effects between activities. Policymakers could therefore focus on providing basic inputs, such as funding for research, and expect proportional returns in terms of economic growth and job creation.

The ecosystem approach reveals the limitations of this linear thinking. Innovation emerges from complex interactions among researchers, entrepreneurs, investors, established companies, and supporting institutions. Knowledge spillovers occur continuously as people move between organisations, collaborate on projects, and share information through formal and informal networks. The value created by these interactions often exceeds what any individual participant could achieve independently, suggesting that the whole system generates more value than the sum of its parts.

This systemic view draws heavily from endogenous growth theory, which emphasises that knowledge and innovation are key drivers of long-term economic growth. Unlike physical capital, knowledge exhibits unique economic properties: it can be shared without diminishing its value, it generates increasing returns when widely applied, and it creates cumulative advantages for regions that successfully attract and retain innovative activities.

The characteristics of knowledge capital means that innovation ecosystems can become self-reinforcing, with success attracting more talent, investment, and activity in a virtuous cycle.

The economic implications are significant. Rather than simply funding individual research projects or providing tax incentives to isolated firms, policymakers must consider how to strengthen the overall system of relationships and interactions. This requires attention to network effects, knowledge flows, and the quality of connections between different types of economic actors. It also means recognising that innovation ecosystems require time to develop and that early investments may not generate immediate measurable returns.

Cluster Economics and Competitive Advantage

Michael Porter's research on industrial clusters provides crucial insights into how innovation ecosystems generate competitive advantages. To Porter, clusters create economic value through four main mechanisms: increased productivity due to specialisation and efficiency gains; enhanced innovation through knowledge spillovers and collaboration; improved access to specialised suppliers, services, and talent; and reduced transaction costs through geographic proximity and trust-based relationships.

The productivity gains from clustering stem from the economies of scale and scope that emerge when related activities concentrate in specific locations:

- Suppliers can develop more highly specialised capabilities when they serve multiple nearby customers.
- Workers can develop industry-specific skills that command premium wages.
- Support services, such as legal, financial, and consulting firms, can build deep expertise in particular sectors.

These productivity advantages can be expected to compound over time as the cluster becomes more efficient and attractive to new entrants.

Knowledge spillover benefits arise from the rapid circulation of ideas, techniques, and market intelligence within clusters.

- *Face-to-face interactions* facilitate the transfer of tacit knowledge that cannot easily be codified or transmitted over long distances.
- *Joint projects* between firms and research institutions accelerate the development and commercialisation of new technologies.
- *Competition for talent and customers* incentivises continuous improvement and experimentation.
- *The concentration of innovative* activity attracts venture capital investors and other sources of risk financing.

Cluster advantage also flows from enhanced access to specialised inputs. When related firms locate near each other, they collectively create a greater demand for highly specific goods and services that would not be economically viable in dispersed markets. These goods and services include, for example, precision manufacturing equipment, testing and measurement equipment, industry-specific software platforms, specialised legal and IP services, and venture and development capital investors.

Trust and reduced transaction costs emerge from repeated interactions and shared institutional frameworks within clusters. That is:

- Formal contracts become less necessary when parties have ongoing relationships and reputations to uphold.

- Information flows more freely when people share common technical languages and professional networks.
- Collaboration on pre-competitive activities, such as training, research, and infrastructure, becomes possible when firms trust each other and recognise mutual benefits.

Public Goods and Market Failures

Innovation ecosystems depend heavily on public goods and collectively shared resources that markets alone cannot efficiently provide. Research infrastructure, educational institutions, transport networks, and regulatory frameworks all contribute to ecosystem performance but require coordination and public investment.

Understanding market failures in the delivery of public and collective goods and services is essential for designing appropriate policy interventions.

Basic research is the most obvious public good in innovation systems. The knowledge generated by universities and research institutions creates benefits that extend far beyond any individual organisation's ability to capture. While firms may fund applied research directed towards specific commercial applications, they will systematically underinvest in fundamental research whose benefits are uncertain and difficult to appropriate. Public funding for research is, therefore, essential for maintaining the knowledge base that underpins innovative activity.

Education and skills development is also a critical public good. Innovation ecosystems require workers with high-level technical capabilities, creative problem-solving skills, and the ability to collaborate across disciplinary boundaries. Individual firms may provide specific training for their immediate needs, but they cannot efficiently produce the broad-based human capital that ecosystems require. This necessitates public investment in universities, technical colleges, and continuing education programs that serve the entire innovation community.

Physical infrastructure creates network effects that no individual organisation can fully capture. High-speed internet, efficient transport links, quality office and laboratory space, and liveable neighbourhoods all contribute to ecosystem attractiveness. While private developers may provide some of these amenities, public coordination is necessary to ensure adequate provision and appropriate integration across different infrastructure systems.

Regulatory frameworks and institutional arrangements also require public provision. Intellectual property regimes, business registration systems, planning processes, and quality assurance mechanisms all affect the performance of the innovation ecosystem. Markets cannot provide institutional arrangements efficiently or effectively; they require active government interventions to ensure that they support, rather than hinder, innovative activity.

Economic Development Strategy and Resource Allocation

Successful innovation ecosystem development requires strategic choices regarding resource allocation and competitive positioning. Cities and regions cannot excel at everything and must focus their limited resources on areas where they can build a genuine competitive advantage. This necessitates a thorough examination of existing capabilities, market opportunities, and competitive dynamics. This should involve addressing the following issues and questions:

A candid and realistic assessment of local economic assets and capabilities. What research strengths exist in local universities? Which industries already demonstrate competitive performance? What infrastructure advantages can be leveraged? Where do skilled workers currently cluster? This analysis should identify potential building blocks for ecosystem development rather than simply listing aspirational goals.

Market analysis examines global trends and opportunities in specific sectors. Which industries are growing rapidly? Where are new technologies creating opportunities for disruption? What gaps exist in global value chains that local capabilities might fill? How are other regions positioning themselves in these markets? This analysis helps identify sectors where focused development efforts may succeed.

Competitive assessment evaluates how local capabilities compare with those of other regions pursuing similar strategies. What advantages can be sustained over time? Where do other locations have insurmountable head starts? Which activities require such a large scale that only a few global centres can succeed? This analysis helps avoid futile competition and identifies realistic positioning strategies.

Resource allocation decisions must balance the need for focus with the requirement for critical mass. Spreading investments too thinly reduces the likelihood of achieving a competitive advantage in any area. Concentrating resources too narrowly risks creating fragile monocultures that cannot adapt to changing circumstances. The optimal approach typically involves selecting two or three related sectors that can share infrastructure, talent, and knowledge while creating sufficient scale to attract global attention.

Policy Implications and Implementation Challenges

The economic logic of innovation ecosystems has important implications for how policymakers approach economic development. Traditional economic approaches that focus on individual firm incentives or isolated infrastructure projects are unlikely to generate a sustainable competitive advantage.

Instead, policies must address the systemic requirements for ecosystem development while recognising the long-term nature of building competitive clusters.

Investment in shared infrastructure and public goods requires sustained commitment *over decades* rather than just political cycles. Research facilities, educational institutions, and transport networks all require long-term funding commitments that extend beyond individual government terms. This creates political challenges when benefits may not be visible until after the current leaders have left office.

The current and ongoing development of Sydney's Metro demonstrates that long-term government commitments can be made. The Metro is a game-changer for Sydney's innovation ecosystem and the districts and precincts that constitute it.

Coordination between different levels of government and across policy domains becomes essential when ecosystem development spans municipal boundaries and multiple policy areas. Universities may be funded by state governments, while local governments control planning and infrastructure. Industry development may be a federal responsibility, while education remains at the State-level. Effective ecosystem development requires alignment across these different jurisdictions and agencies.

International connectivity increasingly determines ecosystem success as innovation becomes more globally distributed. Access to international talent, markets, and knowledge networks requires policy attention to immigration, trade, and international education. Regions, districts and precincts that remain insular or focus solely on domestic activity are unlikely to achieve global competitiveness in knowledge-intensive industries.

The economic principles underlying innovation ecosystems provide a foundation for more effective development strategies; however, translating these principles into practice requires sustained effort, strategic focus, and coordination across multiple stakeholders.

Success depends on understanding local assets, identifying realistic opportunities, and consistently committing resources over time to build a genuine competitive advantage.

7. Business: Commercial Logic and the Architecture of Investment

Business remains the essential mechanism for transforming ideas into economic value within innovation ecosystems. Without viable commercial pathways, even the most promising research discoveries and elegant urban planning remain unrealised potential. The business dimension of innovation ecosystems encompasses far more than individual companies pursuing profit maximisation. It is a complex web of commercial relationships, investment flows, and collaborative arrangements that determine whether innovation districts succeed or fail.

Understanding this dimension requires examining how business thinking has evolved, how commercial logic shapes ecosystem development, and why the architecture of investment demands careful integration between public objectives and private motivations. The most successful innovation ecosystems demonstrate that commercial viability and public benefit can be mutually reinforcing when properly structured.

The Evolution of Business Thinking in Innovation Ecosystems

The relationship between business and innovation ecosystems has undergone a dramatic transformation over the past three decades. During the 1990s, business strategy primarily focused on achieving competitive advantage through proprietary technologies, protected intellectual property, and market dominance. Companies viewed innovation as an internal capability that set them apart from competitors. Business plans emphasised capturing value through exclusivity and maintaining strategic advantages through information asymmetries.

This paradigm worked well in stable markets with clear boundaries between industries and predictable technology trajectories. However, the accelerating pace of technological change and the increasing complexity of innovation challenges have rendered this approach insufficient. Contemporary innovation requires capabilities that no single organisation can develop or maintain independently.

Today's business environment demands collaborative approaches to innovation. Companies recognise that their success increasingly depends on external partnerships, shared infrastructure, and collective capabilities. This shift is a fundamental change in how businesses perceive value creation and competitive positioning.

The emergence of platform business models exemplifies this transformation. Rather than competing solely through product differentiation, companies now compete by participating in ecosystems that generate value through network effects and collaborative innovation. For example, Amazon's cloud computing services support countless start-ups and established companies, creating value by enabling others rather than merely protecting internal capabilities.

The New Collaborative Paradigm

Modern businesses understand that innovation success requires access to diverse capabilities that extend far beyond their organisational boundaries. Companies work with universities and public research organisations for research, with government agencies for regulatory support and infrastructure investment, with financial institutions for capital, and with community organisations for social licence and local knowledge.

This interdependence is not a retreat from commercial principles; it involves securing a deeper understanding of how value creation occurs in complex environments.

Companies that effectively participate in innovation ecosystems maintain a strong commercial focus while recognising that their success depends on the health of the ecosystem.

In the pharmaceutical industry, for example, drug development requires basic research from universities, regulatory approval from government agencies, clinical trial capabilities from healthcare systems, and patient access through insurance mechanisms. No single company can control all these elements; however, success depends on coordinating them effectively.

Similarly, cleantech companies require collaboration among research institutions developing new materials, manufacturers scaling production processes, utilities integrating new technologies into existing infrastructure, and policymakers creating supportive regulatory frameworks. The most successful companies in these sectors participate in ecosystem development rather than simply pursuing individual market opportunities.

This collaborative approach extends to the processes of company formation and growth. Traditional venture capital models assumed that entrepreneurs could develop businesses independently with the appropriate financial support. Contemporary start-up ecosystems recognise that new companies require access to specialised facilities, mentoring networks, regulatory guidance, and market development support that cannot be provided through financial investment alone.

Programs like MIT REAP and the UK's Catapult Centres demonstrate how systematic ecosystem development can accelerate business formation and growth. These initiatives create infrastructure that reduces the cost and risk of innovation for individual companies while generating collective benefits that strengthen entire sectors.

Commercial Logic and Investment Architecture

Every innovation ecosystem requires substantial private investment to achieve its development goals. Developers must finance construction projects, companies must commit to leasing arrangements, investors must provide growth capital, and entrepreneurs must invest their time and resources in building new businesses. However, the commercial logic driving these decisions often conflicts with public policy objectives.

Private investors focus on generating returns that compensate for risks while maintaining sufficient liquidity to respond to changing market conditions. They evaluate opportunities based on projected cash flows, exit possibilities, and portfolio effects. Public agencies, conversely, pursue broader objectives, including social inclusion, environmental sustainability, and long-term economic development, that may not generate immediate financial returns.

These different objectives create tensions that can undermine ecosystem development if they are not addressed systematically. Private investors may favour developments that maximise short-term returns at the expense of community integration or environmental sustainability. Public agencies may impose requirements that reduce project viability without providing offsetting benefits or risk mitigation.

Successful innovation districts resolve these business tensions through structured agreements that align private incentives with public objectives. These arrangements recognise that sustainable ecosystem development requires both commercial viability and the achievement of public benefits.

The architecture of business investment in successful ecosystems typically includes several key elements:

- *Value capture mechanisms* that link private returns to the delivery of public benefits, ensuring that investors benefit from the community improvements they help create.
- *Performance-based approvals* that tie development permissions to specific outcomes, creating incentives for private developers to achieve public objectives.
- *Risk-sharing agreements* that distribute development risks between public and private partners based on their respective capabilities and interests.
- *Community benefit agreements* that establish clear commitments to affordable housing provision, local employment creation, and community facility development.

These agreements are structured as binding obligations rather than aspirational goals, creating certainty for both private investors and community stakeholders.

Mixed-use development strategies integrate commercial, residential, and community functions to create diverse revenue streams that support ecosystem sustainability. Successful districts develop balanced portfolios that can withstand market volatility while serving diverse community needs, rather than relying solely on premium office rents or high-end residential sales.

Alignment Strategies and Implementation

Creating effective alignment between commercial logic and public objectives requires systematic planning rather than ad hoc and unconnected negotiations. Successful ecosystems develop comprehensive frameworks that outline how public and private interests will be balanced throughout the development, delivery and operational phases.

These frameworks begin with a clear articulation of ecosystem objectives that encompass both commercial and public goals. Successful districts identify synergies between economic development and community benefit rather than seeing them as competing goals. For example, affordable housing requirements can be structured to support local workforce development, benefiting companies seeking skilled employees.

Financial structuring, such as tax deductions and offsets, land value capture, and development impact fees, creates mechanisms for private development to fund public infrastructure and community benefits. These tools distribute costs and benefits among multiple stakeholders while maintaining project viability.

Governance arrangements establish ongoing mechanisms for coordinating public and private decision-making throughout the ecosystem development process.

Rather than limiting interaction to the initial planning phases, successful districts establish institutional structures that facilitate continuous collaboration and problem-solving as conditions evolve.

Performance measurement systems track both commercial and public outcomes to ensure that ecosystem development achieves its intended objectives. These systems provide feedback that enables corrections and demonstrates accountability to both private investors and community stakeholders.

Common Pitfalls and Mitigation Strategies

Many innovation districts fail to achieve their potential because they treat private sector engagement as consultation rather than genuine partnership. This approach limits funding options, reduces private sector commitment, and often leaves public agencies bearing disproportionate financial and operational risks.

Consultation-based approaches typically involve public agencies developing plans independently and then seeking input from the private sector on implementation details. While this maintains public control over planning processes, it fails to leverage private sector expertise in project design and risk management. It reduces private sector ownership of outcomes and commitment to long-term success.

It is also possible to underestimate the complexity of aligning public and private interests. Many districts assume that shared economic development goals will automatically create alignment, without recognising that different stakeholders operate under fundamentally different constraints and incentives.

> *Genuine partnership requires involving private sector stakeholders in planning decisions while maintaining public accountability for ecosystem outcomes.*

This approach recognises that private sector participants bring essential capabilities, including market knowledge, financial resources, and operational expertise that public agencies cannot replicate.

Addressing this complexity requires a systematic analysis of stakeholder interests and careful design of governance and financial arrangements that create genuine mutual benefit. Success depends on structuring relationships that enable each stakeholder to achieve their core objectives while contributing to the collective success of the ecosystem.

Building Sustainable Commercial Foundations

It follows that the business dimension of innovation ecosystems succeeds when commercial logic and public objectives reinforce each other. This means recognising that sustainable ecosystem development depends on creating genuine value for all stakeholders rather than just redistributing costs and benefits among them.

Successful innovation districts demonstrate that commercial viability and public benefit represent complementary, rather than competing, objectives. When properly structured, private investment can deliver community benefits while generating appropriate returns, and public investment can strengthen commercial opportunities while achieving social and environmental goals.

This alignment creates positive feedback loops that strengthen ecosystem development. Commercial success generates resources that support ongoing public investment in infrastructure and community development. Public investment in education, transportation, and quality of life improvements attracts private investment and supports business growth.

Aligning requires a mature understanding of both commercial dynamics and public objectives, along with the institutional capacity to design and implement complex partnership arrangements. However, the results justify this investment by creating innovation ecosystems that can adapt to changing conditions while maintaining their core development objectives.

The business dimension ultimately determines whether innovation ecosystems can sustain themselves through changing economic conditions and evolving community needs. Districts that integrate commercial logic with public objectives create a foundation for long-term prosperity that benefits all stakeholders while advancing innovation and economic development goals.

8. Governance: Orchestrating Innovation Through Institutional Design

Governance possibly is the most complex and politically sensitive dimension of developing innovation ecosystems. Unlike the relatively straightforward challenges of physical placemaking or economic analysis, governance touches on fundamental questions of power, legitimacy, and institutional design that lie at the heart of democratic societies. In innovation districts, these challenges are amplified by the need to coordinate across multiple sectors, jurisdictions, and stakeholder groups, each with its own priorities, timelines, and accountability mechanisms.

That said, the understanding of innovation ecosystem governance has come a long way from its rudimentary origins in national innovation systems research. What began as a focus on network building and policy coordination has evolved into a recognition that successful ecosystems may require entirely new forms of institutional architecture, particularly institutional structures that can orchestrate collaboration without relying on traditional command-and-control mechanisms.

The evolution of thinking about governance reflects a deeper understanding that innovation ecosystems operate according to different logics than either markets, hierarchies, or networks, requiring approaches that are both more collaborative and complex, and new approaches in public administration.

From Coordination to Orchestration: Redefining Leadership in Innovation

The transition from coordination to orchestration is a fundamental shift in our understanding of leadership within innovation ecosystems. Traditional coordination assumes that different actors can be brought together through better communication, shared planning processes, and well-defined roles and responsibilities.

While these elements remain important, orchestration recognises that innovation ecosystems require a more nuanced form of leadership that creates conditions for collaboration rather than directing specific outcomes. Leadership must create an environment of trust and partnership, allowing parties to build long-term relationships

The orchestration metaphor, taken from Peter Drucker's notion of the jazz combo, captures something essential about how innovation ecosystems function.

> *Like jazz musicians, ecosystem participants must be skilled in their own domains while remaining responsive to the creative contributions of others.*

The orchestrator's role is to establish the rhythm, suggest themes, and create space for improvisation rather than dictating every note. This requires leaders who understand the technical aspects of innovation while possessing the interpersonal skills to build trust across diverse stakeholder groups.

This shift has major implications for how governance institutions are designed. Rather than focusing primarily on formal authority structures and reporting relationships, orchestration-based governance emphasises relationship-building, knowledge sharing, and the creation of platforms for ongoing collaboration. This includes developing new roles such as innovation brokers, system integrators, and boundary spanners who can work across traditional sectoral divides to identify opportunities and resolve conflicts.

The emergence of these bridging roles reflects a deeper recognition that innovation ecosystems require specialised forms of institutional knowledge. Understanding how universities, businesses, government agencies, and community organisations can work together effectively requires expertise that spans traditional professional boundaries. These roles cannot be filled simply by assigning additional responsibilities to existing staff; they require dedicated positions with appropriate training, resources, and institutional support.

Institutional Infrastructure: Building Capacity for Complex Coordination

It follows that the development of effective governance for innovation ecosystems requires substantial investment in *institutional* infrastructure. This goes well beyond establishing new organisations or reorganising existing agencies. *It involves creating new capabilities for strategic planning, stakeholder engagement, project management, and performance evaluation that can operate across traditional institutional boundaries.*

The complexity of this challenge becomes apparent when looking at the multiple levels of government that typically influence the development of innovation ecosystems. In the Australian federal system, the national government controls research funding and immigration policy, while state governments oversee higher education accountability and reporting, as well as major infrastructure investments. Multiple local governments are responsible for zoning, local transportation, and community development.

Each level of government operates according to its own political cycles, funding mechanisms, and accountability requirements, making coordinated action extraordinarily difficult.

In addition, the governance structures within these levels of government typically involve multiple stakeholder groups in decision-making roles. This extends beyond advisory committees and consultation processes to encompass shared authority over budget allocations, strategic planning, and project selection. Such arrangements reflect different power dynamics and decision-making processes that bring different perspectives and capacity for timely action.

This coordination and collaboration challenge might be addressed through the creation of dedicated management organisations with clear mandates, adequate resources, and genuine political support. These organisations could function as "institutional entrepreneurs", working to align incentives, resolve conflicts, and maintain momentum across multiple political, administrative, and economic cycles. They could also serve as repositories of institutional memory, ensuring that lessons learned from previous initiatives inform future decision-making.

Effective governance also requires well-developed approaches to performance evaluation that can assess success across multiple dimensions simultaneously. Traditional public sector evaluation focuses on efficiency and compliance, while private sector evaluation emphasises financial returns.

Innovation ecosystems require evaluation frameworks that can capture social, economic, and environmental outcomes while recognising that many benefits may not become apparent for years or even decades.

Managing Complexity: Integration Across Domains and Timelines

As discussed, one of the most significant governance challenges facing innovation ecosystems is the need to integrate activities across domains that operate according to distinct logics and timelines. Urban development projects may span decades, while economic development strategies typically focus on five- to ten-year horizons. Commercial investments often seek returns within three to five years, and political cycles may last only two to four years.

Creating governance systems that can maintain strategic coherence across these different governments and timeframes requires institutional capabilities that few cities possess. The problem is worse in Australia, which, apart from Brisbane and Canberra, does not have metropolitan governance systems, which contrasts starkly with the UK, North America and Europe.

This integration challenge is compounded by the fact that different professional domains approach their tasks in distinct ways:

- *Urban planners* focus on design quality, environmental sustainability, and community engagement.
- *Economic developers* emphasise job creation, tax revenue generation, and business attraction.
- *Commercial developers* prioritise financial returns and market responsiveness.
- *Political leaders* have their own priorities and must balance these considerations while remaining accountable to electoral constituencies that may have limited patience for long-term investment strategies.

As noted above, a successful innovation district will address these challenges through governance systems that create genuine shared accountability across different domains. This involves developing planning processes that consider multiple objectives from the initial stages of project development, rather than attempting to coordinate separate plans developed according to different logics. It also requires financial mechanisms that align incentives across different sectors, such as public-private partnerships that share both risks and rewards based on comprehensive performance measures.

The development of shared accountability requires institutional arrangements that go beyond simple coordination.

It involves governance structures in which different stakeholders have genuine decision-making authority and where success or failure in any domain affects all participants.

This might include governance boards with shared authority across multiple agencies, financing mechanisms that tie public investment to private sector performance, or community benefit agreements that link development approvals to specific social and economic outcomes.

The Political Economy of Stakeholder Management

The governance of innovation ecosystems operates within broader political and economic contexts that significantly influence what is possible and how different approaches are likely to be received by key stakeholders. Understanding these contexts is essential for developing governance strategies that can maintain political support and adapt to changing circumstances.

Political considerations include the need to demonstrate progress within electoral cycles, maintain support across different political parties and levels of government, and respond to changing public priorities. Economic considerations include the availability of public resources, private sector investment patterns, and broader economic conditions that may affect the viability of different development strategies.

Effective ecosystem governance requires effective stakeholder management that recognises the different motivations, constraints, and success criteria of various participants.

Universities may prioritise research excellence and student outcomes; businesses may focus on market access and talent acquisition; government agencies may emphasise job creation and tax revenue; and community organisations may prioritise affordable housing and local business development.

Creating governance systems that can accommodate these different priorities while maintaining strategic coherence requires institutional capabilities for negotiation, conflict resolution, and ongoing relationship management. This includes the development of formal and informal mechanisms for ongoing communication, the creation of shared platforms for problem-solving, and the establishment of processes for addressing conflicts as they arise.

Democratic Legitimacy and Inclusive Governance

Innovation ecosystems raise questions about democratic legitimacy and representation that cannot be resolved solely through technical or economic analysis. Decisions about ecosystem development affect land use, public investment priorities, and community character in ways that significantly impact existing residents and businesses. The governance of these systems must, therefore, address questions of power, representation, and accountability that go-to the heart of democratic governance.

Innovation districts that receive significant public investment, or risk contributing to gentrification, face acute challenges around democratic legitimacy. Securing *social licence* and community support is therefore essential. This requires developing clear narratives about why the innovation ecosystem is needed, how it will develop, and what benefits it will deliver.

When communities understand that the objective is building long-term resilience and prosperity for both people and planet, they are more likely to support the use of public resources for these initiatives.

Traditional democratic processes, focused on electoral representation and public consultation, may not ensure that ecosystem development serves the broader community's interests rather than simply benefiting already advantaged groups.

Addressing democratic challenges requires governance approaches that go beyond traditional consultation to include meaningful community participation in strategic decision-making.

This might involve community representation on governance boards, participatory budgeting processes that allow residents to influence investment priorities, or community benefit agreements that ensure local residents share in the economic benefits of ecosystem development.

The development of inclusive governance also requires attention to capacity building within community organisations. Meaningful participation in complex governance processes requires technical knowledge, time, and resources that many community groups may lack.

Future Directions: Adaptive Governance for Dynamic Systems

Innovation ecosystems are inherently dynamic, evolving in response to technological change, market conditions, and shifting social priorities. The governance systems that oversee them must therefore be designed for adaptation and learning rather than simply implementing predetermined strategies.

This requires governance approaches that emphasise experimentation, evaluation, and continuous improvement. Adaptive governance focuses on creating institutional capabilities for sensing change, experimenting with new approaches, and scaling successful innovations while discontinuing unsuccessful initiatives.

The development of adaptive governance capabilities requires investment in monitoring and evaluation systems that can provide ongoing feedback about ecosystem performance, stakeholder engagement processes that can identify emerging needs and opportunities, and institutional arrangements that enable rapid responses to changing circumstances.

> *As innovation ecosystems continue to evolve, their governance systems must develop the maturity to match their complexity while maintaining the democratic legitimacy and community engagement that ensure they serve broader social purposes rather than narrow economic interests.*

This is one of the most significant institutional challenges facing cities and regions that seek to develop successful innovation ecosystems in the coming decades.

9. Behind the Buzz: How Innovation Ecosystems Deliver Value

The idea of innovation ecosystems has become central to contemporary policy, city planning, and university strategy for good reason. When well-designed, these ecosystems generate genuine economic, industrial, and public value that far exceeds the sum of their individual parts. Yet the key challenge remains: How can we clearly demonstrate their value, and what evidence truly convinces both policymakers and the public?

This chapter aims to provide some answers. It explores how innovation ecosystems create the conditions for productivity growth, capability building, and enduring competitiveness. Rather than simply assembling assets, successful ecosystems mobilise talent, capital, and ideas across organisational boundaries, enabling knowledge to flow and new opportunities to emerge.

Drawing on international research and the case studies, the chapter argues that innovation ecosystems should be recognised as vital productivity infrastructure, not just vehicles for start-up creation or research commercialisation. Their real power lies in nurturing collaboration, accelerating diffusion, and building resilience across sectors and regions.

With policy and investment decisions under increasing scrutiny, this analysis sets out a clear and actionable framework for measuring ecosystem performance—ensuring that future strategies are grounded in real impact, not rhetoric, and deliver lasting public benefits.

From Concept to Consequence: What Do Innovation Ecosystems *Really* Do?

Innovation ecosystems are not merely clusters or innovation precincts. They are adaptive socio-technical systems involving universities, firms, entrepreneurs, intermediaries, government agencies, capital providers, and communities. These actors interact through knowledge flows, joint ventures, worker mobility, and institutional relationships.

Research and comparisons of ecosystems increasingly show that innovation ecosystems produce better results than merely adding up the contributions of each individual part. This happens not just because the actors are in the same place, but also because of the strong and meaningful connections they have with each other across different organisations, industries, and fields.

This is not due to the mere co-location of actors but rather to the density and quality of their interactions across organisational boundaries, sectors, and disciplines. These interactions foster what Metcalfe (1995) terms a "distributed innovation process,", wherein knowledge creation and diffusion are not linear but are shaped by relational dynamics, institutional trust, and co-evolution.

For example, research by Autio et al. (2018) demonstrates that ecosystems characterised by high levels of actor interdependence and institutional alignment outperform loosely coordinated networks in sustaining innovation over time. Similarly, the OECD's analysis of knowledge triangle policies (2013) finds that ecosystems with deliberate coordination across education, research, and innovation functions exhibit greater adaptive capacity and higher rates of knowledge diffusion into markets.

In short, ecosystem value lies in the ability of universities, firms, and government actors to interact productively, share risks, build collective intelligence, and co-invest in shared futures.

Productivity as the Central Policy Rationale

In the current policy milieu, the decisive question is whether innovation ecosystems demonstrably contribute to productivity growth, which is the principal channel through which innovation drives prosperity, competitiveness, and fiscal sustainability. Without such assurance, governments will hesitate to prioritise ecosystem development in constrained budgetary environments.

Well-designed ecosystems contribute to productivity growth through multiple mechanisms:

- *Reducing Innovation Friction*: Ecosystems decrease time to market by streamlining experimentation, regulatory navigation, and technology diffusion.
- *Boosting Firm Capability*: By providing firms, especially SMEs. with access to high-quality knowledge, talent, and collaborative R&D, ecosystems enhance managerial and operational efficiency (Bloom et al., 2019).

- *Accelerating Spillovers*: Proximity and dense interconnections enable tacit knowledge spillovers, often resulting in stronger productivity effects than R&D inputs alone (Audretsch & Feldman, 2004).
- *Facilitating Technological Convergence*: Ecosystems enable cross-sector breakthroughs, such as robotics in agriculture and genomics in manufacturing, which underpin future productivity frontiers.
- *Supporting Workforce Adaptation*: Integrated ecosystems can incorporate training, reskilling, and informal learning, which increase absorptive capacity and, in turn, labour productivity.
- *Aligning Capital and Coordination*: Ecosystems can consolidate fragmented public and private investments into coordinated capability building platforms, enhancing long-run economic efficiency.

This framing demands that innovation ecosystems be recognised as strategic productivity infrastructures, not just enablers of entrepreneurial activity or R&D commercialisation.

Of course, the spatial and market realities must be acknowledged. In Sydney, for example, many innovation-intensive firms, both emerging and established, have demonstrated a preference for locating outside the formal *Tech Central Innovation District*. Instead, organic agglomerations are emerging in areas such as Surry Hills, Pyrmont, Walsh Bay, Barangaroo, North Ryde, Liverpool, and other dispersed industrial and retail districts.

These locations may offer lower land and rental costs, better logistics, and established industry connections. This suggests that ecosystem value may reside less in *branded precincts* and more in *distributed networks of capability and interaction* across a city or a larger region.

Innovation policy must adapt accordingly, supporting the often-overlooked connective tissue of ecosystem health—metropolitan workforce pipelines, shared testing infrastructure, peer networks, trusted intermediaries, and place-based coordination platforms. They should certainly not be seen as flagship property developments.

The Value Innovation Ecosystems Create

The economic value created by innovation ecosystems can be captured under three headings.

Economic Value: Innovation ecosystems drive firm creation, job growth, investment attraction, and export expansion. More importantly, they provide the *adaptive capacity* to maintain competitiveness in fast-changing environments. Eindhoven's transition from a single-firm dependency on Philips to a globally recognised deep tech ecosystem illustrates how ecosystems can renew economic capacity after industrial declines (Trippl et al., 2015).

Industrial Value: Ecosystems support the scaling and upgrading of industry capabilities through knowledge spillovers, workforce mobility, and shared R&D. Germany's Fraunhofer model demonstrates how embedded applied research institutions can diffuse advanced technologies into SMEs, thereby increasing sector-wide productivity (Eickelpasch & Fritsch, 2005). Similarly, Canada's MaRS district facilitates sector-wide platform building in health, cleantech, and enterprise technology.

These dynamics underpin the creation of industrial commons—shared talent pools, research platforms, and institutional capabilities that drive long-term competitiveness (Pisano & Shih, 2009).

Public and Societal Value: Ecosystems also generate public goods—green technologies, inclusive education models, digital public health tools, and culturally enriching activities. Barcelona's 22@ and Montréal's Quartier de l'Innovation illustrate how civic ecosystems can enhance urban quality of life while fuelling innovation. In Australia, this dimension is less developed. Strategies often overemphasise commercialisation metrics while undervaluing civic engagement, inclusive innovation, and social outcomes.

Core Attributes of a Well-Functioning Innovation Ecosystem

If innovation ecosystems are to be taken seriously in policy and budget processes, their advocates must go beyond rhetoric and clearly demonstrate the presence and effectiveness of specific attributes. A categorisation of attributions, rationale, evidence, indicators, and metrics might be constructed as follows.

Attribute	Rationale, Explanation	Evidence, Indicators, Metrics
1. Frequent and Diverse Knowledge Exchange	Actors share insights, data, and experiences across organisational, disciplinary, and sectoral boundaries—enabling new ideas to emerge and spread.	Evidence includes co-publications, co-patents, and cross-sector projects.
2. Low Friction for Collaboration and Experimentation	The ecosystem lowers barriers—legal, financial, and cultural—to partnerships and pilot projects.	This is demonstrated by the speed and frequency of collaborative ventures, cross-licensing agreements, and regulatory testing sandboxes.
3. Shared Infrastructure and Intermediary Platforms	Physical and institutional resources (labs, testbeds, accelerators, data platforms) are accessible and actively utilised by multiple actors.	Evidence includes utilisation rates, case examples, and cross-institutional service agreements
4. Talent Mobility and Circulation	People move between research, business, government, and intermediaries, transferring know-how and building relational capital.	Trackable through secondments, alumni networks, job placement flows, and joint appointments.
5. Supportive Intermediaries and Brokers	Entities such as accelerators, translational institutes, and innovation hubs actively connect supply and demand, de-risk innovation, and build trust.	Effectiveness is demonstrated through user satisfaction, funding leverage, and deal conversion rates.
6. Institutional Trust and Strategic Alignment:	Organisations trust one another enough to share risks, data, and decision-making. Strategic goals align across institutions.	Measurable through joint strategies, partnership longevity, or stakeholder surveys.
7. Dynamic Capability Building	Actors and institutions enhance their capacity to innovate, scale, and adapt over time.	Evidence of capability lifts includes improved innovation maturity, digital adoption rates, and upskilling programs.
8. Relational Density and Informal Networks	Informal ties complement formal partnerships, enabling rapid problem-solving and knowledge flow.	Tracked through social network analysis, collaboration patterns, or stakeholder mapping exercises
9. Coordinated Governance and Feedback Loops	Governance bodies coordinate actions, resolve tensions, and adapt strategies based on feedback and performance.	Indicators include adaptive policy frameworks, meeting transparency, and changes made in response to evaluations.

Attribute	Rationale, Explanation	Evidence, Indicators, Metrics
10. Inclusive Participation and Social Embeddedness	Broad participation across firm sizes, demographics, and communities strengthens legitimacy and diffusion.	Demonstrated through diversity metrics, engagement with underserved groups, and community partnerships
11. Mission-Driven or Purpose-Led Activities	In mature ecosystems, actors align around societal challenges or strategic missions (for example, decarbonisation and food security).	Demonstrable via mission statements, challenge-led funding, and cross-sector alliances tackling public problems
12. Spatial and Sectoral Coherence	The ecosystem displays coherence either through spatial proximity (precincts, campuses) or sectoral specialisation (for example, healthtech, clean energy).	Measurable through value chain mapping, locational clustering, or sectoral intensity.
13. International Connectivity	Well-functioning ecosystems tap into global knowledge and capital flows.	Demonstrated through international research collaboration, foreign direct investment, export data, and participation in global innovation networks.
14. Anchored Public and Private Investment	Sustained co-investment from public, private, and philanthropic sources demonstrates confidence and commitment.	Evidence includes capital commitments, matched funding, and long-term investment strategies.
15. Observable Economic and Innovation Outcomes	*Visible economic outputs* include new firms, increased exports, new intellectual property, and productivity gains. *R&D absorption*: capability upgrading, supply chain anchoring *Public Value*: Emissions reductions, health and education access, cultural participation.	Tracked through standard metrics and case examples

These attributes can be used to develop ecosystem health diagnostics, baseline assessments, longitudinal performance tracking, and benchmarking frameworks.

Conclusion: Towards Strategic Productivity Ecosystems

The policy case for innovation ecosystems rests heavily on their capacity to enhance long-term productivity, industrial renewal, and civic wellbeing, rather than simply creating start-ups or generating patents.

Their true value lies in what they make possible: upgraded capabilities, resilient industries, adaptive cities, and public value. For this value to be visible and sustained, governments must invest in strategic patience, systems-based measurement, and distributed, place-aware governance models. Innovation ecosystems should not be viewed as discretionary investments for economic boom times; they are essential infrastructure for national resilience and productivity renewal.

This issue is considered further in Part II, which follows.

PART II: CREATING VALUE IN INNOVATION ECOSYSTEMS

Part II turns from theory and framing to the practical question of how innovation ecosystems actually generate value. It explores the processes through which assets, knowledge, capital, and relationships are mobilised to deliver measurable outcomes for economies, institutions, and communities. The chapters demonstrate that successful ecosystems do more than co-locate research, infrastructure, and businesses: they build cumulative advantage by aligning physical, financial, intellectual, and social capital.

The discussion begins with how ecosystems emerge from existing assets and capabilities, showing that place-based innovation is rarely created from scratch. Particular attention is given to Australia's health, medical, and biomedical precincts, which have achieved international recognition through strong anchors and translational research capacity. Lessons from this sector are used to identify what manufacturing and technology ecosystems might adopt to improve scaling performance.

Later chapters address the importance of collaboration, coordination, and governance as drivers of value creation, culminating in a perspective on how Australia might craft an innovation-led industrial strategy. Throughout Part II, the emphasis is on practical insights: what success looks like, how it is achieved, and what policy, institutional, and business choices sustain performance over time. Together, these chapters provide a roadmap for turning promising ideas and assets into enduring economic and social outcomes.

1. Building on Assets: How Innovation Ecosystems Emerge and Grow

Innovation ecosystems emerge through remarkably diverse pathways that challenge simplistic assumptions about building on seemingly existing strengths. While successful economic and industry clusters often leverage research organisations and established corporations as anchors, ecosystem development frequently involves transformation, adaptive reuse, visionary leadership, and the recognition of latent potential, rather than a straightforward continuation of existing advantages.

The observations in this chapter draw heavily on the case studies prepared as part of the project and referenced in Part VII.

Pathways to Ecosystem Development

Adaptive Reuse and Urban Transformation represent perhaps the most common pathway for contemporary innovation districts. Many celebrated ecosystems occupy former industrial sites, abandoned docklands, or declining commercial areas that offered affordable space and urban regeneration opportunities rather than obvious sectoral advantages.

Barcelona's 22@ district, with its textile and manufacturing heritage, left behind interesting buildings and connectivity; however, the economic activities that created these assets had largely disappeared. The innovation ecosystem that emerged leveraged the physical legacy while creating entirely new activities around digital media and technology.

Similarly, London's King's Cross and Dublin's Docklands have transformed former railway lands into a major knowledge district by attracting universities and technology companies looking for modern facilities in an accessible location.

Response to Industrial Decline constitutes a second pathway, where ecosystems emerge as deliberate responses to economic disruption. The Pittsburgh and Ruhr Valley Innovation Districts transitioned from coal and steel to technology and environmental innovation, requiring fundamental changes while selectively preserving valuable elements, such as engineering expertise and manufacturing capabilities. This pathway reveals both opportunities and challenges - declining industries create assets like skilled workers and infrastructure, but also potential liabilities including economic lock-in and resistance to change.

Visionary Leadership and Institution Building involve catalysts who recognise potential where others see problems. Singapore's transformation into an innovation hub reflects the strategic choices made by government leaders who positioned the city-state at the intersection of global trends in biotechnology, financial technology, and advanced manufacturing. Similarly, university-led ecosystems emerge when academic leaders recognise opportunities to translate research excellence into economic development, as demonstrated by MIT in Kendall Square, Stanford's role in Silicon Valley, and Cambridge University's contribution to the Cambridge technology cluster.

Property-Led Development and Government Partnership involve commercial developers recognising market opportunities for high-end commercial spaces designed to attract knowledge workers and innovative companies, working closely with governments seeking economic development. This pathway typically features master planned developments that create comprehensive physical infrastructure and urban environments intended to attract innovative activity. Developers provide modern facilities, mixed-use environments, and attractive amenities, while governments contribute land, infrastructure investment, planning facilitation, and often direct funding.

Examples include many newer innovation districts that emerged as real estate development projects with innovation themes, creating a physical and social environment designed to attract companies and talent. The challenge lies in ensuring property development generates genuine innovation outcomes rather than simply attractive commercial space, hosting technology companies.

Opportunistic Connections and Serendipitous Development acknowledges that many ecosystems emerge through somewhat unexpected interactions rather than deliberate planning. Stockholm's gaming ecosystem developed around a combination of technical education, digital culture, successful early companies like Mojang, and supportive government policies. While each element was important, their combination was not obviously predictable, and the ecosystem's character emerged through interactions between these elements.

Research Organisations and Corporate Anchors consistently serve as important drivers of ecosystem growth. Universities attract talent, conduct multidisciplinary research, maintain long-term location commitments, and often possess substantial property holdings. However, their impact depends on the willingness to engage with external partners rather than pursuing academic excellence in isolation.

Toronto's MaRS Discovery District demonstrates how research organisations can anchor development while enabling transformation. Building on the University of Toronto's research strengths and hospital networks, the ecosystem has expanded to encompass financial technology, artificial intelligence, and digital services well beyond the original health focus. The research base provides credibility and expertise while creating space for new activities and unexpected connections.

Established corporations anchor ecosystems by providing market knowledge, technical expertise, and commercial credibility. Their most valuable contributions often involve sharing knowledge and creating collaboration platforms while remaining open to learning from ecosystem partners.

Sectoral Specialisations

Analysis reveals seven distinct specialisation patterns that can emerge through any development pathway, reflecting intersections of opportunity, capability, and strategic choice.

Health and Biomedical Innovation appear most frequently across global ecosystems - from Boston's Kendall Square to Sydney's Camperdown precinct. This reflects the requirement for close collaboration between medical schools, category 4 hospitals, medical research institutes, pharmaceutical companies, and technology developers in medical innovation. The specialisation encompasses medical devices, biotechnology, pharmaceutical development, and emerging digital health convergence. Success often depends on navigating complex regulatory environments and building relationships with healthcare purchasers.

Advanced Manufacturing and Deep Technology typically emerge in regions with established industrial capabilities transitioning towards higher-value activities. Brainport Eindhoven exemplifies this pattern, building upon Philips' engineering heritage for semiconductor technology and precision manufacturing. The specialisation involves intersections of traditional manufacturing excellence with digital technologies, advanced materials, and sophisticated engineering systems.

Defence and Aerospace Technology develops around government facilities, military installations, or major manufacturers. Adelaide's Defence Precinct leverages submarine construction programs, the Henderson Precinct in Perth builds on naval shipbuilding, and Seattle's aerospace ecosystem builds on Boeing's presence. These ecosystems involve complex interactions between government agencies, prime contractors, and specialised suppliers.

Financial Technology and Services Innovation develop in established financial centres, leveraging deep domain expertise. The fintech ecosystems in London, Singapore and Sydney have built on their global financial centre status while nurturing innovation in payment systems and regulatory technology. The specialisation requires the intersection of financial domain knowledge with digital capabilities and regulatory innovation.

Creative and Digital Industries reflect the convergence of cultural creativity with digital technologies. Barcelona's 22@ district and Stockholm's gaming ecosystem demonstrate how technical capabilities can intersect with creative culture. This specialisation encompasses digital content creation, design, architecture with digital tools, and digital marketing, combining creativity with data analytics.

Clean Technology and Sustainability address environmental challenges through technological innovation. Copenhagen's Ørestad builds upon wind energy leadership, while Australian regional centres develop renewable energy capabilities. The specialisation encompasses renewable energy technologies, environmental technologies, and circular economy innovations.

Agricultural Technology and Food Innovation emerge in regions with strong primary production capabilities. Wagga Wagga-Riverina demonstrates how premium agriculture intersects with technology innovation. The specialisation combines traditional agricultural expertise with precision agriculture, food technology, and sustainable farming practices.

The most dynamic ecosystems often combine multiple specialisations, creating convergence innovation opportunities at sectoral boundaries. Toronto's MaRS District brings together health innovation, financial technology, and artificial intelligence, enabling solutions that span traditional industry boundaries. This convergence recognises that contemporary challenges require multi-domain expertise while creating unexpected innovation opportunities.

Summary of Observations from the Case Studies

Specialisations

From the 96 case studies covered in the study, and briefly profiled in the Appendix (on page 490, and observations drawing on the discussion in this chapter, two broad categories of specialisation are revealed:

Sectoral Concentration. Advanced manufacturing, defence/aerospace, and health/biomedical account for more than 60 per cent of the districts. Their dominance reflects the digital transformation of traditional industries and the convergence of life sciences with enabling technologies such as digital health, AI, and quantum. University–

hospital partnerships seem to be especially effective in creating biomedical clusters with global recognition.

Geographic Specialisation. Information from the 96 case studies reveals the following general observations:

- *Australia* shows a strength in health/biomedical precincts, for example, Parkville, Clayton (Melbourne) Camperdown, Randwick, Liverpool/Westmead (Sydney), Boggo Road, Herston (Brisbane), Floreat (Perth), Adelaide BioMed, and in defence/aerospace (e.g. Henderson, Lot 14, Bradfield Aerotropolis, Fishermans Bend).
- *Europe* is strong in advanced manufacturing technologies, clean technology and sustainability-focused districts, strongly influenced by policy priorities (for example, Hammarby Sjöstad, Ruhr, Aspern Seestadt).
- *North America* is very strong in defence and aerospace, entrepreneurial software, AI, and venture-driven models (e.g. Kendall Square, Silicon Valley, Pittsburgh Robotics Corridor).
- The *Asia-Pacific* shows success in creative/digital clusters (Tokyo's Bit Valley, Pangyo Techno Valley, Shenzhen) and in large-scale planned hubs (Jurong, One North, Zhongguancun, HITEC Hyderabad).
- *Latin America and Africa* are represented by emerging districts (Medellín, Porto Digital, Lagos, Kigali, Konza Technopolis) that highlight the diffusion of the model but also its dependence on governance and investment maturity.

The findings reinforce the discussion that successful specialisation stems from building on institutional anchors and sectoral strengths, rather than creating competencies from scratch.

Development Pathways

The distribution of ecosystem development and growth among the case studies reveals:

- *Research organisation and corporate anchors* dominate with 35 districts, particularly strong in Australia (22 districts), where university-industry collaboration is a well-established policy. Health and biomedical precincts cluster heavily in this category, reflecting the

natural partnership between university health, medical and clinical schools, affiliated medical research institutes and hospitals.
- *Property-led development and Government partnership* showcase in 22 districts, concentrated in Asia-Pacific planned developments and some European sustainable city projects. This pathway appears most common where governments actively direct spatial and economic development.
- *Adaptive reuse and urban transformation* are prevalent in 11 districts, notably in European contexts, where industrial heritage provides opportunities for creative repurposing. Australian examples often involve former institutional sites (prisons, hospitals).
- *Response to industrial decline* encompasses 14 districts, distributed across regions experiencing manufacturing transition. These represent deliberate strategies to replace declining industries with knowledge-based alternatives.
- *Opportunistic/serendipitous hubs (14 cases).* Exemplified by Kendall Square, Cambridge Cluster, Silicon Valley, Cyber Valley (Stuttgart-Tübingen) and Bellevue, which emerged through cumulative advantages in talent, culture, urban transport, and investment emerging into public-private research partnerships and institutional collaborations.

The dominance of research-anchored approaches suggests that institutional assets are critical for innovation ecosystem development.

However, the diversity of successful pathways suggests that context-specific approaches, which leverage existing strengths and address particular challenges, are more effective than applying universal models.

Sectoral Focus

The sectoral analysis reveals a number of significant strategic insights.

- *Dominant sectors*: Advanced manufacturing and deep technology, combined with health and biomedical, account for 61% of all innovation districts globally. This dominance reflects two major trends: the digital transformation of traditional manufacturing industries and the increasing convergence of life sciences with technology platforms.
- *Geographic strengths:* Australia demonstrates particular excellence in health and biomedical innovation (14 districts), leveraging strong

university-hospital partnerships and integrated health systems. Europe leads the development of clean technology, reflecting policy priorities centred on sustainability and circular economy principles.
- *Underrepresented opportunities*: Agricultural technology is represented in only three districts globally (all in Australia), suggesting significant untapped potential given the growing food security challenges and the advancement of precision agriculture technologies. Similarly, defence and aerospace technology represent just four districts despite substantial government investments in these sectors.
- *Cross-border potential:* The single cross-border example (Medicon Valley spanning Denmark-Sweden) suggests opportunities for international collaboration, particularly relevant for smaller nations or regions seeking to achieve critical mass in specialist sectors.
- *Capital convergence applications*: The sectoral concentration aligns with the capital convergence framework observations, discussed in the next chapter, as health/biomedical logically brings together knowledge capital (universities), physical capital (hospitals), and social capital (clinical networks). Meanwhile, advanced manufacturing leverages existing industrial infrastructure and corporate research capabilities.

The implications of the analysis of sector focus suggest that regions, districts, and precincts should focus on sectors where they already possess existing anchor institutions and capabilities, rather than attempting to build entirely new sectoral competencies.

The success of university-hospital partnerships in health suggests similar anchor-based approaches could work in other sectors with appropriate institutional foundations.

Policy Implications

These perspectives suggest that policymakers and ecosystem builders should recognise different ecosystem development pathways rather than assume natural emergence around obvious strengths:

- The adaptive reuse pathway requires attention to underutilised urban assets with repurposing potential
- Industrial transformation demands substantial interventions, including retraining programs and infrastructure investments
- Visionary leadership suggests supporting catalytic individuals and organisations rather than directing development from above

- Opportunistic pathways imply creating conditions for serendipitous connections while remaining open to unexpected outcomes

Across all pathways, patient capital and sustained commitment remain essential. Building relationships, capabilities, and collaborative infrastructure requires time and consistent effort.

The most successful approaches appear to be a focus on creating enabling conditions rather than directing specific outcomes, recognising that ecosystems emerge through complex interactions and that valuable innovations often occur at boundaries between sectors and disciplines.

The policy challenge involves supporting this complexity while maintaining a focus on outcomes that benefit the broader economy and community.

Success depends on understanding how different types of assets—whether physical, institutional, or human—can be activated and connected through infrastructure investments and collaborative frameworks.

2. Capital Convergence: Knowledge, Physical, Financial and Social

Innovation ecosystems thrive when various forms of capital collaborate synergistically, rather than operating in isolation. Knowledge, physical, financial, and social capital each have unique characteristics and institutional logics; however, their true potential is realised through strategic convergence, which enhances collective capabilities beyond those of individual assets.

This convergence, however, does not necessarily occur organically. Universities prioritise academic metrics, corporations emphasise competitive advantage, and government agencies focus on risk mitigation. Simultaneously, elevated property costs, misaligned funding cycles, and uniform networks can disrupt ecosystems, even in instances of apparent proximity.

The challenge for policymakers and ecosystem builders is to facilitate convergence while preserving the constructive tension between different institutional logics.

This necessitates an understanding of how spatial design, governance structures, and relationship-building frameworks can facilitate the productive intersection of these forms of capital.

This chapter explores how successful innovation districts can achieve this equilibrium by nurturing dynamic systems in which diverse assets converge to generate sustained innovation and economic transformation.

Knowledge Capital as the Foundation

Knowledge capital is the foundational element of innovation ecosystems. It comprises both *explicit knowledge*, such as research outputs, patents, and technical expertise, and *tacit knowledge*, including skills, experience, and institutional wisdom. The value of knowledge capital does not come from its existence in books, papers and patents: value comes from its circulation and recombination across institutional boundaries.

The circulation of knowledge may require the adoption of enabling mechanisms that amplify spatial proximity. In successful districts, knowledge capital is integrated through collaborative infrastructures that facilitate the exchange of academic rigour, commercial application, and public policy requirements.

It is the case, however, that knowledge capital can encounter substantial obstacles to convergence. The presence of siloed institutional logics, where universities prioritise publication metrics, corporations concentrate on proprietary advantages, and government agencies focus on risk mitigation, results in fragmentation, even within densely populated environments. Intellectual property regimes and confidentiality considerations contribute to isolation, thereby hindering the knowledge spillovers that are crucial for fostering innovation.

Resolution of these barriers necessitates building "trust infrastructure": neutral spaces, boundary-spanning roles, and the long-term relationship-building initiatives that facilitate secure knowledge exchange.

Trust infrastructure requires distinguishing between areas designated for open collaboration and those intended for the protection of intellectual property, thereby enabling diverse institutions to engage effectively while upholding their fundamental missions.

Physical Capital as Enabling Infrastructure

Physical capital within innovation ecosystems encompasses buildings and equipment, spatial arrangements, technological infrastructure, and urban design elements that facilitate or impede innovative activity. Physical capital derives its strength from its ability to enhance interaction, accessibility, and the "collision density" essential for fostering serendipitous encounters.

A feature of physical capital is that it is shared among research organisations and industry, which provides for scalability and multiple uses. Organisations like CSIRO have invested heavily in shared infrastructure, as have universities through public-private partnerships between universities, businesses, and government.

Districts such as 22@Barcelona exemplify the convergence of physical capital with other forms of capital through mixed-use development, enhanced walkability, and integrated transportation systems. These spatial configurations promote socioeconomic diversity and facilitate a range of interactions, encompassing both formal collaborations and informal knowledge exchanges.

The physical environment serves as a platform that enhances the value of knowledge, social, and financial capital by facilitating opportunities for their intersection.

The effectiveness of physical capital may be diminished when overshadowed by speculative property development or inflexible zoning practices that separate different functions. Elevated real estate prices, automobile-oriented designs, and rigid spatial configurations can displace the diverse and creative individuals essential for fostering innovation, while simultaneously restricting opportunities for interaction.

Strategic approaches to physical capital convergence encompass participatory planning models that involve a wide range of stakeholders, affordability mandates designed to sustain ecosystem diversity, adaptive reuse of existing structures that uphold community character, and the development of inclusive public spaces that cater to both innovation activities and the broader needs of the community.

Financial Capital as the Catalyst

Financial capital functions as both a catalyst and a determinant within innovation ecosystems, influencing the allocation of resources for idea development while also shaping the risk tolerance and timeframes associated with innovative endeavours. The alignment of incentive structures across institutional boundaries heavily influences the integration of financial capital with other forms of capital.

Exemplary institutions such as the Fraunhofer Institutes illustrate how blended funding strategies can effectively align academic excellence with commercial relevance and objectives for the public and collective good. These hybrid financing mechanisms encourage genuine collaboration, as opposed to transactional partnerships that arise solely to meet funding compliance requirements. Innovation vouchers and risk-sharing mechanisms facilitate access to knowledge capital for smaller enterprises, which can mitigate barriers to collaboration.

Financial capital may contribute to ecosystem *divergence* flowing from the misalignment of metrics and short-term funding cycles. When academic institutions are incentivised by publication and citation metrics, corporations prioritise quarterly returns, and government agencies emphasise risk mitigation, the potential for collaboration remains untapped.

The convergence of financial capital with other capital components necessitates the creation of funding frameworks that incentivise authentic cross-sector collaboration, encourage patient capital for long-term innovation initiatives, and ensure the availability of diverse funding sources to support various types and stages of innovative activities.

Financial convergence encompasses public investment in shared infrastructure, private venture capital for high growth enterprises, and hybrid mechanisms that integrate diverse institutional logics.

Social Capital as the Binding Force

Social capital, comprised of networks of relationships, trust, and shared norms that facilitate collective action, emerges as one of the most essential yet delicate components in the convergence of innovation ecosystems. Dense, diverse, and trust-based networks function as the *circulatory system* that disseminates knowledge, opportunities, and resources throughout the ecosystem.

In districts like Kendall Square, social capital is reflected in cross-sectoral mobility, inclusive event curation, and network facilitators who actively connect diverse communities. These relationships establish informal channels that facilitate the emergence of formal collaborations, thereby promoting rapid information flow and the identification of opportunities. Social capital creates the resilience necessary for ecosystems to adapt to external shocks and changing circumstances.

Nevertheless, social capital can also contribute to divergence by creating homophilous networks that exclude diverse perspectives, creating demographic barriers that restrict participation, and limiting cross-sectoral mobility, thereby reinforcing institutional silos. Certain factors can delineate how historical mistrust among sectors, cultural differences in communication styles, and unexamined assumptions regarding value creation may establish unseen barriers to collaboration.

The development of convergent social capital may necessitate a deliberate investment in relationship-building infrastructure. This includes the implementation of structured mobility programs across various sectors, the establishment of community building platforms that cater to a diverse range of participants, and the creation of inclusive governance mechanisms that empower various stakeholder groups.

Social capital cannot be rapidly produced; it must be developed through ongoing engagement and the creation of mutual value.

The Dynamics of Capital Convergence

Capital convergence is a dynamic process that necessitates ongoing management and adaptation. Effective innovation ecosystems maintain a constructive balance between alignment and diversity, structure and flexibility, as well as strategic direction and emergent creativity. They acknowledge that excessive convergence may result in homogenisation and diminished potential for innovation, whereas inadequate convergence can lead to fragmentation and missed opportunities.

Governance structures serve as essential intermediaries in the convergence of capital. Independent, multi-stakeholder governance models that effectively balance strategic direction with operational flexibility are capable of managing the intricate interactions among various forms of capital. These governance mechanisms must be sufficiently robust to address the varying logics, time horizons, and success metrics of diverse stakeholders, all while prioritising collective innovation outcomes.

Furthermore, capital convergence must be contextually aware, acknowledging the historical, cultural, and institutional factors that influence the trajectory of each ecosystem. Path-dependent influences, such as relational capital derived from previous collaborations and the adaptive reuse of industrial heritage, can serve to establish trust and distinctiveness or present obstacles to contemporary integration.

Sustainable innovation ecosystems achieve capital convergence by supporting the dynamic equilibria that transform tensions into sources of creative energy, rather than aiming to eliminate the tensions among various forms of capital.

This, in turn, requires a comprehensive understanding of the interactions among knowledge, physical, financial, and social capital, along with adaptive governance strategies that can respond to emerging patterns while ensuring strategic coherence.

In this context, innovation ecosystems are characterised as complex adaptive systems in which the convergence of various forms of capital cultivates collective capabilities that surpass the aggregate value of individual assets, thereby fostering conditions for sustained innovation and economic transformation that benefit the broader public interest.

Conclusion

The integration of knowledge, with physical, financial, and social capital, is a significant opportunity and a complex challenge for the development of innovation ecosystems. The discussion in this chapter suggests that successful convergence cannot be achieved solely through spatial proximity; it requires the intentional coordination of governance frameworks, funding mechanisms, and relationship-building infrastructures that connect institutional boundaries while respecting diverse organisational logics.

The evidence also suggests that sustainable ecosystems preserve a dynamic equilibrium between alignment and diversity, thereby preventing both homogenisation, which hinders creativity and fragmentation, which diminishes collective potential. Governance approaches must be attuned to historical and cultural contexts while remaining adaptable to emerging opportunities and external pressures.

For policymakers and ecosystem professionals, the chapter points to intervention strategies that address multiple forms of capital concurrently, rather than treating them as isolated policy areas. Establishing neutral spaces for knowledge exchange, designing inclusive physical environments, implementing blended funding models, and investing in cross-sectoral relationship-building are essential interconnected components of an effective ecosystem strategy.

Ultimately, effective innovation ecosystems will cultivate collective capabilities that advance broader public interests, thereby transforming urban and regional economies and addressing economic and societal challenges through continuous collaborative innovation.

---

3. Australia's Innovation Success Story: Health, Medical, and Biomedical Innovation Precincts

Australia's health, medical, and biomedical innovation ecosystem is one of the nation's most significant innovation success stories. It combines world-class research, strong institutional frameworks, and a distinctive regulatory environment to create clusters that are internationally competitive. These precincts are anchored by Category 4 teaching hospitals, universities with strong medical schools, independent research institutes, and pharmaceutical and medical technology companies.

The clustering effect is powerful. Proximity enables collaboration, accelerates translational research, and provides fertile ground for the emergence of start-ups, particularly in medical technology. However, while the formation of med-tech start-ups is prolific, scaling them into large, globally competitive companies remains limited.

This imbalance between start-up vitality and scale-up fragility is one of the most pressing issues for the next stage of innovation ecosystem development.

This chapter explores the origins, strengths, and current trajectory of Australia's biomedical innovation ecosystem, analyses the role of clustering in driving innovation, and outlines the scale-up challenge that must be addressed to sustain momentum.

The Geography of Biomedical Clustering

Australia's leading biomedical innovation precincts cluster around teaching hospitals, medical schools, and associated research institutes, which provide the intellectual and clinical backbone of the ecosystem. Industry players, from pharmaceutical companies to emerging med-tech firms, gravitate towards these hubs to access expertise, facilities, and patient populations. Notable examples include:

- *Parkville Biomedical Precinct, Melbourne:* one of the largest biomedical clusters in the Southern Hemisphere, integrates the University of Melbourne, the Royal Melbourne Hospital, the Walter and Eliza Hall Institute, and CSL.

- *Monash-Clayton Health Precinct:* built around Monash University's Faculty of Medicine and the Monash Medical Centre.
- *Camperdown (Tech Central, Sydney):* anchored by the Royal Prince Alfred Hospital, The University of Sydney's medical school, and several medical research institutes, including the Centenary Institute and the Brain and Mind Centre.
- *Randwick Health and Innovation Precinct:* centred on the Prince of Wales Hospital, UNSW Medicine, and associated research institutes.
- *Liverpool and Westmead, Western Sydney:* large health districts with growing translational capacity linked to both Western Sydney University and The University of Sydney.
- *Herston Health Precinct, Brisbane:* connecting the Royal Brisbane and Women's Hospital, QIMR Berghofer Medical Research Institute, University of Queensland Centre for Clinical Research, Herston Imaging Research Facility, and Herston Biofabrication Institute.
- *Floreat Biomedical Precinct, Perth:* emerging around the Harry Perkins Institute of Medical Research, the University of Western Australia, and Perth's teaching hospitals.
- *Adelaide BioMed City:* a major cluster centred around the new Royal Adelaide Hospital, the University of Adelaide, and the South Australian Health and Medical Research Institute.
- *Gold Coast Health and Knowledge Precinct:* a collaboration between Griffith University, Gold Coast University Hospital, and translational research infrastructure.
- *TropiQ Health & Knowledge Precinct, Townsville:* a Tropical medicine and health research cluster anchored by James Cook University focusing on Indigenous health and tropical disease innovation.

These precincts provide density and interaction.

Co-location allows clinicians, researchers, entrepreneurs, and students to work side by side, sharing facilities such as biobanks, imaging equipment, and clinical trial centres.

International evidence consistently demonstrates that such clustering enhances research productivity, accelerates translation, and increases the likelihood of successful commercialisation.

Building the Ecosystem: Institutional Foundations

Australia's biomedical strength rests on a century of institutional development. The Walter and Eliza Hall Institute (established in 1915),

CSL (1916), and the Garvan Institute (1963) have established reputations for research excellence that garnered global recognition. These institutions produced Nobel Prize winners, breakthrough discoveries, and public health innovations that laid the foundation for today's ecosystem.

Cochlear (formed in 1981) emerged from university and hospital research and became a global pioneer in hearing implants. Pro Medicus (1983) developed software to digitise medical practice administration and later radiology imaging workflow and has scaled internationally. ResMed (1989) has grown into an international leader in respiratory devices.

These firms demonstrated that Australian biomedical research could underpin world-leading commercial enterprises. CSL's privatisation in 1994 transformed a small government laboratory into a multinational biotechnology leader.

Government policy also played a catalytic role. The R&D Tax Incentive, the establishment of technology transfer offices at universities, and the strengthening of intellectual property frameworks in the 1990s created an environment conducive to commercialisation. AusBiotech, the national industry body, provides advocacy and sector coordination. The Therapeutic Goods Administration (TGA) refined a regulatory approach that balanced safety and efficiency, providing Australian firms with credibility in global markets.

By the early 2000s, precincts were being developed more strategically. Melbourne, Sydney, and Brisbane invested in co-location, shared facilities, and the physical integration of hospitals, universities, and institutes. The Victorian Science, Technology and Innovation (STI) initiative, the Queensland Smart State Research Facilities Fund, and Atlantic Philanthropies made significant investments in research infrastructure in Melbourne and Brisbane.

The Medical Research Future Fund (MRFF), announced in 2014 (legislated in 2015), created a perpetual $20 billion endowment that provides stability and long-term vision for medical research. The Biomedical Translation Fund, introduced in 2016, enhanced the capacity for later-stage financing.

Expansion and Global Recognition

Since 2018, biomedical innovation has accelerated. The $650 million Sydney Biomedical Accelerator, now under construction, exemplifies a new generation of translational infrastructure designed to move discoveries from bench to bedside. Melbourne's $6 billion precinct expansion consolidates its status as a global hub, featuring research platforms in cancer, infectious diseases, neuroscience, child health, and ageing.

The COVID-19 pandemic highlighted both the strengths and vulnerabilities of the system. Australia rapidly mobilised research capacity and clinical trial participation while attracting mRNA manufacturing from Moderna and BioNTech. These investments, backed by government partnerships, cemented Australia's reputation as a trusted location for advanced biomedical activities.

Clinical trial capacity has also grown substantially. Around 1,000 new trials are initiated each year, representing an investment of about $1 billion. Australia ranks near the top globally for early-phase trials, aided by regulatory efficiency, seasonal counter-cyclicality, and a diverse population.

The commercial outcomes are also impressive. In September 2025, four ASX-listed pharmaceutical and healthcare equipment companies had a combined market capitalisation of $214 billion: CSL ($101.5 billion), Cochlear ($19.5 billion), ResMed ($61.7 billion), Pro Medicus ($31.16 billion) are all global leaders with Australian headquarters.

Mid-tier firms such as Telix Pharmaceuticals ($5.1 billion), Orthocell ($316.3 million), and Imugene ($78.4 million) are growing strongly, while digital health innovators like Artrya ($326.7 million) and Volpara (acquired by Lunit in May 2024) signal a diversification into new domains.

Distinctive Features of the Australian Model

Several characteristics give Australia's biomedical innovation ecosystem international distinctiveness.

- *Stable, long-term funding*: the MRFF's perpetual endowment provides continuity beyond political cycles.
- *Institutional anchoring*: precincts are structured around teaching hospitals, universities, and institutes that combine clinical care, research, and education.

- *Regulatory credibility*: the TGA offers efficient, internationally aligned pathways recognised in global markets.
- *Clinical trial advantages*: a diverse population, seasonal differences, respected data standards and well-regulated and trusted processes, with competitive pricing attract international pharmaceutical and device firms.
- *Precinct integration*: Australian cities have strategically invested in co located biomedical precincts, creating scale and density.
- *International market presence:* leading firms have expanded globally while retaining their Australian base, sustaining domestic capability.
- *State government commitment to public health, public hospitals, and medical research:* these commitments are long-standing and have wide community support. Medical research also attracts significant philanthropy from corporate foundations and the community.

Med-Tech Start-ups: A New Generation of Innovators

Perhaps the most vibrant feature of Australia's biomedical ecosystem today is the steady flow of med-tech start-ups emerging from hospitals, universities, and research institutes. These firms often arise from clinician-led insights into unmet medical needs or from novel engineering and digital health solutions that intersect with life sciences.

The precinct environment is critical. Access to clinicians and patient cohorts within teaching hospitals provides start-ups with real-world testing grounds. Universities provide expertise in engineering, computer science, and biomedical research, while incubators and accelerators within precincts offer workspace, mentoring, and investor connections.

A new generation of companies illustrates the breadth of activity.

- *Synchron* (Parkville, Melbourne) is developing a brain–computer interface that allows people with paralysis to control digital devices through neural signals. It is one of the few Australian companies to gain FDA approval for human trials in the United States.
- *Optalert* (Melbourne) is building wearable alertness monitoring devices for fatigue management, which are used in both medical and industrial applications.
- *4DMedical* (Melbourne and Brisbane) is commercialising advanced imaging technology that provides dynamic, four-dimensional insights into lung function. The company was listed on the ASX in 2020 and has expanded into US markets.

- *CureXsys* (Adelaide BioMed City) is pioneering exosome-based therapies with applications in regenerative medicine and oncology.
- *Artrya* (Perth) is using artificial intelligence to detect vulnerable plaque in coronary arteries, aiming to transform the early detection of heart disease.
- *Vaxxas* (Brisbane) is developing a needle-free vaccine patch technology that was spun out of the University of Queensland. It has attracted major international partnerships with pharmaceutical companies and the US Biomedical Advanced Research and Development Authority.
- *Inventia Life Science* (Sydney) specialises in 3D bioprinting platforms for cell biology and drug discovery. Its RASTRUM system is gaining global traction.
- *Ferronova* (Adelaide) is developing nanoparticle tracers for improved cancer diagnostics and is currently progressing through clinical trials.
- *Lumos Diagnostics* (Melbourne) is commercialising rapid diagnostic platforms for infectious diseases and point-of-care testing.
- *EpiAxis Therapeutics* (Canberra) is advancing epigenetic therapies for cancer treatment, with translational links to hospital-based oncology services.

These firms represent the diversity of the med-tech pipeline: neural interfaces, imaging, diagnostics, AI-driven health technologies, vaccine delivery platforms, and regenerative medicine. They also demonstrate how precincts create an environment conducive to breakthroughs and enable entrepreneurs to transform ideas from the laboratory into prototypes.

The start-up pipeline is healthy, supported by programs such as MTPConnect's BioMedTech Horizons, the Biomedical Translation Fund, and state government initiatives like Victoria's LaunchVic and NSW Health's Medical Devices Fund. University accelerators (e.g., UNSW Founders, the University of Melbourne's TRaM program) and precinct-based incubators (e.g., Jumar Bioincubator in Parkville) are nurturing new ventures.

The limiting factor, however, is what happens next. While seed and early-stage funding is available, and many companies successfully progress to ASX listing, the capacity to fund late-stage trials, scale production, and build global distribution networks remains constrained.

Without greater access to growth capital and scale-up infrastructure, Australia risks losing many of these promising ventures to offshore acquisition or relocation.

The Start-up–Scale-up Gap

The most significant challenge for the ecosystem is the lack of scalable pathways for med-tech ventures. Start-ups are plentiful—spun out of universities, hospitals, or research institutes. They benefit from seed funding, accelerator programs, and translational initiatives.

As these ventures progress to late-stage clinical trials and prepare for international market entry, financing needs escalate dramatically. Australia's venture capital market is relatively small, and superannuation funds have been reluctant to commit to higher-risk, growth-stage investments.

Moreover, while precincts excel at supporting discovery, translational research, and incubation, there are limited facilities for advanced biomanufacturing, large scale GMP production, and device prototyping at a commercial scale. The establishment of the National Vaccine and Therapeutics Laboratory is an important step, but broader investment in scale-up capacity is required.

The result is an ecosystem that excels in research and invention, performs strongly in start-up formation, but struggles to create the conditions for large scale, globally competitive med-tech firms to emerge and remain headquartered in Australia.

Future Directions and Policy Priorities

To sustain momentum and address structural weaknesses, several priorities stand out:

- *Growth-stage capital:* expand the depth of the domestic venture market, incentivise superannuation funds to participate, and extend co-investment programs to support late-stage ventures.
- *Scale-up infrastructure:* invest in advanced biomanufacturing, GMP cleanrooms, and device prototyping centres within precincts.

- *Talent retention:* ensure career pathways and funding continuity for researchers and entrepreneurs to reduce the outflow of talent to larger markets.
- *Regulatory innovation:* establish sandboxes for digital health, AI-driven discovery, and next-generation therapies to maintain alignment and flexibility.
- *National integration*: strengthen networks between precincts to avoid duplication, build critical mass, and facilitate cross-country collaboration.
- *Global engagement:* deepen partnerships with Asia-Pacific health markets to expand trial populations, diversify supply chains, and access larger manufacturing capacity.

Conclusion

Australia's biomedical innovation ecosystem is a strong example of how deliberate investment, institutional anchoring, and regulatory credibility can stimulate global competitiveness. The clustering of activity around teaching hospitals, medical schools, and research institutes has created fertile ground for innovation. The success of firms such as CSL, Cochlear, and ResMed demonstrates the potential of the model.

The ecosystem's next test is whether it can support the scale-up of med-tech ventures. Addressing the capital and infrastructure gaps is essential if Australia is to capture more value domestically and sustain the growth of a new generation of globally competitive firms.

The lesson is that precincts work when they bring together research, clinical care, education, and industry in close proximity, supported by stable funding and credible regulation.

Australia has built a system that performs well by international standards. Its challenge now is to consolidate success and ensure that start-up energy translates into scale-up outcomes that secure lasting economic and social benefits.

4. Can Manufacturing and Technology Learn from the Health-Medical-Biotech Engine?

Australia's health, medical, and biotech sectors have delivered reliable pipelines from discovery to trials and into global markets. Manufacturing and broader technology have pockets of excellence; however, the system design has been less coherent, more stop-and-start, and weaker at the translation and scale-up stages. The contrast offers practical lessons for policy, institutions, finance, and execution.

What the Health-medical-biotech Sectors Got Right

Long-horizon public funding, secured by statute and renewed through published investment plans, has created predictability. The Medical Research Future Fund is a permanent endowment now reported as a $24 billion long-term investment, guided by rolling 10-year plans and administered through competitive programs. This sits alongside the National Health and Medical Research Council's grant engine, featuring a transparent pipeline and a published funding calendar. Together, they anchor the sector's research base and signal continuity to industry and investors.

Australia built a clinical translation "machine" that companies can actually use. The Therapeutic Goods Administration's Clinical Trial Notification (CTN) pathway places primary responsibility on Human Resource Ethics Committees (HREC)s allowing trials to commence without waiting for separate TGA approval. For sponsors, this means that the time to first patient is often faster and administrative costs are lower.

The ecosystem is connected by purpose-built, shared translational infrastructure. National facilities supported through NCRIS, including Therapeutic Innovation Australia, Bioplatforms Australia, the National Imaging Facility, and others, provide specialised laboratories, platforms, biobanks, and services that reduce technical and financial risk along the translation pathway.

The government anchored patient capital at the translation pinch point. The Biomedical Translation Fund was established as a matched, commercially managed co-investment vehicle to bridge the risk associated with early clinical stages. It signalled that public capital would take on translation risk alongside private VCs under clear governance, which catalysed follow-on investment and capability in specialist fund managers.

The sector benefits from dense, place-based networks around teaching hospitals, medical schools, and MRIs. This proximity supports clinician-led innovation, rapid ethical review, recruitment, and adoption. It is supported by nationally recognised anchors with global manufacturing footprints, such as CSL, Cochlear, and ResMed, that provide career ladders, supply chain gravity, and credibility for exports.

Policy messaging has remained consistent. The life sciences growth agenda has spanned multiple governments with stable vehicles, published strategies, and regular reporting. Credible, predictable rules reduce friction in clinical trials, R&D tax claims, and institutional collaboration.

Where Manufacturing and Broad Technology Fell Short

Manufacturing's most significant break in continuity was the end of domestic automotive assembly in 2016–2017, which erased a deep industrial foundation for processes, tooling, supplier capabilities, and vocational pathways. The closures were preceded by years of policy churn and contested assistance, followed by transition programs that helped many firms survive, yet could not replace the lost integrator role of OEMs. The outcome left a thinner base for advanced manufacturing scale-ups.

The broader technology economy has grown quickly in terms of job creation and enterprise development, but the architecture for scaling up remains incomplete. Australia has set a goal of 1.2 million tech workers by 2030 and has been tracking its progress. However, the platform for patient growth capital, mission-oriented demand, and late-stage commercialisation is uneven outside of a handful of firms.

Business R&D investment has risen again in absolute dollars after several years of stagnation but remains flat as a proportion of GDP. Public support still depends heavily on a tax-based instrument and less on coordinated translational programs that mitigate risk down the pipeline. The latest ABS release shows business R&D at $24.4 billion in 2023–24, which is welcome; however, it does not, by itself, build shared test beds, pre-commercial manufacturing lines, or integrator institutions.

Institutional continuity has also been weaker. The Industry Growth Centres were setup by the Commonwealth with a finite five years of funding to become self-sufficient. They were subsequently given a one-year extension. MTPConnect with their model and securing multiple

MRFF programs has become self-sufficient. The AMGC continues to operate reasonably well, as is AusCyber which folded into Stone+Chalk.

Nonetheless, Manufacturing has lacked an enduring equivalent of a Fraunhofer-style network with delegated authority and multi-year funding to run applied programs, operate shared infrastructure, and engage in procurement.

Finally, the demand signals for domestic scale-up have been weak. Health benefits from trial activities are linked to hospitals and the PBS/MBS environment, which accelerates evaluation and adoption. By comparison, manufacturing scale-ups have faced fragmented lead markets and limited strategic use of public procurement or standards to convert prototypes into production runs domestically.

Why the Difference Matters

The health, medical, and biotech system combines five reinforcing features: predictable long-term funding, streamlined regulation for trials, shared translational infrastructure, mission-aligned patient capital, and place-based anchors with global reach. This bundle transforms research advantages into investable pipelines.

Manufacturing and broad technology have strong research, energetic founders, and growing talent pools. What they have not had to the same degree as health, medical, and biotech is an integrated translation-to-scale blueprint that binds finance, facilities, demand, and institutions over long horizons.

What Manufacturing and Technology Can Learn and What Should Be Done

A lesson from health is the value of permanent structures that outlast political cycles. Endowments with investment mandates, published multi-year plans, and a cadence of competitions reduce uncertainty for firms and investors. Making translation easier than delay is a policy choice. The TGA's CTN pathway demonstrates how regulation can empower ethical oversight while accelerating activity.

Translational facilities funded under NCRIS and the MRFF demonstrate how shared platforms reduce capital costs for firms and accelerate the time to pilot. Patient co-investment at the most fragile stage attracts private capital without attempting to pick winners program by program. Precincts with real anchors reduce search and coordination costs.

These insights suggest several policy trajectories:

- *Lock in long-term, translational funding for manufacturing and technology.* The National Reconstruction Fund and the Future Made in Australia framework should have a clear translational spine: multi-year streams that reduce scale-up risk in priority platforms such as advanced materials, sensors, robotics, power electronics, and bioprocessing equipment. Facilities funding should be contestable yet persistent, with published 10-year roadmaps, similar to the MRFF model.
- *Create an industrial trials pathway that mirrors the principles of CTN.* Establish a national "industrial test and trials" regime for non-medical technologies, with a single-entry point for ethics, safety, standards, and liability frameworks. The goal would be to achieve speed with assurance, allowing companies to run pre-commercial pilots on national testbeds without navigating duplicative approvals. The CTN precedent shows that this can be done with the right allocation of responsibility.
- *A permanent network of applied institutes with a shared kit is absolutely essential.* Australia requires a stable, arm's-length network of translational institutes for manufacturing and industrial technology, with mandates to operate open-access pilot lines, demonstrators, metrology, and certification services. The network should operate in NCRIS-like settings and integrate with state facilities, such as advanced manufacturing research centres. This raises the floor for SMEs and reduces Capex for scale-ups.
- *Use public procurement to create lead markets.* Health adoption incentives are effective because they involve a buyer with established evaluation frameworks. Governments can take a similar approach for industrial technology by establishing outcome-based specifications in major programs, publishing standards roadmaps, and offering first customer contracts that require domestic production at early volumes when feasible.
- *Hard-wire patient capital at the late-stage translation.* Establish a standing, commercially managed co-investment sleeve for manufacturing scale-ups analogous to the Biomedical Translation Fund. It should take minority stakes alongside private funds in pilot-to-first-factory stages with strict governance and recycling of returns. The BTF

experience demonstrates that this crowd-in model can enhance capabilities in fund managers and firms.

- *Tie precincts to production, not just to property.* Health precincts are effective because they integrate clinical demand with translational assets. Integrator firms, shared lines of supply, export logistics, skilled labour pipelines, and contracts should anchor manufacturing precincts. Zoning and buildings are necessary, but insufficient without the production system.
- *Measure what matters.* Shift performance metrics from inputs and venture counts towards conversion rates, focusing on the time from grant to pilot, pilot to first sale, and first sale to repeat export. ABS will continue to track business R&D, but policymakers should require translational KPIs for all major industry programs and publish them annually.
- *Support the workforce with earn-and-learn pathways.* The goal for the tech workforce is useful. It needs more integrated industrial apprenticeships and mid-career conversion programs that combine wages, credentials, and hands-on experience with real equipment across a national network of applied institutes and partner precincts.

A Closing Observation

Australia did not "get lucky" in health. It built an enabling architecture and maintained it long enough to compound. It also worked, and the reason the MRFF started is the *social licence.* Health innovation is understood and appreciated by the public, unlike many other deep tech and digital initiatives.

Manufacturing and broader technology can achieve the same goals if there is a commitment to long-term collaboration, shared infrastructure, patient financing, and demand that rewards building to produce.

The policy work is to design those features as permanent capabilities of the system, rather than a sequence of short-run initiatives.

5. The Australian Way: How Rural Innovation Became a National Strength

Australia's rural innovation system is one of the nation's most distinctive policy achievements. For over a century, this system has transformed Australian agriculture from a struggling colonial experiment into a competitive global industry while pioneering institutional arrangements that other countries study but have found difficult to replicate.[1]

A Distinctive Partnership Model

The rural innovation system combines institutional elements that distinguish it internationally. At its centre lies a partnership between farmers and the government, channelling approximately $800 million annually through 15 Rural Research and Development Corporations (RDCs) funded by industry levies matched by government investment. This co-investment model, established in 1989, has generated estimated returns of $11 for every dollar invested, according to independent economic assessments.

The system's distinctiveness extends beyond funding arrangements. While most innovation systems concentrate in capital cities, Australian agricultural research has developed strong regional centres from Wagga Wagga to Armidale and Narrabri, creating knowledge ecosystems that integrate universities, research institutes, and farming communities.

This geographic distribution reflects practical necessity and institutional design: rural innovation requires proximity to actual farming conditions and regional expertise.

The RDCs operate under the Primary Industries and Energy Research and Development Act 1989, creating stable funding through research and development levy components on commodities, with matching government funding up to 0.5% of industry gross value of production. This creates direct connections between industry investment and

---

[1] This Chapter draws on the Rural Innovation System Project for the Rural Industries Research and Innovation Committee of the Primary Industry Ministers Council and extended in subsequent project advisory work.
See: Acton Institute for Policy Research and Innovation (2018). *Australia's Rural Innovation Future: Performance Review of The Rural Innovation System*, for the Rural Industries Research and Innovation Committee of the Primary Industry Ministers Council.

research priorities, ensuring farmer demand influences research directions.

Origins in Colonial Necessity

The development of the system traces to existential agricultural challenges. When the first wheat planted at Sydney Cove in 1788 failed, it became clear that Australian agriculture could not simply adopt European practices. The 1889 rust epidemic, which cost the industry £3 million and forced grain imports, catalysed the first coordinated research efforts.

William Farrer's work exemplified the emerging Australian approach. Beginning wheat breeding experiments in the 1880s near present-day Canberra, Farrer developed rust-resistant varieties suited to Australian conditions. His Federation wheat, released in 1903, tripled Australia's wheat harvest over twenty years and established a template for science-based agricultural improvement tailored to local conditions.

Farrer's breakthrough demonstrated principles that would define Australian rural innovation: systematic, science-based approaches addressing local constraints and the potential for substantial productivity gains when research aligns with practical farming needs.

The establishment of CSIRO in 1949 institutionalised this approach, creating a distributed research network that addresses regional farming system variations across the continent. Agricultural problems dominated the research agenda from the beginning, with early successes including the biological control program that eradicated prickly pear across millions of hectares of grazing land.

Rural and Regional Innovation Ecosystems

Australia's rural innovation system has created a geography of knowledge production that differs from other innovation sectors. As mentioned, rural research is anchored in regional centres, creating innovation ecosystems that integrate research, education, and commercial applications.

Wagga Wagga illustrates this model. It hosts Charles Sturt University's agricultural and veterinary programs and the NSW Department of Primary Industries' research facilities. The University employs over

1,000 people in Wagga Wagga, making it one of the region's largest employers and creating knowledge spillovers that benefit surrounding farming communities.

Narrabri has developed as a centre for cotton research excellence through the Australian Cotton Research Institute, operated by NSW Department of Primary Industries and The University of Sydney. The institute has developed expertise in cotton breeding, integrated pest management, and sustainable production systems that have influenced cotton industries internationally while supporting Australia's position as a premium cotton producer.

The University of New England demonstrates this integration through its SMART Farms, which cover 3,650 hectares and feature the SMART Farm Innovation Centre as a technology testbed with advanced connectivity infrastructure, including LoRaWAN and satellite networks. The university's SMART Region Incubator, with locations across Armidale, Tamworth, Moree, and Narrabri, supports start-up development, with 40% of its focus in AgTech.

This regional distribution creates advantages that are difficult to achieve in metropolitan locations. Proximity to commercial farming operations enables field trials to be conducted under realistic conditions, allowing for immediate feedback from farmers.

Regional centres can specialise in research relevant to local farming systems and climatic conditions. Land costs in regional areas facilitate large scale experimental facilities that would be prohibitively expensive in metropolitan areas.

The social dimensions of these regional innovation ecosystems are equally important. Agricultural researchers in regional centres often develop long-term relationships with local farming families, creating trust and knowledge exchange that facilitates the adoption of technology. This social capital enhances both the quality of research and the effectiveness of technology transfer.

Performance and Global Impact

In the agriculture, fisheries, and forestry sector, R&D investment has helped Australian agriculture double its productivity over the past 25 years, creating the foundation for Australia's position as a major food exporter.

Plant breeding achievements extend across multiple commodities. Australian research has established leadership in developing crops suited to marginal environments. The CIMMYT wheat breeding program, which contributed to the Green Revolution, drew extensively on Australian germplasm and breeding techniques. Australian cotton varieties developed through systematic breeding programs have provided the foundation for cotton industries in multiple countries with similar climatic conditions.

Biological control is also an important area of international influence. Following the success of controlling prickly pear, Australian research developed biological control programs for over 30 invasive species, creating methodologies that are now used worldwide. Animal health innovations, including vaccines for diseases such as anthrax and blackleg, combined with systematic breeding programs, have improved livestock productivity across multiple continents.

Contemporary achievements include leadership in precision agriculture. Cotton growers are among the most digitally connected farmers in the world and early adopters of Internet of Things technology. The Cotton Research and Development Corporation has led the Pathways to Digitally Connected Farms project on behalf of all RDCs.

Australian researchers have developed CRISPR applications, including high energy ryegrass with increased digestibility for dairy production, high-protein sorghum, disease-resistant rice, and non-browning potatoes with reduced carcinogen formation.

In 2019, CRISPR-edited plants classified as SDN-1 organisms were deregulated in Australia, facilitating the development of new crop varieties with enhanced nutritional profiles, disease resistance, and climate adaptation.

The Structural Limitation

There has been a downside. By concentrating almost exclusively on-farm-level productivity, the system has limited Australia's ability to develop the broader food and agriculture value chain.

Raw products and minimally transformed exports have comprised most of Australia's agricultural exports from 1988 to 2018. Around 56 cents of every dollar of raw agricultural export value accrues to the agriculture sector itself, with 25 cents going to domestic services and only 7 cents going to domestic industrial sectors.

While Australian farmers excel at producing high-quality commodities, the nation has struggled to capture value through food manufacturing, product differentiation, and brand development. This is a missed opportunity, given Australia's reputation for food safety, quality, and sustainability, combined with its multicultural culinary heritage.

Countries like Denmark, the Netherlands, and New Zealand have developed integrated food and agriculture innovation systems that span from primary production to consumer products. These systems capture a higher-value per unit of primary production through systematic attention to the entire food value chain.

The rural innovation system's focus on commodities has also limited research into food product development, cultural food traditions, and premium food experiences that could command higher market premiums.

Despite the diversity of food cultures and the growing sophistication of consumers regarding food experiences, the innovation system has generated limited research in these areas.

Contemporary Challenges

The rural innovation system faces pressure to adapt to expanding agricultural objectives beyond productivity. Australian agriculture must now consider climate change, biodiversity loss, food-related human health, and the vitality of rural communities, alongside maintaining international competitiveness.

Australian agriculture has become increasingly vulnerable to weather extremes, requiring the development of more resilient production systems. Australian agricultural emissions are expected to slowly decline over the next 15 years; however, agriculture's share of emissions is anticipated to increase from 19% in 2023 to over 30% in 2040 as other sectors decarbonise.

The innovation system is responding through increased cross-sector collaboration. The RDCs are adapting their business strategies for collaboration, with the Cotton Research and Development Corporation's involvement in cross-sectoral research initiatives growing from simple knowledge sharing to increasingly large multiparty co-investment programs. For CRDC, this has supported doubling cross-sectoral research to 40% of its annual investment portfolio.

A Pathways to Digitally Connected Farms project highlighted barriers, including a lack of critical connectivity infrastructure and generally low levels of digital literacy and skills within the industry. Addressing these constraints requires coordination across multiple innovation systems beyond agriculture.

The emergence of alternative protein sources, changing consumer preferences, and sustainability demands create new innovation imperatives.

The system's traditional commodity focus may be inadequate for addressing these challenges, requiring greater integration with food technology, nutrition science, and consumer research.

Future Directions

The future development of Australia's rural innovation system requires addressing the structural limitations created by its commodity focus while preserving the institutional strengths that have generated success in primary production. This means expanding the innovation ecosystem to encompass broader food, fibre, and materials value chains without undermining existing effectiveness.

Developing innovation infrastructure that bridges primary production, and downstream processing requires institutional mechanisms comparable to the RDC model but focused on value creation beyond the farm gate. This could involve establishing Food and Materials Innovation Corporations that apply similar levy and co-investment principles to product development, processing technology, and brand development.

Leveraging Australia's research excellence in fibre science and materials engineering requires systematic investment in advanced applications. The innovation system has underutilised opportunities in technical textiles, smart materials, bio composites, and medical applications, despite having established expertise in wool, cotton, and emerging alternative fibres.

Developing the circular economy potential of agricultural waste streams is an opportunity for additional value creation. Sugarcane bagasse, wheat straw, cotton gin trash, rice hulls, and other agricultural residues could support industries in construction materials, packaging, biochemicals, and bioenergy.

Expanding the bio-derived economy is a significant opportunity to build on existing strengths in animal health and livestock research.

Similar potential exists in developing biomaterials from wool and other animal fibres, pharmaceuticals from animal-derived compounds, and industrial biochemicals from both crop and livestock processing streams.

International engagement strategies should emphasise collaborative processing and value-adding partnerships rather than primarily focusing on commodity export relationships. Collaborations with Southeast Asian partners could establish processing hubs that enhance the value of Australian raw commodities, with premium markets such as Japan and the Republic of Korea as ultimate destinations.

Building on Success

Australia's rural innovation system demonstrates how focused institutional design can achieve strong results within its scope. The system has established Australia as a competitive force in agricultural technology through institutional arrangements that align industry investment with research priorities, distribute knowledge production across regional centres, and maintain a systematic focus on practical applications.

The RDCs mow have the opportunity to show leadership by investing in ecosystems in partnership with key players. They can build on their success beyond primary production to encompass broader food and agriculture opportunities. Contemporary food system challenges, including climate adaptation, sustainability, health and nutrition, and rural community vitality, cannot be addressed through commodity-focused research alone.

Australia's capacity to make this transition while preserving the institutional strengths that generated agricultural competitiveness will determine whether the nation's agricultural advantages translate into broader food system leadership and economic opportunities. The foundation exists in the RDC model's industry-government collaboration, systematic research orientation, and focus on practical application.

The lesson from Australia's rural innovation experience is that focused institutional design can achieve strong results within its scope; however, addressing broader challenges requires expanding that scope without undermining institutional effectiveness.

Australia has demonstrated how to build agricultural competitiveness through innovation. The opportunity now is to extend that approach across the entire food and agriculture value chain.

6. Contacts, Connections, and Collaborations: Creating Value in Innovation Ecosystems

In an era saturated with digital platforms, it is easy to mistake connections for collaborations in innovation ecosystems. A growing list of contacts on LinkedIn or entries in a CRM database may signal reach, but they rarely provide the trust, reciprocity, and shared purpose that underpin effective cooperation and collaboration. For innovation ecosystems, this distinction is critical.

For decades, marketing firms have relied on customer relationship management (CRM) systems to strengthen the connections between businesses and their current and prospective customers. In parallel, LinkedIn has emerged as a platform where professionals connect around shared interests. Some of these connections develop into genuine collaboration, but the dynamics of cooperation extend far beyond these transactional or platform-based interactions.

In innovation ecosystems, the contemporary emphasis is on building more effective relationships between research organisations, businesses, and governments to enhance research, development, and innovation (RD&I) outcomes. LinkedIn is widely used by researchers, innovation managers, and policymakers; however, meaningful collaboration requires more than merely accumulating contacts.

This chapter explores how interactions between individuals and organisations within an ecosystem can evolve from connections to cooperation, and ultimately to collaboration founded on trust-based relationships.

Context

Innovation ecosystems have emerged as the primary framework for understanding how knowledge, talent, and capital come together to create economic value. Despite the numerous mapping exercises, network diagrams, and system analyses presented in policy documents, a significant gap remains between the structural representation of these ecosystems and their actual functioning.

The typical approach to ecosystem thinking has often resembled electrical engineering: identify the components, outline the potential connections, and assume that their proximity and complementarity will naturally create productive relationships.

This mechanistic perspective overlooks an important aspect of human networks: there is a significant distinction between merely having someone's contact information, even on LinkedIn, and genuinely engaging and collaborating with them. The difference can determine whether an innovation ecosystem generates real value or remains merely an appealing diagram.

The Anatomy of Dormant Networks

Many innovation ecosystems are defined more by their *structural potential* than by their actual functionality:

- *Universities* are home to researchers whose work has the potential to transform industries; however, their discoveries often remain confined to academic publications and fail to reach a broader audience.
- *Corporations* have the technical expertise to address urgent social issues, but they often lack connections to community organisations that can help them understand these problems.
- *Government agencies* possess valuable regulatory knowledge that could accelerate innovation, but they often operate independently from the entrepreneurs who require their guidance.

These dormant connections remain because proximity in structure does not necessarily lead to functional interaction. Physical co-location, shared membership in industry associations, and formal partnership agreements can nurture collaboration, but they do not ensure that it will occur.

The ecosystem may be perfectly structured, but without current—whether it be information or knowledge—flowing through its connections, it remains inactive.

The challenge is identifying *what activates these latent relationships*. Research in network theory suggests that activation involves three key elements: recognising opportunities, aligning capacities, and investing in relationships. Opportunity recognition refers to the ability of potential collaborators to perceive the mutual benefits of working together.

Capacity alignment necessitates that organisations possess complementary resources and compatible working styles. Investing in a relationship requires individuals to dedicate time and energy to building trust and shared understanding.

Activation Mechanisms

The *shift from mere contact to genuine connection* starts with interactions that focus on solving problems rather than simply building relationships through networking. Successful ecosystem activation typically focuses on *specific challenges* that require a variety of expertise to address.

Climate adaptation, digital transformation, and health system innovation, for example, serve as focal points that enable various stakeholders to contribute their specialised knowledge while working towards common goals and outcomes.

This challenge-centred approach is effective because it creates an immediate context for collaboration.

Participants focus on specific problems that showcase the benefits of combining their capabilities, rather than engaging in abstract discussions about potential synergies. For example, a clean technology start-up collaborates with university researchers and government regulators on energy storage solutions, discovering synergies through their shared commitment to problem-solving.

The Trust Equation

These collaborations are founded on trust and respect. They develop through repeated interactions focused on specific challenges. There can also be an emphasis on opportunities to create something new or achieve beneficial and lasting change.

Every successful collaboration enhances partners' confidence in each other's capabilities and reliability, laying the groundwork for more ambitious joint ventures. This process is inherently iterative. Trust cannot be imposed or created artificially; it must be earned through demonstrated personal commitment and competence. A common saying is that people perform better when they trust each other.

Management theorist David Maister formalised this in his "trust equation": Trust equals credibility plus reliability plus intimacy, divided by self-orientation[2]. Credibility is linked to one's expertise and past experiences, reliability refers to consistent behaviour, and intimacy involves creating a sense of psychological safety when sharing information. Self-orientation, or the extent to which a person prioritises their own interests over those of others, is a major barrier to building trust.

Trust is fragile, especially when compared to the time it takes to build it. Broken commitments, inconsistent behaviour, and poor communication can undermine months or even years of relationship-building in an instant.

Once trust is damaged, it becomes very difficult to restore. Mitigation involves establishing realistic expectations from the beginning, maintaining honest and transparent relationships, and promptly acknowledging mistakes as they occur.

The role of system integrators is critical in this activation process. Successful ecosystems often include entities that specialise in laying the groundwork for effective collaboration. These integrators can include prominent and visionary leaders from the research, business, government, and community sectors. What sets them apart is their ability to understand and engage with the various languages and incentives that can drive collaboration towards breakthrough outcomes.

From Connections to Collaborations

Active connections are the foundation of deeper collaborative relationships; however, transitioning *to true collaboration requires additional elements.* Collaboration involves shared risk, collective decision-making, and aligned incentives, rather than merely exchanging information or engaging in other forms of transactional relationships.

The key difference is in how participants tackle problems. Connections typically involve organisations preserving their distinct identities and capabilities while sharing information and resources. Collaborations involve creating new hybrid capabilities that none of the partners could develop independently.

---

[2] Maister, D. H., Green, C. H., & Galford, R. M. (2004). *The Trusted Advisor.* The Free Press. (Original work published in 2000)

Collaborative relationships require governance mechanisms that strike a balance between autonomy and coordination. Partners should maintain enough independence to leverage their unique strengths while also embracing a level of interdependence necessary to achieve common goals. Achieving balance on these factors is challenging as participants operate under different incentive structures: academic careers reward publications, corporate success relies on market returns, and government agencies prioritise effective policy and program outcomes.

Successful collaborations establish shared metrics and governance processes that recognise diverse criteria for success while also creating unified measures of progress.

Joint intellectual property agreements, shared risk-reward frameworks, and collaborative decision-making processes indicate a true partnership rather than a mere transactional relationship.

Transactional Traps

The ongoing challenge in ecosystem development is to ensure that connections evolve from being merely transactional to developing into genuine, collaborative relationships or partnerships. Transactional relationships prioritise immediate exchanges, such as providing information and knowledge in return for payment, granting access to data, or offering services for a fee. Although these exchanges provide short-term benefits, they seldom lead to the long-term innovative capacity that characterises successful ecosystems.

Transactional relationships continue when participants perceive collaboration as a zero-sum game rather than a value-creating endeavour.

- *Universities* that prioritise licensing revenue, contracted and commissioned research, and broader research income may overlook opportunities for more meaningful research partnerships.
- *Companies* that view start-ups solely as acquisition targets often miss the opportunity to build long-term, innovative partnerships.
- *Government agencies* that prioritise compliance over collaboration may hinder their capacity to foster genuine innovation.

To move beyond transactional relationships, it is essential to create interactions that ensure shared interests in outcomes rather than just exchanging information.

Joint ventures, collaborative research programs, and shared infrastructure investments align the interests of participants around a common purpose and success, rather than focusing on individual transactions.

Conclusion

Innovation ecosystems often hold significant untapped potential until inactive connections are transformed into active collaborations. The difference between a network diagram and a functioning ecosystem lies in the *flow* of interactions.

In well framed ecosystem, there are shared challenges that create mutual value, repeated interactions that build trust, collaborative governance that aligns the various incentives of organisations, and intentional efforts to counteract biases that exclude relevant organisations and talented individuals.

For policymakers and ecosystem builders, the emphasis should shift from mapping potential connections to actively encouraging collaborations that ensure inclusive participation.

This involves establishing key challenges that encourage diverse stakeholders to contribute their skills, supporting intermediary organisations and systems integrators that promote cross-sector collaboration, and designing incentive systems that reward cooperative behaviour.

The goal should be to ensure that information and knowledge flow in a way that harnesses the full range of available institutional capabilities and human talents to create value for all participants in the ecosystem and the broader community they serve.

Importantly, while ecosystem activation can be encouraged, it cannot be enforced. Successful interventions generate valuable collaborations by creating opportunities for them, allowing relationships to develop organically based on genuine shared interests.

7. Innovation Through Coordination: Australia's Path to Economic Transformation

For many years, Australian governments have invested billions in innovation programs aimed at transforming the economy. Yet, the results remain modest. These observations might indicate that Australia's innovation challenge is not due to a lack of funding or poor program design. Instead, the issue lies in viewing innovation as something that governments can buy rather than facilitate.

Innovation arises from coordinated interactions across various domains working together. The "jazz combo" metaphor, first used by management expert Peter Drucker in 1995, often illustrates the importance of group coordination. No single agency controls all the conditions that enable breakthrough innovations. No single funding program alone can produce the market-ready solutions needed for economic growth. Innovation demands coordination among research capabilities, business investment, talent development, regulatory frameworks, and infrastructure systems—areas where traditional government portfolios often struggle.

Academic, professional, and even popular innovation literature recognises that coordination is a key factor distinguishing countries, regions, districts, industries, businesses, and communities that achieve innovation-led economic transformation from those that merely promote impressive initiatives while importing the technologies developed by others.

The Coordination Challenge: Why Innovation Cannot Be Purchased

Australia's innovation policy relies on a misleading fallacy: that governments can directly buy innovation outcomes through research funding, business support programs, and institutional development. This input-driven approach has led to significant public investment focused on measuring research outputs, partnership numbers, and program compliance, rather than on creating the conditions necessary for innovation to genuinely emerge.

The Cooperative Research Centres program illustrates this pattern. Despite producing impressive research outputs and forming partnerships, doubts persist about whether these programs generate the cross-sector coordination necessary to transform breakthrough innovations into market-ready solutions. The R&D Tax Incentive boosts research activity but doesn't necessarily create the collaborative environment that transforms research into economic value.

This approach underestimates what innovation policy can realistically achieve. Research funding supports the creation of knowledge. Business programs assist individual enterprises. Infrastructure investment develops physical capabilities. None of these inputs will automatically coordinate with one another to produce innovative outcomes that depend on multiple capabilities working together.

Innovation is coordination made visible.

It is well known that the Internet did not come about from a single innovation initiative. It required defence research working alongside telecommunications infrastructure, advances in computer science aligned with university-industry collaboration, and regulatory decisions coordinating network interconnection over the decades. Multiple institutions aligned their efforts towards complementary goals without any one agency controlling the outcome.

Modern breakthrough innovations follow similar coordination patterns. Electric vehicles require advances in battery technology, the development of charging infrastructure, alignment of emissions regulations with consumer incentives, and manufacturing investments across the automotive, energy, environmental, and transportation sectors. The iPhone combines telecommunications, computing, manufacturing, design, and software capabilities that no single organisation could develop on its own.

This systems approach requires a fundamental shift in thinking: from managing innovation as an input that can be controlled to coordinating the conditions that allow innovation to emerge as an outcome.

Domains Requiring Coordinated Alignment and Collaborative Commitment

From a policy perspective, and with some integrative thinking, it is possible to identify numerous technology and knowledge systems that interact and require coordination and collaborative commitment to achieve innovative outcomes rather than optimising any single area. Seven such systems are outlined below.

1. *Business Innovation Investment* encompasses venture capital regulation, R&D incentives, foreign direct investment, competition policy, and business formation regulations that must be coordinated to support the scaling of innovation.
2. *Talent development* encompasses education policy, skilled migration, labour market regulation, and professional development programs that must coordinate to provide innovation capabilities.
3. *Regulatory Architecture* encompasses intellectual property systems, regulatory sandboxes, government procurement, standards setting, licensing, certification, and approval processes that must be coordinated to enable the adoption of innovation.
4. *Financial markets* coordinate capital market regulation, superannuation investment policies, banking regulations, and co-investment mechanisms to support innovative financing.
5. *International engagement* aligns trade policy, research collaboration agreements, foreign investment frameworks, and technology transfer arrangements to connect domestic innovation with global opportunities.
6. *Physical infrastructure* coordinates land use planning, urban development, innovation precincts, transport networks, and energy systems to support the implementation of innovation.
7. *Information infrastructure* integrates digital government systems, data governance frameworks, cybersecurity policies, interoperability standards, and computational capabilities that enable cross-domain coordination.

Countries that achieve superior innovation outcomes have developed institutional mechanisms for coordinating and collaborating across these and many other systems. Those that struggle with innovation typically have fragmented approaches that optimise individual system elements without ensuring overall ecosystem coordination and integration.

Denmark's global wind energy leadership stems from the coordination of systems covering energy policy, industrial growth, environmental regulations, export loans, and grid infrastructure investments across various agencies over the course of twenty years. Danish companies now produce 50% of the world's wind turbines through coordinated alignment rather than through isolated excellence in any single domain.

Why Traditional Approaches Predictably Fail

Three decades of research, consultation, and observation reveal predictable failure patterns that persist because they reflect institutional biases towards departmental control rather than system coordination.

Single-portfolio ownership assigns innovation responsibility to a single minister or department, fragmenting naturally interconnected processes. Even when the Department of Industry leads innovation policy, the portfolios of education, skills development, regulatory reform, foreign direct investment, and international trade remain poorly coordinated, despite all being essential for innovation outcomes.

Funding-centric design focuses resources on research or business support without ensuring that these investments connect to other system elements. The Australian Research Council funds world-class research, while the Department of Education provides block grants to universities. Individual universities make independent decisions about how to spend their substantial revenue from international students. The Department of Industry supports business research, separate from the Department of Health, which supports health research, and the Department of Agriculture, which supports agricultural research. There are limited mechanisms to coordinate this activity to achieve common objectives.

Best practice replication attempts to transplant successful policies without understanding how they interact with different institutional, economic, and cultural contexts. Attempts to replicate Silicon Valley consistently fail because they copy visible elements, such as venture capital, university research, and entrepreneurship programs, while missing the coordination mechanisms that make these elements effective within broader metropolitan innovation ecosystems.

Sector-specific fragmentation develops separate innovation approaches for different industries, rather than recognising that breakthrough innovations increasingly emerge at the intersections of sectors. For example, health innovation requires coordination among medical research, regulatory approval, manufacturing capability, and digital infrastructure across multiple uncoordinated portfolios.

These patterns persist for many reasons, but government policy advisers and program managers are incentivised to advance their careers by delivering programs within their portfolio rather than achieving outcomes that require cross-portfolio collaboration and systems integration. The incentive structure rewards departmental success over system coordination.

Mission-Oriented Coordination in Practice: An Example of Net Zero Emissions

The Net Zero Emissions Mission strategy exemplifies innovation as coordination thinking at scale. Achieving net zero emissions demands unprecedented coordination across the policy systems mentioned above.

Temporal coordination aligns all systems with net zero deadlines.

- *Cross-system integration* requires smart grid systems to effectively coordinate renewable energy generation, trained operators, supportive regulations, research backing, and investment finance. Success requires simultaneous progress rather than sequential development.
- *Feedback loops* enable adaptive coordination through real-time emissions monitoring systems as conditions change, and technologies emerge. When offshore wind costs drop faster than expected, coordinated responses across multiple areas can accelerate deployment while ensuring grid stability and facilitating workforce transitions.
- *Mission coherence* ensures that each system's contributions reinforce rather than contradict efforts in other systems. Education programs training renewable energy technicians must align with industry investment incentives, regulatory approval timelines, and infrastructure development schedules.

This coordination challenge reveals why traditional innovation policy approaches often fail for complex missions.

No single agency can deliver net zero outcomes; success requires more than just good intentions regarding whole-of-government approaches.

The Path Forward: Coordination and Integration Architecture

Effective innovation policy requires institutional frameworks that prioritise system coordination over program delivery. This does not mean abandoning existing programs; it means redesigning how they connect and reinforce one another.

- *A Cross-Portfolio Innovation Coordination Office* should be established as a small, high-level unit reporting to the Prime Minister's department, with the authority to coordinate innovation-relevant policies across portfolios. Unlike traditional departments, this office would focus on alignment mechanisms rather than program delivery, with success measured by system-wide innovation outcomes.
- *Mission-Based Policy Pilots* should select 2-3 specific innovation challenges as coordination pilots, requiring relevant agencies to develop integrated approaches with shared success metrics and conduct regular cross-portfolio reviews.
- *Innovation Impact Assessment* should require major policy proposals across all portfolios to assess their innovation implications, making coordination needs visible and creating incentives for departments to consider system-wide effects.
- *System Performance Measurement* should evaluate innovation system coordination effectiveness rather than individual program compliance, developing metrics that assess interaction quality between system elements.
- *Adaptive Policy Architecture* should enable continuous adjustments based on evidence and changing circumstances rather than fixed implementation over electoral cycles, with built in learning mechanisms that improve coordination effectiveness.

Addressing Implementation Challenges

The coordination approach faces numerous genuine obstacles that require systematic solutions, including, for example:

Obstacle	Solution
Ministers gain political visibility by announcing programs within their portfolios rather than coordinating with other departments. Career progression depends on delivering measurable outcomes within departmental boundaries. Coordination success is harder to attribute and claim credit for than program delivery	Design coordination mechanisms that create visible wins for participating ministers and departments, ensuring clear attribution of success to coordination efforts.
Current performance management systems reward departmental achievement rather than cross-portfolio collaboration. Budget processes allocate resources to agencies, rather than coordinate	Pilot performance metrics that reward coordination outcomes, along with shared budgets for mission-based initiatives

Obstacle	Solution
outcomes. Risk aversion discourages the experimentation that effective coordination requires.	requiring multiple agencies to succeed together.
Innovation outcomes emerge over longer timeframes than political and administrative cycles. System coordination effects are harder to measure than program outputs. Attribution challenges make demonstrating coordination effectiveness difficult for treasuries and audit offices.	Develop intermediate coordination metrics that demonstrate progress towards innovation outcomes, along with regular independent reviews assessing system integration.

The Innovation Choice

Australia faces a crucial decision. We can continue announcing innovation programs, measuring research outputs, and hoping for economic change, even though the results of innovation remain out of reach. Alternatively, we can adopt coordination-based strategies that view innovation as a natural outcome of well-organised system components.

After thirty years of input-driven methods yielding only modest outcomes despite substantial investments, the case for a coordination-centric innovation policy has become compelling.

The real question is whether we will begin generating innovation outcomes through coordination and teamwork or continue relying on the hope that purchasing innovation inputs will somehow lead to results.

This choice will shape our economic future and determine how effectively every dollar spent on innovation inputs is utilised. Australia has the institutional strength and policy expertise to lead with coordination-based approaches that other middle powers facing similar challenges can adopt.

Innovation springs from coordination and collaboration. The only remaining question is whether we will act on this common knowledge or continue to trust that purchasing innovation will somehow bring about the change that coordination can truly achieve.

---

8. Towards an Australian Innovation-Led Industrial Strategy: A Public Administration Perspective

For nearly 30 years, Australian innovation policy advisers and analysts have been calling for a comprehensive "industrial policy" to transform the nation's economic foundation. There is now an even greater imperative for Australia to transition from its traditional reliance on resource extraction and rural commodity production towards the knowledge-intensive industries of the future—advanced manufacturing, clean technology, biotechnology, artificial intelligence, and digital services.

This transformation is essential for long-term competitiveness and resilience in an era of technological disruption and geopolitical uncertainty. Proponents point to successful industrial strategies in Europe and Asia as evidence that targeted government interventions can steer an economy towards a more competitive and innovative path, with higher-value industries and emerging technologies.

The aspiration for an industrial transformation in Australia confronts a fundamental challenge that is often underestimated or overlooked. The issue is not the absence of ambition: it is more about the unique and highly complex institutional and administrative framework in which an industrial strategy must operate.

The core of the problem lies in the deep mismatch between the concept of a unified industrial policy, typically conceived for unitary states, and the political, administrative, and economic reality of Australia as a vast continental federation composed of six independent sovereign states, each with distinct economic specialisations, institutional structures, and political cultures.

This challenge is far more complex than it initially appears. Australia is not Denmark, Finland, Sweden, or Singapore—the small, cohesive economies that often serve as exemplars of successful industrial coordination. Australia is a large continent with a population comparable to that of a medium-sized European country, but it is spread across vast distances with varying economic specialisations and governance structures among its states and regions.

This reality complicates the pursuit of a coherent national industrial strategy, though it does not render it impossible. The task is to acknowledge this structural complexity and develop an approach to industrial transformation that aligns with Australia's size, diversity, and federal system of governance, rather than working against it.

The Appeal and Limits of Industrial Policy for Australia

Industrial policy is attractive because it offers a sense of unity and consistency. It connects investments, skills, research, and infrastructure to accelerate significant changes. The well-known cases of Japan's Ministry of International Trade and Industry (MITI) in the years after World War II and South Korea's organised approach to industrial change show that governments in unitary governance systems can influence market outcomes and direct resources towards emerging industries. China, as a command economy, has also embarked on implementing a top-down industrial strategy.

In Western economies the intellectual foundation of industrial policy draws on national *innovation systems* (NIS) thinking developed in the late twentieth century (Freeman, 1987; Lundvall, 1992; Nelson, 1993). These frameworks assume that innovation is integrated within institutions and that national governments play a key role in coordinating collaboration among businesses, universities, and public agencies. They are based on the experiences of small, closely knit European economies where national governments can operate similarly to regional development authorities.

The Australian industry policy challenge lies in these assumptions not aligning with the realities of a large, decentralised, and geographically diverse nation. A national industrial policy relies on the notion of a unified economy that can be guided by a single set of priorities.

Australia's economic landscape is shaped by diverse state and regional systems, each influenced by distinct conditions of geography, climate, natural resource endowments, historical legacies, and political cultures.

Some people have claimed that the answer would be to "Abolish the States", but in truth, this is merely a political and constitutional illusion. Federalism incorporates mechanisms of checks and balances that help prevent the concentration of power, a characteristic often found in unitary and centrally governed countries.

Australia's Continental and Federal Reality

Spatial scale. In Denmark, the road trip from Copenhagen to Aarhus takes less than three hours. In Finland, the drive from Helsinki to Tampere takes two hours. The flight from Sydney to Perth takes five hours and covers a distance of 3,300 kilometres, which is more comparable to international collaboration between Europe and North America than to exchanges within a single country. The challenge of distance continues to be a significant aspect of Australia's economic situation.

Economic diversity. Western Australia's economy, which relies heavily on mining, differs greatly from Victoria's blend of manufacturing and services, as well as Queensland's focus on agriculture and resources. Sydney has emerged as a centre for finance and technology, whereas Adelaide is establishing itself in the fields of defence and space. Hobart and Darwin encounter unique industrial challenges influenced by their geographic locations, sizes, and historical backgrounds.

To conflate these State differences into one "national" framework risks favouring certain industries while neglecting others or creating an overly extensive list that loses its significance.

Federalism. Section 51 of the Australian Constitution provides the Commonwealth Government with many levers relevant to industrial strategy. Powers over trade and commerce (s 51(i)), taxation (s 51(ii)), corporations (s 51(xx)), defence (s 51(vi)), and external affairs (s 51(xxix)), along with powers over patents and intellectual property, communications, statistics, and bounties. Additional provisions such as Section 90 (customs and excise), Section 92 (free interstate trade), and Section 96 (conditional grants to States) all extend Commonwealth influence.

While the Commonwealth powers are significant, the States retain substantial residual authority in areas central to industrial policy, including land use planning, development control, infrastructure, education and training, natural resources, and industry development.

These responsibilities give States substantial influence over the location and growth of industries, regional specialisation, and workforce capability. States also design their own industrial and innovation strategies, reflecting their unique priorities and perceived opportunities.

The result is a shared, uneven and highly complex policy space where State-level initiatives play a decisive role in shaping industrial outcomes within their jurisdictions, sometimes complementing federal priorities as well as competing with other States.

In practice, the Commonwealth's principal policy lever is its control over money, which allows it to create broad guidelines and develop *funding programs* built around grants of financial assistance that are subject to stringent conditions and controls. These programs have not always been well-designed or coordinated among ministers. This issue is currently being addressed by the Strategic Examination of Research and Development (SERD).

The States also leverage Commonwealth research, development and innovation (RD&I) programs, such as the Cooperative Research Centres program. Even so, the actual implementation and delivery of industrial policy is largely a shared Commonwealth-State responsibility. Efforts to implement centralised strategies usually face strong resistance from State, Territory and local governments—unless they are mainly about money.

Financial relationships between the Commonwealth and the States are seldom marked by smooth coordination; instead, they are influenced by negotiation, competition, and varying levels of capability.

However, as COVID demonstrated in the realm of public health, both federal and state governments can collaborate to achieve national objectives.

A Constitutional and Administrative Challenge that Cannot be Ignored

These structural characteristics indicate that achieving a cohesive national industrial policy is particularly challenging. Several points illustrate why.

- *Fragmented institutions.* Universities are funded federally but embedded regionally. Vocational education is supported by both Commonwealth and State funding. Some industry programs function on a national scale, but infrastructure delivery and planning are managed at the State-level, with significant authority delegated to 537 democratically elected local governments, each with substantial statutory planning and infrastructure responsibilities.
- *Conflicting priorities.* Industrial policy requires a clear and achievable set of national goals. However, the complexities of federalism and regional identity create challenges in reaching a consensus. A Commonwealth Government that favours defence industries in South Australia, for example, may distance itself from Queensland and Western Australia. Efforts to elevate one region's specialisation as a "national priority" invite pushback from others.
- *Cycles of dependency.* Federal fiscal dominance leads states to seek financial support from Canberra. Instead of collaboratively developing strategies, States actively and competitively seek money from "funding programs". The outcome is a lack of integrated capability, competitive bidding processes, and duplication, rather than a unified transformation of industrial capability.
- *The Political economy of short cycles.* Industrial policy requires long horizons. Australia's federal political landscape is influenced by brief electoral cycles, regular alterations in the Administrative Arrangements Order, and a competitive parliamentary opposition. Strategies labelled as "industrial" may also be pejoratively labelled by the press as "picking winners" in specific sectors or regions, which weakens their legitimacy.

Thus, the concept of "one national industrial policy" is difficult to grasp. The more Canberra attempts to centralise, the more the reality of interstate diversity and the administrative and political economy of federalism undermine the effort.

International Comparisons

Other large federal economies illustrate the points made above.

The *United States* does not have a national industrial policy in the European sense. Instead, it has a mix of federal programs: defence procurement, research foundations and institutes, energy funding, and the CHIPS and Science Act, which provide platforms and missions, while states drive execution. Silicon Valley, the life sciences hub in Boston, and the technology sector in Austin result from local and state environments rather than national industry plans.

The federal role has, however, been catalytic: funding research, underwriting infrastructure, and creating markets through procurement, principally via large and continuous government contracting, particularly in defence and aerospace.

In *Germany,* which is also a federal republic, industrial policy is coordinated but decentralised. Berlin defines priorities, funds research notably through the Max Planck and Fraunhofer Institutes, and builds platforms, while states (Länder) and cities, and districts and precincts within them, initiate and adapt ecosystem strategies to their strengths. In this way, Germany combines national frameworks with strong regional execution.

The federal government establishes missions in energy transition, digitalisation, and advanced manufacturing, such as Industrie 4.0, while also leading in education, skills, and industry support. Institutions such as the Fraunhofer Society provide applied research capacity across regions.

Canada balances federal ambition with strong provincial control. Ottawa establishes wide-ranging guidelines by implementing initiatives such as the Global Innovation Clusters and the Strategic Innovation Fund, collaborating in fields like digital technologies, protein industries, and ocean economies. Ontario focuses on automotive and artificial intelligence, British Columbia emphasises clean technology, and Quebec specialises in aerospace.

Federal funding creates missions and platforms, while provinces adapt and deliver. Industrial policy is thus collaborative and place-based, anchored in regional ecosystems while being supported by national coordination and EU-style cluster initiatives.

Spain also implements its industrial policy within a decentralised framework, where the national government establishes overarching strategies, such as energy transition, digitalisation, and green industry, that are frequently aligned with European Union missions and funding. Autonomous regions play a significant role in service delivery by customising support for local specialisations in sectors such as automotive (Catalonia), aerospace (Andalusia), and renewable energy (Navarre, Basque Country).

National frameworks, European Union cohesion funds, and recovery plans provide guidance and resources, while implementation takes place at the regional level. The outcome is a multilevel industrial strategy influenced by both Brussels and Madrid, grounded in local territorial ecosystems.

Chile does not have a unified national industrial policy. Instead, its strategy integrates national programs with robust sectoral and regional initiatives. Santiago has prioritised digital and financial services, whereas regions such as Antofagasta concentrate on mining technologies and renewable energy.

National agencies like CORFO (Corporación de Fomento de la Producción) offer funding, incentives, and cluster programs, often in partnership with international co-financiers. The approach focuses on mission-oriented sectors such as green hydrogen, lithium value chains, and AgTech; however, execution is tailored to regional and ecosystem-based contexts, reflecting the diverse geography and resource endowments of Chile.

Brazil implements its industrial policy through national missions, which are executed in a decentralised manner across its states. The federal government supports financing, procurement, and research initiatives in sectors such as aerospace, agribusiness, energy, and digital technologies through agencies like BNDES and EMBRAPII.

States such as São Paulo, Minas Gerais, and Paraná support specialised clusters in the automotive, biotechnology, and AgTech sectors, while regional development banks and industrial parks enhance federal programs. The outcome is a hybrid system in which Brasília determines priorities and allocates funding for platforms, while states and metropolitan regions tailor strategies to align with their economic strengths and institutional capabilities.

In *China*, which has a unitary governance system, the delivery of industrial policy is highly decentralised. Provinces and municipalities tailor central directives to fit their unique contexts, frequently competing to attract investment and talent. Special Economic Zones like Shenzhen demonstrate how local experimentation, backed by central incentives, facilitates transformation.

These examples demonstrate that large, federated, and diverse countries do not function under a unified industrial strategy. They operate within a governance system comprising multiple levels, where national governments establish goals, provide incentives, and create frameworks, while regional authorities develop and execute strategies that align with their resources and organisations.

Policy Implications for Australia

For Australia, these cases suggest that striving for a comprehensive and cohesive, nationally directed, and controlled industrial policy may be impractical.

What is needed is a method that recognises Australia as a federation comprised of diverse State and regional economies, each with its own unique assets, industries, institutions, and political and administrative cultures.

A national industrial strategy should direct efforts by setting long-term goals, funding research and development, building regional information and AI data platforms at the innovation district and precinct levels, creating demand through procurement, and making strategic investments to strengthen national capabilities.

The approach should enable States and metropolitan areas to comprehend and execute these missions in a manner that aligns with their national economic frameworks, capabilities, and sociopolitical contexts.

This approach would embrace Australia's federalism instead of opposing it. This would result in New South Wales, Victoria, Queensland, Western Australia, South Australia, Tasmania, and the two Territories adopting tailored strategies that leverage their strengths, ranging from advanced manufacturing and defence to energy, artificial intelligence, mining technologies, and agribusiness.

The Commonwealth's role will remain pivotal in defining missions such as net zero, health resilience, and digital security; investing in platforms like research infrastructure, data, and digital networks; and ensuring that procurement and regulation stimulate markets for innovation. States and regions should facilitate delivery by developing innovation ecosystems and regional specialisations that integrate into national and global value chains.

The takeaway from international comparisons is that industrial policy in a large federation is effective when it operates at multiple levels, is adaptable, and is grounded in local context.

For Australia, this means moving beyond the concept of a singular national plan and embracing a cohesive strategy for industrial transformation.

This approach leverages the benefits of federation to boost the economy's long-term resilience and competitiveness.

Conclusion: From Illusion to Adaptation

Australia's quest for an industrial policy should transcend the illusion of centralised control and instead align with the realities of its federal structure. The country's diversity is not a challenge to be eliminated, but a fundamental strength to be embraced.

A practical framework will regard industrial transformation as a collaborative endeavour: the Commonwealth outlines missions, sets up platforms, and supports market growth through procurement and regulation, while States and regions tailor their approaches to align with their economic resources and institutional capabilities.

This approach is consistent with international practices in large federations, where policy coherence arises from collaboration across multiple levels rather than from central mandates.

What is important is having a clear national purpose and the flexibility to execute plans at the local level. Investments in research, skills, and physical and digital infrastructure should aim to enhance distinct specialisations rather than weaken them. Federal support should promote evidence-based strategies, encourage collaboration across jurisdictions, and consider distributional effects to ensure that industrial transformation enhances inclusivity and legitimacy.

By anchoring industrial policy in missions that extend beyond electoral cycles and focusing on place-based strategies that enhance resilience and competitiveness, Australia can transform federalism from a limitation into an asset. The alternative of adhering to a unitary model that is ill suited to Australia's scale will only exacerbate fragmentation and result in missed opportunities.

9. What Innovation Ecosystem Success May Look Like

Innovation Ecosystem Success refers to the capacity of an innovation ecosystem *to generate sustained, transformative, and inclusive economic, technological, and societal value* through dynamic interactions among knowledge institutions, entrepreneurs, industry, government, and civil society.

Based on the research and analysis undertaken for this Handbook, a successful innovation ecosystem might be characterised by its ability to:

- Integrate knowledge, entrepreneurship, industry, and public purpose on a metropolitan-scale.
- Create and scale breakthrough innovations with economic, technological, and societal impact.
- Extend opportunities and participation across diverse communities and geographies.
- Adapt and renew itself in response to external technological, environmental, and economic shifts.
- Build long-term institutional, human, and infrastructural capabilities beyond short-term market fluctuations.

Taking these characterisations further and drawing on the Acton Institute Research Report outlined in Part VI, success can be understood across at least six integrated dimensions:

- Systemic Convergence and Integration
- Innovation Output and Transformation
- Inclusivity and Societal Engagement
- Resilience and Adaptability
- Sustainability and Long-Term Capability Building
- Global Connectedness and Strategic Positioning

The table below summarises these dimensions and provides example metrics.

Dimension	Dimensions of success	Example metrics
1. Systemic Convergence and Integration	Deep, dynamic connectivity among universities, start-ups, corporations, investors, government, and communities. Knowledge, capital, and people flow fluidly across sectors and disciplines.	Number of formal university-industry partnerships, public-private R&D consortia, cross-sector collaboration projects.

Dimension	Dimensions of success	Example metrics
2. Innovation Output and Transformation	Creation and scaling of globally competitive products, platforms, technologies, and services Significant contributions to emerging deep tech, civic tech, healthtech, climate solutions, and digital industries.	Number and quality of deep tech start-ups, scale-ups, unicorns, transformative technologies commercialised, societal impact innovations launched.
3. Inclusivity and Societal Engagement	Broad-based participation in innovation processes across socioeconomic, cultural, and geographic divides. Civic innovation platforms that link technological advancement to public value creation.	Participation rates in innovation programs by underrepresented groups; geographic distribution of start-ups and R&D centres across metro regions.
4. Resilience and Adaptability	Capacity to reinvent, diversify, and renew the ecosystem in the face of shocks (for example, technological disruption, geopolitical shifts, economic downturns).	Ecosystem diversification index (sectors beyond core specialisations); rate of new cluster emergence following external shocks.
5. Sustainability and Long-Term Capability Building	Investment in foundational infrastructures: education, research, urban connectivity, entrepreneurial capabilities, and public institutions. Innovation is embedded within sustainable urban and environmental development strategies.	Investment in education, research, innovation infrastructure as a share of regional GDP; longevity and evolution of core institutions.
6. Global Connectedness and Strategic Positioning	Integration into global knowledge networks, R&D consortia, venture ecosystems, and policy leadership circles. Ability to influence and respond to global innovation trajectories.	Number of active international R&D partnerships, venture investment sources, global market expansion successes by local firms.

Innovation districts and precincts are much more than mere spatial concentrations of research and business activity or simple instruments of economic policy. The argument developed in this, and previous chapters demonstrates that they represent complex adaptive systems, whose success depends on a deep understanding and management of multiple interacting forces.

The Australian experience offers particularly valuable lessons. Sydney's Tech Central exemplifies the coordination challenges inherent in large scale urban innovation initiatives, where multiple stakeholders and regulatory authorities must find common ground. Canberra's transformation demonstrates the power of patient, bottom-up ecosystem development. Regional Queensland demonstrates how non-metropolitan areas can leverage their distinctive assets for knowledge-intensive growth.

Three overarching principles emerge for the successful development of innovation districts:

- *Balance matters more than maximisation.* The goal is not to eliminate divergence in favour of complete convergence, which would produce stifling homogeneity. Rather, successful districts maintain a productive tension between alignment and diversity, structure and flexibility, as well as strategic direction and emergent creativity. This "generative tension" becomes a source of innovation rather than an impediment.
- *Context shapes everything.* While international best practices provide valuable insights, they cannot be transplanted wholesale. Each district must develop approaches suited to its unique historical legacies, institutional relationships, and cultural contexts. What works in Kendall Square may not work in Camperdown; what succeeds in Barcelona may fail in Brisbane.
- *Systems integration is essential.* Leading edge physical infrastructure and generous funding cannot overcome institutional fragmentation. Districts require dedicated mechanisms, whether organisations, roles, or processes, that bridge boundaries and enable diverse actors to work productively together. Without such integration, proximity generates competition rather than collaboration.

With these in mind, the imperative for effective innovation districts will only intensify. As Australia navigates technological disruption, economic transformation, and mounting environmental challenges, our capacity for innovation becomes increasingly critical.

This demands a fundamental shift in how we conceive and develop innovation districts. Rather than viewing them primarily through the lenses of property development or research commercialisation, they should be understand as vital civic infrastructure—environments where diverse communities come together to address shared challenges and create collective futures.

The path ahead requires patience, maturity, and sustained commitment. Innovation ecosystems are cultivated, not constructed. They emerge from countless acts of connection, collaboration, and creative recombination. By understanding and thoughtfully managing the forces that shape these dynamics, we can build innovation districts that fulfil their promise as engines of transformation, generating economic returns and fostering more prosperous, sustainable, and inclusive societies for all Australians.

PART III: CONCEPTS AND CONTEXTS – SPATIAL, SECTORAL, AND SYSTEMIC DIMENSIONS

Part III examines the contexts in which innovation ecosystems take shape and operate. While Parts I and II addressed the conceptual foundations and value creation dynamics of ecosystems, this part explores their geographical, sectoral, and cultural dimensions. It investigates how place, policy, and practice interact to shape innovation outcomes and why these contextual factors cannot be treated as neutral backdrops.

The chapters examine the spatial logic of innovation, exploring how ecosystems derive strength from geography, infrastructure, and connectivity, while also highlighting the tensions between regulatory certainty and the need for adaptive planning. They analyse sectoral drivers of innovation, including the interplay of culture, creativity, and science, as well as the often-overlooked contributions of the humanities and social sciences. Attention is also given to hidden innovation, cultural meaning, and the distributional effects of who is included in and excluded from ecosystem participation.

Together, these discussions provide an understanding of the systemic factors that either enable or constrain innovation. They reinforce the view that successful ecosystems depend on aligning spatial, economic, business and social dimensions in ways that generate inclusive growth, cultural vitality, and long-term resilience.

1. The Geographical Foundation of Innovation

This chapter draws on a draft working paper prepared for the Australian 2030 Innovation Strategy, 2017

The intellectual foundations for understanding geographical proximity in innovation trace back to Alfred Marshall's pioneering work in 1920. Marshall identified what he termed "industrial atmosphere" and agglomeration effects, recognising how knowledge spillovers and innovation could flourish within geographically concentrated industrial districts. His observation that "the secrets of industry are in the air" captured something essential about how tacit knowledge transfers between firms and individuals when they share physical space.

This insight about the significance of place has been picked up and developed by successive generations of economic geographers, sociologists, and strategy economists. Michael Porter popularised the cluster concept in the 1990s, emphasising how geographical proximity can amplify competitive advantage through enhanced knowledge exchange, collaboration, and innovation within industry-specific groupings. Richard Florida's work on creative cities added another dimension by highlighting how places themselves could become magnets for talent and innovation.

These various strands of thinking share a common recognition that innovation is fundamentally a social process that benefits from face-to-face interaction, shared cultural understanding, and the kind of serendipitous encounters that occur when people with complementary knowledge and interests occupy the same spaces.

Industrial Milieu and Innovative Environments

The concept of an innovative milieu, introduced by the Groupe de Recherche Européen sur les Milieux Innovateurs in 1986, provided a more nuanced understanding of how geographical spaces support innovation. Rather than focusing purely on economic transactions, the milieu concept emphasises the importance of informal social relationships, shared cultural identity, and collective learning processes within defined geographical areas.

An innovative milieu encompasses cooperation and information exchange among regional actors, repeated face-to-face contacts, engagement across different economic sectors, and a shared sense of belonging to a coherent regional culture. Critically, these environments must remain open to the outside world to access new information and resources while maintaining their internal coherence and identity.

This framework helps explain why some places consistently produce innovation while others with similar resources struggle to generate dynamic economic activity. The difference lies in the quality of relationships, the depth of trust networks, and the presence of what might be called an industrial atmosphere.

Clusters and Economic Agglomeration

Clusters represent a particular manifestation of geographical innovation thinking, characterised by their organic emergence and flexible governance structures. Unlike more formally planned innovation environments, clusters typically develop through market forces and policy initiatives rather than through deliberate spatial planning.

Porter's cluster theory identifies several key advantages that emerge from geographic concentration. Companies within clusters can take advantage of agglomeration economies, observe competitors directly, benefit from collective knowledge and network effects, develop stronger social interactions, and gain early access to new technologies and market opportunities. These benefits arise from shared infrastructure, specialised labour markets, and the kind of knowledge spillovers that Marshall first identified.

The origins of clusters are diverse. Some emerge from university research, as seen in Massachusetts with MIT and Harvard University. Others develop from geographical advantages, such as the Dutch transportation clusters that benefit from the Netherlands' central European location. Some arise from the need to solve particular problems, such as Finland's environmental cluster that emerged in response to pollution challenges. Others begin with a single innovative firm that stimulates the growth of related companies.

The biotechnology sector provides compelling examples of cluster development. Early biotechnology clusters like Cambridge, Heidelberg and San Diego demonstrate the "virtuous cycle" that can emerge when a strong science base, available funding, robust industrial infrastructure and supportive contextual factors combine effectively. Melbourne's Bio21 precinct is an Australian example of this cluster development, though it remains in an evolutionary phase compared to more established international examples.

The commercial appeal of clusters extends beyond academic theory. Property developers, investors, and real estate agents recognise the value proposition that proximity creates through knowledge spillovers, specialised labour pools, and supplier networks.

This understanding has driven significant property premiums in established cluster locations and has encouraged speculative development in emerging regions.

Science Parks and Technology Infrastructure

Science parks emerged as a deliberate policy intervention to strengthen collaboration among universities, industries, and governments. Over many years, governments, universities, and property developers have invested heavily in these developments, though with mixed results. The performance of science parks has been inconsistent, with many struggling to achieve sustained renewal without continuous government support.

The Adelaide Multi-function Polis, announced in 1987, exemplifies both the ambition and limitations of large-scale technology park development. Intended to be a futuristic metropolis built in partnership with Japan, the MFP was designed to accommodate up to 250,000 people and become a centre for biotechnology, computing, education, health and tourism. The South Australian Government terminated the initiative in 1998, leaving only the Mawson Lakes Technology Park as its legacy.

Similarly, the Australian Technology Park, created in 1995 by the New South Wales Government in partnership with three Sydney universities, struggled to meet commercial expectations. The universities appeared to prefer conducting innovation activities closer to their campuses, near their staff and students. While ATP did produce the successful Cicada Innovations incubator, the broader park concept struggled to achieve the scale and integration seen in prominent international examples.

More recently, the Sydney Science Park, announced in 2015, is best understood as a large-scale property development structured around the science park model. While it promises research, education, and innovation-led employment linked to the Western Sydney Aerotropolis, its delivery has been slow and contentious. The project blends urban development with a smart city vision, but scepticism remains about whether its ambitions will translate into tangible outcomes. In effect, it exemplifies the tension between property-driven urban growth and the genuine development of innovation ecosystems.

Research on science park performance suggests that success requires more than simply co-locating research institutions and companies. The single most critical factor appears to be the ability to create new firms continuously, enabling ongoing renewal in both existing and emerging sectors. This requires what Audretsch terms a critical mass of scientific talent, but also an entrepreneurial culture and nearby financial support.

Successful parks typically feature a large research university or national laboratory at their core, strong public-private partnerships, sustained long-term public financial and policy support, and services that extend well beyond real estate provision.

These services include shared laboratory and prototyping facilities, business incubation, intellectual property advice, and assistance with early-stage capital.

Industrial and Innovation Districts

The evolution from industrial districts to innovation districts is a mature understanding of how planning and organic development can work together. Industrial districts historically emerged through two pathways: organic clustering driven by agglomeration economies and shared knowledge, and deliberate planning interventions that designated specific areas for industrial use.

Innovation districts represent a more integrated approach that recognises both planned and emergent dimensions of economic development. While planning continues to involve strategic land use designation and infrastructure development, the ecosystem dimension emphasises network effects, knowledge flows, and adaptive capacity that allow for organic growth within planned frameworks.

This is what might be called "designed emergence", where planning creates conditions for ecosystem development rather than attempting to control outcomes directly. Modern innovation districts operate across multiple scales simultaneously, from regional positioning within broader innovation networks to neighbourhood-scale social spaces and community building.

Barcelona's 22@ district exemplifies this integration. Beginning with a deliberate urban transformation of an old industrial area, it succeeded by incorporating key ecosystem elements: innovation capabilities, human capital development, connectivity infrastructure, place quality, and governance capacity.

The planning created the platform, but ecosystem thinking ensured sustainable innovation dynamics that could adapt and evolve over time.

Innovation Precincts and Hubs

Innovation precincts represent a middle ground between large scale districts and small-scale hubs. Typically covering 8 to 80 hectares, these developments focus on specific innovation activities with planned spatial arrangements designed to encourage collaboration. They feature shared resources and infrastructure, including laboratories, maker spaces, and meeting areas, with coordinated management of tenant relationships and activities.

Several Australian universities have excelled at creating innovation precincts around their campuses, including the University of Technology Sydney, Macquarie University, the University of Melbourne, Monash University, RMIT University, Deakin University, and the University of Newcastle. Many successful precincts began as traditional science parks but evolved through deliberate transformation strategies to embrace more dynamic and integrated approaches.

This transformation involved creating active ground floors with public-facing businesses, extending operating hours beyond standard work schedules, implementing governance structures that include diverse stakeholders, and measuring success through innovation outputs rather than merely occupancy rates.

Successful precincts actively manage their tenant mix to create complementary capabilities and host regular events that bring together different sectors and disciplines.

Innovation hubs operate on an even smaller scale, often within a single building or campus covering less than eight hectares. These highly concentrated environments feature specialised equipment, expertise, and programs that enable intense collaboration and knowledge exchange. Examples include Melbourne's biosciences hub in Parkville and various university-based initiatives across Australia.

Learning Regions and Knowledge Infrastructure

The concept of learning regions, initially promoted by Richard Florida, describes geographical areas that are continuously engaged in innovation and knowledge generation through learning processes. These regions prioritise lifelong learning, digital literacy, inclusivity, and rapid skill adaptation, enhancing their resilience and responsiveness to global challenges.

Learning regions function as integrative knowledge ecosystems that involve educational institutions, businesses, and government agencies. They serve as collectors and repositories of knowledge and ideas, providing the underlying infrastructure that facilitates the flow of knowledge, ideas, and learning across different sectors and organisations.

During the 1990s and early 2000s, cities worldwide competed to identify themselves as "knowledge cities," reflecting concentrations in tertiary education and educated workforces. While some policy initiatives emerged from this trend, it often became primarily a strategy for regional government branding and investment attraction.

Universities have always played a particularly important role as "public spaces" within learning regions.

Beyond their traditional research and education functions, universities provide venues for open-ended conversations about industry development pathways, as well as new technological and market opportunities.

These public spaces include university-hosted meetings and conferences, standard-setting forums, investor forums, business plan competitions, industrial liaison programs, alumni networking activities, and industry-practitioner committees.

Co-working Spaces and New Work Patterns

Co-working spaces have emerged as vibrant innovation hubs that accommodate changing work patterns in service-dominated economies. These environments create collaborative, flexible, and entrepreneurial settings that are crucial for start-up ecosystems and innovation in the gig economy.

The co-working model emphasises shared values and social interaction, based on the understanding that working alongside other companies and individuals in friendly social spaces creates synergies that lead to new opportunities. Many co-working spaces cater specifically to high-technology and creative industries, often integrating with university-based incubators and accelerators to create innovation clusters.

Co-working is both a cultural phenomenon linked to start-up culture and a practical response to the high costs of urban office accommodation.

The focus on community within shared spaces reflects contemporary innovation patterns where collaboration across organisational boundaries becomes increasingly important.

Regional Innovation Systems

Regional Innovation Systems frameworks evolved from broader national innovation system thinking, emphasising regional characteristics in innovation processes. These models incorporate territorial embedding, networking among regional actors, and linkages with broader national and international systems.

Three main types of regional innovation systems can be identified based on their connections, production structures, and institutional arrangements.

- *Territorially embedded systems* encourage networking among small and medium-sized enterprises in industrial districts, building a competitive advantage through localised learning processes.
- *Regionally networked systems* aim to enhance public-private cooperation by strengthening institutional infrastructure.
- *Regionalised national systems* combine regional production structures and institutional infrastructure with national or international innovation systems.

The flexibility and dynamism of these systems, along with their openness and the extent and quality of their interactions, provide important evidence of the health of the innovation ecosystem. Regional facilitators who act as gatekeepers between different networks play crucial roles in enabling strong regional partnerships.

Urban Innovation Ecosystems

Urban innovation ecosystems represent complex networks of interconnected actors, institutions, and resources working together within cities. These encompass universities, start-ups, established companies, government agencies, funding organisations, and physical spaces like innovation districts.

For multiple precincts to function as cohesive urban innovation systems, several additional elements become necessary. These include:

- Consistent innovation planning across precincts
- Physical and digital connections linking different areas
- Cohesive branding that positions precincts as parts of a larger whole

- Activities that encourage movement and collaboration between precincts
- Funding mechanisms are accessible throughout the ecosystem.

Metropolitan innovation systems operate at an even larger scale, positioning substantial metropolitan areas as global centres of innovation. These feature multiple innovation districts and precincts within connected frameworks, signature specialisations that define global reputation, comprehensive innovation policies at the state and metropolitan levels, and major globally recognised research institutions.

Australian Context and Implications

Australia's innovation ecosystem development reflects both international trends and local conditions. With only a handful of true innovation districts emerging in cities like Canberra, Sydney's Tech Central, and Westmead, Australian planners and policymakers have opportunities to learn from international experiences while adapting their approaches to local governance structures and market conditions.

The integration of planning and ecosystem thinking is a maturation in understanding how cities can deliberately encourage economic transformation while remaining responsive to the organic dynamics that drive innovation and growth. This becomes particularly critical as innovation increasingly depends on cross-sector collaboration, rapid adaptation to technological change, and the ability to attract and retain diverse talent.

Success in creating sustainable innovation environments requires neither pure planning nor pure emergence alone, but rather careful orchestration of both approaches in service of broader economic and social objectives. This integration becomes essential in contemporary contexts where innovation demands cross-sector collaboration, rapid technological adaptation and diverse talent attraction and retention.

The evolution from industrial districts to innovation ecosystems advances the understanding of how cities can deliberately encourage economic transformation while remaining responsive to the organic dynamics that ultimately drive innovation and growth. This balanced approach offers the best prospects for creating innovation environments that can adapt and thrive in an increasingly complex and rapidly changing global economy.

2. Innovation Ecosystems and Economics: Innovation Policy Meets Regional Development Strategy

The transformation of regional economies through innovation presents one of the most compelling challenges in contemporary innovation policy development. At its core lies a fundamental tension between two distinct approaches to economic growth: the emergent, network-driven dynamics of innovation ecosystems and the structured frameworks of traditional regional development.

Understanding this intersection has become increasingly crucial as regions worldwide grapple with technological transformation while maintaining economic stability.

Theoretical Foundations of Divergent Approaches

The intellectual foundations of innovation ecosystem theory lie in complexity science and evolutionary economics. This perspective views regional development as an adaptive system characterised by non-linear interactions, emergence, and self-organisation. Innovation ecosystems operate through knowledge spillovers, network effects, and complex feedback loops that resist traditional planning approaches. The theory emphasises the role of institutional depth, particularly in research, knowledge absorption capacity in business, and social capital in fostering innovation.

Traditional economic development theory, grounded in neoclassical economics and industrial organisation, conceptualises regional growth through production functions, industry value added, agglomeration economies, and linear value chains. This framework emphasises factor conditions, market structures, and infrastructural foundations as primary drivers of economic development. The approach prioritises quantifiable inputs and outputs, seeking to establish clear causal relationships between policy interventions and economic outcomes.

These divergent theoretical perspectives generate distinctly different analytical frameworks.

Innovation ecosystem analysis employs network mapping, institutional analysis, and knowledge flow tracking to understand regional dynamics.

Traditional economic development relies on input-output analysis, location quotients, and multiplier effects to guide policy decisions. The methodological implications ripple through every aspect of policy design and implementation.

Policy Mechanisms and Implementation Frameworks

Contemporary policy mechanisms reflect these theoretical divisions. Innovation ecosystem policies typically deploy infrastructure investments focused on research and development, collaborative platforms, and institutional capacity building. These include initiatives to stimulate knowledge transfer, cooperative research programs, and innovation brokers who facilitate the flow of knowledge between actors.

Traditional economic development employs more direct intervention mechanisms, such as capital grants, tax incentives, and investments in physical infrastructure. These tools aim to influence firm location decisions, reduce production costs, and enhance market access. The policy toolkit includes enterprise zones, investment allowances, and workforce development programs with clearly defined outcomes.

Emerging hybrid mechanisms aim to bridge these approaches. Innovation districts combine the development of physical infrastructure with network building activities. Innovation funds blend traditional grant-making with objectives for ecosystem development. Skills partnerships integrate workforce development with innovation capacity building.

The Reality of Practice: Beyond Theoretical Frameworks

The practice of economic development often diverges significantly from theoretical ideals, instead following a pragmatic, process-driven approach. While rarely articulated in theoretical terms, this methodology dominates much of contemporary practice. It typically progresses through familiar stages: environmental scanning, document review, issue identification, stakeholder consultation, consensus building, and formal reporting.

This atheoretical approach offers distinct advantages. Its flexibility allows practitioners to adapt to local contexts and immediate needs without theoretical constraints. The emphasis on stakeholder consultation and consensus building can create broad-based support for development initiatives. The iterative nature of the process enables continuous adjustments based on emerging insights and changing circumstances.

However, this pragmatic approach has its limitations. Without theoretical grounding, practitioners may overlook crucial systemic relationships or underlying causal mechanisms. The focus on achieving stakeholder consensus can lead to "lowest-common-denominator" solutions that fail to address fundamental development challenges. The process-driven nature may emphasise short-term, visible outcomes over longer-term structural changes.

Perhaps most critically, the atheoretical approach struggles to differentiate between correlation and causation in developmental outcomes.

Without theoretical frameworks to guide analysis, practitioners may misinterpret successful outcomes in one context as universally applicable solutions, leading to inappropriate policy transfers between regions with different underlying conditions.

Methodological Tensions and Implementation Realities

The interplay between theoretical approaches and pragmatic implementation creates distinctive tensions in policy development. Traditional metrics struggle to capture the dynamics of innovation ecosystems, while process-driven approaches may overlook systemic relationships. These challenges extend beyond measurement to fundamental questions about how to understand and evaluate regional progress.

The temporal dimension adds further complexity. Innovation ecosystem development demands longer horizons than traditional approaches, while pragmatic methodologies often prioritise immediate, demonstrable outcomes. This temporal mismatch challenges policymakers in balancing short-term needs with structural transformation.

Governance structures complicate this dynamic even further. Structured and siloised organisational approaches may clash with the essential premise of knowledge networks in innovation ecosystems, while consensus-driven processes may dilute transformative initiatives. This necessitates a fundamental rethinking of institutional arrangements and decision-making frameworks.

Emerging Frameworks for Integration

Integration frameworks address a fundamental challenge in regional development: how to merge traditional economic planning with innovation ecosystem approaches. These emerging frameworks, discussed earlier in the book, move beyond simple combinations to create new analytical tools that recognise the inherent complexity of regional and place-based innovation ecosystems.

Smart specialisation offers a structured approach to integration by linking entrepreneurial discovery processes with regional economic planning. This framework helps regions identify and build on existing strengths while addressing innovation imperatives. It provides practical tools for aligning traditional industry development with the cultivation of innovation ecosystems.

The *triple helix framework* provides integration mechanisms through a systemic analysis of university-industry-government relationships. By examining how these institutions interact, the framework can reveal practical pathways for coordinating traditionally separate policy domains. It identifies specific intervention points where economic and innovation objectives naturally align.

Innovation networks facilitate organic knowledge flows while providing structured pathways for resource sharing and collaborative development. By combining formal institutional arrangements with informal relationship-building, innovation networks create flexible integration mechanisms that can adapt to evolving regional needs. This dual capability can make them effective as integration tools, allowing regions and places to maintain the necessary structure while creating the adaptability required for innovation driven growth.

Innovation intermediaries constitute critical mechanisms for bridging traditional economic planning with the development of innovation systems. Intermediary organisations transcend transactional approaches to technology transfer by developing partnerships and alliances that perform vital translation functions between different institutional domains. They interpret, recombine, and repurpose knowledge across boundaries, enabling integration between formal economic structures and informal innovation networks.

Innovation system integrators perform distinct yet complementary roles in creating cohesion across innovation ecosystems. These specialised entities work at multiple levels simultaneously, connecting individual firms with research capabilities, aligning industry clusters with knowledge infrastructure, and linking regional innovation systems with global networks. Their effectiveness comes from a deep understanding of both structured economic processes and fluid innovation dynamics.

Future Directions in Policy Development

The evolution of regional development approaches points towards an increasingly refined integration of innovation systems and traditional economic development frameworks. This integration requires ongoing attention to theories of innovation and economic development, policy design, evaluation, measurement, and governance.

The future effectiveness of regional development policy will likely depend on the ability to transcend traditional dichotomies between innovation and economic development approaches.

This suggests the need for new theoretical frameworks that can better accommodate the complexity of modern regional development while maintaining practical applicability.

Policy experimentation is becoming increasingly important in this context. Regions must develop capabilities for systematic learning and adaptation, creating feedback mechanisms that can inform ongoing policy development and implementation.

Conclusion

The integration of innovation ecosystems and regional economic development approaches is a crucial frontier in policy development. Success requires a comprehensive understanding of the strengths and limitations of both approaches, combined with the ability to create new frameworks that can effectively bridge these different perspectives.

The path is in developing more nuanced frameworks that can accommodate both structured and emergent development patterns. This integration demands a continued evolution in our thinking about regional development, moving beyond traditional dichotomies to embrace more complex, adaptive approaches that can effectively address the challenges of modern economic development.

The most effective regional development strategies will likely emerge from a thoughtful synthesis of these different approaches, creating frameworks that can maintain the rigour of traditional economic development while embracing the dynamic, networked nature of innovation ecosystems. This synthesis is a practical necessity for regions seeking sustainable prosperity in an increasingly complex economic landscape.

3. The Urban Planning Paradox: Balancing Regulatory Certainty with Strategic Agility

Innovation districts across Australia face a fundamental contradiction. These knowledge-intensive areas require the regulatory certainty that statutory planning provides; yet they must simultaneously adapt to the fluid, experimental nature of modern innovation ecosystem design. This tension between structure and agility is more than an administrative inconvenience; it is compromising our capacity to compete in the global knowledge economy.

The challenge lies at the intersection of two essential but often conflicting planning paradigms. *Statutory urban planning* ensures compliance, manages risk, and provides the market certainty that attracts investment in specialised facilities. *Strategic economic planning* drives innovation, builds ecosystems, and responds to technological shifts. Innovation districts cannot succeed without both; however, current governance frameworks struggle to reconcile the different logics of these two approaches.

The Regulatory Foundation: Why Structure Matters

Statutory planning provides the essential framework for innovation districts. Zoning laws and built form standards shape the physical environments where knowledge workers collaborate, ensuring that mixed-use developments, public interaction spaces, and walkable streetscapes create conditions for knowledge exchange.

Regulatory frameworks provide developers and investors with the certainty they need to commit capital to specialised infrastructure, such as laboratory spaces, advanced manufacturing facilities, and technology demonstration environments.

The accountability mechanisms built into statutory planning are equally important. Independent planning authorities insulate decision-making from political interference, thereby maintaining credibility with the diverse stakeholders on which innovation districts depend—universities, businesses, and surrounding communities.

This regulatory consistency becomes particularly crucial when innovation districts push boundaries through experimental building uses, adaptive reuse projects, and unconventional space programming that challenge traditional planning categories.

The strength of statutory planning, with its focus on regulatory compliance and physical form, can become a weakness when applied to knowledge-intensive environments. Traditional planning frameworks lack mechanisms to consider the innovation dynamics that drive district success: the knowledge spillovers between co-located firms, the informal collaborations that emerge from carefully designed public spaces, and the entrepreneurial networks that require flexible, adaptive environments.

The Innovation Imperative: Why Agility is Essential

Strategic planning addresses the innovation dynamics that statutory frameworks often overlook. Innovation districts require governance that can quickly adapt to evolving technological capabilities, emerging industry clusters, and changing collaboration patterns. The knowledge spillovers that make these districts valuable occur through both programmatic initiatives and physical proximity; however, they cannot be mandated solely through zoning laws.

Strategic planning recognises that innovation districts succeed through clustering and agglomeration effects. Knowledge-intensive firms benefit from co-location due to shared infrastructure and because of the concentration of talent and informal knowledge exchanges that occur within defined innovation geographies.

These benefits require active cultivation through policies that bring together universities, research institutions, businesses, and creative industries in ways that statutory planning cannot orchestrate.

The economic development focus of strategic planning also enables the cross-boundary collaboration that innovation requires. Many technological opportunities transcend administrative borders, necessitating flexible institutional structures that can coordinate among multiple jurisdictions, sectors, and stakeholder groups. Strategic planning can leverage research investments, talent attraction programs, and innovation incentives alongside physical development controls to create comprehensive innovation strategies.

Where Planning Paradigms Collide

The institutional separation between these planning functions creates predictable yet problematic tensions. Multiple agencies oversee development approvals, strategic innovation priorities, and infrastructure delivery without unified frameworks, creating confusion for stakeholders in innovation districts who must navigate fragmented decision-making processes.

More damaging is the temporal mismatch between planning paradigms. Statutory processes prioritise thorough assessment and regulatory compliance, which can take months or years. Innovation paradigms operate on much shorter cycles, where delays in approving transformative projects can undermine competitive positioning in fast-moving knowledge sectors. By the time approvals are granted, technological opportunities may have shifted, or competitor regions may have captured first-mover advantages.

The mismatch extends to priorities and metrics. Urban design and zoning frameworks often focus solely on physical form, inadvertently restricting the flexibility, mixed-use integration, and adaptive spaces that innovation ecosystems require. When statutory planners evaluate proposals against traditional criteria such as traffic impacts or height limits, they may overlook the knowledge dynamics that actually determine the success of innovation districts.

Lessons from Success and Failure

International examples illuminate both the potential and pitfalls of various approaches. Kendall Square in Cambridge demonstrates how delineating regulatory and strategic functions, while ensuring coordination, can support knowledge-intensive economic transformation while maintaining governance stability.

Kendall Square maintains clear regulatory standards for development while enabling strategic innovation initiatives that attract leading technology firms and research institutions.

Australia's own experiences provide cautionary lessons. The Sydney Harbour Foreshore Authority's struggle to reconcile its dual function as both a regulatory planning body and a strategic developer illustrates how blurring statutory and strategic roles can create institutional instability. When the same organisation attempts to regulate development and drive economic strategy, conflicts of interest emerge that undermine both regulatory credibility and strategic effectiveness.

Canberra's development history offers another instructive example. The historical tension between the National Capital Development Commission, which focused on implementing strict design principles, and the Department of Urban and Regional Development, which emphasised economic development and growth strategies, mirrors the challenges that many innovation districts face today.

The capital's current statutory and strategic planning challenges, including excessive red tape, lack of transparency, and bureaucratic complexity, exemplify how fragmentation between statutory and strategic planning functions can undermine the effectiveness of innovation districts by slowing approvals and creating uncertainty that deters investment in knowledge-intensive facilities.

A Framework for Integration

Successful innovation districts require hybrid governance models that maintain essential regulatory functions while embedding mechanisms for coordination with strategic innovation objectives. This integration cannot happen by accident; it requires deliberate institutional design that acknowledges the legitimate but distinct roles of statutory and strategic planning.

Innovation regions and districts require dedicated coordination units to ensure alignment between statutory requirements and strategic innovation priorities, particularly for specialised facilities such as research laboratories, technology demonstration spaces, and mixed-use innovation environments. These units should have clear authority to resolve conflicts between planning paradigms and the resources to facilitate ongoing dialogue between regulatory and strategic agencies.

The separation of responsibilities must remain clear, even as coordination improves. Regulatory bodies should focus on compliance, risk management, and maintaining public trust, while innovation-focused entities drive strategic clustering and knowledge-based development initiatives.

When these roles become confused, both regulatory integrity and strategic effectiveness suffer.

Performance measurement frameworks must evolve to capture both compliance outcomes and the impacts of innovation ecosystems. Traditional planning metrics, such as development approval timeframes and building compliance rates, need to be supplemented with innovation indicators that measure knowledge exchange, entrepreneurial activity, and collaboration patterns. This dual measurement approach helps ensure that governance decisions consider both regulatory requirements and innovative outcomes.

Stakeholder governance structures should provide universities, industry partners, entrepreneurs, and community representatives with structured input into both statutory and strategic frameworks. Innovation districts depend on these stakeholders for their success, and governance arrangements that exclude them from decision-making processes miss crucial insights into how policies affect innovation dynamics.

The Path Forward

Australia's innovation districts require statutory frameworks that maintain regulatory integrity while enabling the adaptability necessary for knowledge economies. Performance-based zoning and innovation district overlays offer promising approaches that allow for greater responsiveness to emerging technologies and collaboration models without compromising regulatory consistency.

The challenge is not choosing between regulatory certainty and innovation agility; both are essential for the success of innovation districts.

> *The challenge is to design governance frameworks that harness the strengths of both planning paradigms while minimising their conflicts. This requires moving beyond administrative convenience to institutional arrangements that reflect the complex, interconnected nature of innovation ecosystems.*

Getting this balance right matters for more than just individual innovation districts. As Australia seeks to strengthen its position in the global knowledge economy, the effectiveness of its innovation districts will increasingly determine its competitive advantage. The planning paradox that currently constrains these districts is a solvable challenge—but only if we commit to the institutional innovation that genuine solutions require.

4. The Personality Science of Start-up Success: Policy Insights for Australia's Innovation Economy

Paul X. McCarthy, Xian Gong, Fabian Stephany, Fabian Braesemann, Marian-Andrei Rizoiu, Margaret L. Kern, 8 August 2025[3].

Recent research analysing over 26,000 start-ups globally has revealed that founder personalities are not just quirky anecdotes but instead they're statistically significant predictors of company success. Using artificial intelligence to analyse social media data and a set of *Big Five* personality traits, researchers can now distinguish entrepreneurs from employees with 82.5% accuracy based purely on personality characteristics. Companies founded with multiple founders who combine certain specific personality types are up more than twice as likely to achieve successful exits through acquisition or public listing.

For Australian policymakers seeking to strengthen our innovation economy, these findings offer unprecedented insights into the human factors that drive start-up success. Rather than focusing only on funding, infrastructure, or regulatory frameworks, this research suggests personality diversity and team composition may be equally critical levers for fostering entrepreneurial success.

The Science Behind Entrepreneurial Personality

The study, led by researchers from the University of Technology Sydney, UNSW Sydney, University of Oxford, and the University of Melbourne, is the largest-scale analysis of entrepreneurial personality traits ever conducted. By combining data from Crunchbase (the world's largest start-up database) with Twitter personality analysis, the research team identified distinct psychological profiles that separate successful founders from the general workforce.

---

[3] This chapter draws on a paper titled *The impact of founder personalities on start-up success*, available at https://www.nature.com/articles/s41598-023-41980-y.pdf
The citation is: McCarthy, Paul X., Xian Gong, Fabian Braesemann, Fabian Stephany, Marian-Andrei Rizoiu, and Margaret L. Kern. "The impact of founder personalities on start-up success." Scientific Reports 13, no. 1 (2023): 17200.

The Adventurous Core

The most significant differentiator is *adventurousness,* which is a preference for variety, novelty, and starting new things. This trait, part of the Big Five personality framework's "Openness" domain, showed the largest effect size in distinguishing entrepreneurs from employees. Successful founders also showed notably lower levels of modesty and higher activity levels, suggesting they're comfortable being the centre of attention and maintaining high energy output.

Critically, the research debunks the myth of a single "entrepreneur type." Instead, it identifies six founder personalities:

- *Fighters* (pure hackers): Spontaneous and impulsive, tough, sceptical, and uncompromising
- *Operators* (operations-focused hackers): Highly conscientious, detail-oriented, and orderly
- *Accomplishers* (technology-focused hustlers): Extraverted and conscientious
- *Leaders* (pure hustlers): High openness, strong in emotional interests and altruism
- *Engineers* (hipsters): Highest in imagination and intellect
- *Developers* (product-focused hackers): Balanced "middle child" personality

The Power of Personality Diversity

Perhaps most significantly for policy consideration, the research shows larger, personality-diverse founding teams dramatically outperform solo founders. Start-ups with three or more founders are more than twice as likely to succeed as solo-founded companies. Teams combining complementary personality types and such as the "Hipster, Hacker, Hustler" trinity, show the highest success rates.

This finding challenges common assumptions about the lone genius entrepreneur and suggests that fostering collaboration between different personality types should be a priority for innovation policy.

Implications for Australian Innovation Policy

Rethinking Incubator and Accelerator Design

Traditional start-up support programs often focus on business planning, technical skills, and access to capital. The personality research suggests these programs should also emphasise team formation and personality compatibility assessment. Australian incubators could implement personality profiling to help founders identify complementary co-founders and build more balanced teams.

The research indicates that teams with specific combinations, such as a leader with two developers, or an Engineer, Leader, and Developer, achieve significantly higher success rates. Incubators could facilitate "founder dating" sessions based on personality complementarity rather than just sector knowledge or technical skills.

Immigration and Talent Attraction Strategies

Australia's start-up visa program and Global Talent Independent Program could benefit from incorporating personality insights. Rather than focusing exclusively on educational qualifications or business experience, immigration assessments could consider personality traits that predict entrepreneurial success.

The research shows that certain personality combinations are powerful when brought together. This suggests Australia might benefit from attracting diverse personality types rather than optimising for any single "ideal entrepreneur" profile. A policy framework that recognises the value of different founder personalities could help Australia build stronger start-up ecosystems.

Regional Development and Innovation Hubs

The study confirms that location significantly affects start-up success, with firms in certain countries and regions showing markedly higher success rates. For Australian regional development policy, this suggests that creating the right environment for personality-diverse teams to form and collaborate may be more important than traditional infrastructure investments alone.

Regional innovation hubs should be designed not just as physical spaces but as social ecosystems that help with the connection of complementary personality types. This might involve programming that brings together technical developers, business-focused leaders, and design thinking engineers from across geographic boundaries.

University Entrepreneurship Programs

Australian universities could redesign their entrepreneurship education to account for personality diversity. Rather than trying to make all students into "entrepreneurs," programs could help students understand their own personality strengths and identify the types of co-founders they need to succeed.

This approach could be valuable in addressing the cultural barriers to entrepreneurship that exist in some Australian contexts. Students who do not fit the stereotypical "hustler" mould might discover they have essential "hacker" or "hipster" qualities that are equally valuable in start-up teams.

Addressing Bias and Expanding Participation

Gender and Diversity Challenges

The research reveals about gender disparities in start-up funding and success rates. Female solo founders received 20% less funding than male counterparts, while female co-founded teams received less than half the funding of all-male teams. However, the study also found that successful female founders show personality profiles more similar to successful male founders than to the general female population.

This suggests that personality traits, rather than gender *per se,* drive success, but systemic bias in funding and support prevents many potentially successful female entrepreneurs from accessing resources. Australian policy could address this through initiatives such as:

- *Targeted funding programs* that use personality assessment rather than traditional business metrics for early-stage evaluation
- *Mentorship programs* that match founders based on complementary personality types rather than demographic characteristics
- *Investor education* about unconscious bias and the importance of personality diversity in founding teams

Cultural and Background Diversity

The research's global scope reveals that start-up success patterns vary significantly across countries and cultures. For Australia's multicultural innovation ecosystem, this suggests that founder teams combining different cultural backgrounds and personality types may have unique advantages.

Policy frameworks should recognise that different personality types may be more common in different cultural communities, and that fostering cross-cultural founding teams could enhance overall ecosystem performance.

Practical Policy Suggestions

Innovation Metrics and Measurement: Government innovation programs should expand beyond traditional metrics (patents, funding raised, jobs created) to include measures of team personality diversity and founding team composition. This would provide better predictive indicators of long-term success and more effective allocation of government support.

Procurement and Government Partnerships: Government procurement programs aimed at supporting start-ups could incorporate personality-based team assessment as a selection criterion. This would help identify start-ups with the personality diversity associated with higher success rates, potentially delivering better outcomes for public investment.

Education and Skills Policy: The research suggests that entrepreneurship cannot be taught as a universal skill but requires an understanding of individual personality strengths and team dynamics. Educational curriculums should shift from generic "entrepreneurship" training to personality-aware team formation and collaboration skills as well as the practical mechanics of establishing and growing successful start-ups.

Research and Development Collaboration: Government R&D programs could use personality insights to optimise team formation for research commercialisation. Academic researchers with strong technical skills (often "Fighter" or "Developer" personalities) could be systematically matched with business-oriented partners ("Accomplisher" or "Leader" personalities) to improve knowledge transfer success rates.

Future Considerations and Limitations

While this research provides valuable insights, policymakers should recognise several limitations. The study focuses primarily on technology start-ups and may not apply equally to all sectors of the innovation economy. The reliance on social media data also introduces potential bias towards younger, digitally native entrepreneurs.

The research also highlights temporal factors and that start-up success patterns change over time, and what works in one economic context may not apply in another. The *Science of Start-ups* research is ongoing with more research underway that exploring whether when adjusted for age whether personality changes by generations. Are people born among Millennial and Gen Z generations as entrepreneurial as those in Gen X and Baby Boomers for example?

Other *Science of Start-ups* research is exploring the macroeconomic impacts of personality diversity. Is any of the variance in GDP explained by personality diversity at a national level for example and does this account for any variance beyond other known factors like immigration and the strength of institutions say.

Conclusion

The *Science of Start-ups* offers Australian policymakers a new lens for understanding and supporting entrepreneurship. Rather than treating founders as interchangeable units or focusing only on traditional support mechanisms, policy frameworks should recognise the critical importance of personality diversity and team composition.

The research suggests that Australia's innovation success depends not just on creating more entrepreneurs but on fostering the formation of personality-diverse founding teams. This requires policy approaches that move beyond individual founder support to ecosystem-level interventions that help with the connection of complementary personality types.

For a country seeking to strengthen its position in the global innovation economy, understanding the personality science of start-up success provides a competitive advantage. By incorporating these insights into immigration, education, regional development, and innovation support policies, Australia can build a more sophisticated and effective approach to fostering entrepreneurial success.

The message for policymakers from the science of start-ups is that personality is a critical economic factor that deserves systematic policy attention. The question is whether Australia will be among the first to leverage these insights for a competitive advantage in the global innovation race.

5. Culture and Creativity: Driving Forces in Innovation

Seventeen years ago, when the National Innovation Review (the Cutler Review, 2008) was underway, the Australian innovation landscape was dominated by a science and technology paradigm that viewed research and development (R&D) as the primary engine of economic progress. The creative industries were seen as peripheral to serious innovation policy, relegated to cultural portfolios rather than recognised as drivers of economic competitiveness. Today, that perspective has transformed.

Many people involved in urban and regional innovation ecosystems recognise culture and creativity as essential components of sustainable competitive advantage in a globally connected, digitally enabled economy, rather than as nice-to-have additions.

This chapter examines how our understanding of innovation has evolved in this domain, explores the contemporary role of culture and creativity in driving innovation, and considers the policy implications for Australia's continued economic prosperity.

The Evolution from Linear to Ecosystem Thinking

The 2008 perspective on innovation was largely linear: basic research led to applied research, which led to development, and ultimately to commercialisation. This model failed to capture how innovation actually occurs in practice, through the serendipitous encounters, cross-cultural pollination, and the creative friction that arises when technologists grab coffee next to street artists, when start-up founders attend gallery openings, and when diverse communities share ideas in informal settings.

Today's ecosystem thinking recognises innovation as emerging from dynamic interactions between multiple actors, institutions, and knowledge domains—interactions that increasingly occur in mixed-use cultural precincts rather than in isolated business parks. The emergence of platform economics, digital transformation, and the creator economy has altered how value is created. Increasingly, this creation occurs in spaces that blur the boundaries between work, culture, and community.

Contemporary innovation ecosystems are characterised by:

- *Place-based clustering*: The concentration of diverse creative and technical talent in culturally rich neighbourhoods
- *Cross-cultural pollination:* Innovation emerging from interactions between different cultural communities and knowledge traditions
- *Live-work-play integration*: Spaces where professional collaboration extends seamlessly into cultural and social engagement
- *Community-driven innovation:* Bottom-up innovation emerging from local communities rather than top-down institutional programs

Arts and Culture Precincts as Innovation Infrastructure

Modern innovation theory recognises arts and culture precincts as essential innovation infrastructure in the form of physical and social spaces that enable the creative collisions and cultural mixing that drive breakthrough thinking. This is a fundamental shift from viewing cultural districts as spaces for consumption to understanding them as environments for the production of new ideas, products, and services.

The Tech Central Model: Integration in Practice

Sydney's Tech Central precinct exemplifies this new model. Rather than creating a sterile "technology park," the development integrates start-up accelerators with artist studios, corporate offices with live music venues, and research facilities with community gardens and multicultural food courts. This design recognises that innovation requires more than technical expertise; it needs the cultural vibrancy and diversity that attract and retain creative talent.

The precinct's success is evident in the innovations that emerge from unexpected collaborations. When game developers work alongside installation artists, new forms of interactive storytelling emerge. When fintech start-ups are located near migrant communities, they can potentially develop more culturally appropriate financial products. When biotech researchers frequent venues showcasing Indigenous artists, they can gain new perspectives on systems thinking and environmental design.

Multicultural Innovation Networks

Australia's multicultural cities provide unique advantages in the global innovation economy. Arts and culture precincts that celebrate this diversity, such as Ethiopian coffee houses next to Vietnamese galleries, Lebanese restaurants alongside Korean tech meetups. They create environments where different knowledge systems and market insights naturally cross-pollinate.

In Melbourne's Collingwood precinct, this multicultural mixing has driven innovations including new food technologies developed by companies serving diverse dietary requirements, translation software created by start-ups embedded in multilingual communities and sustainable fashion emerging from designers who combine traditional textile techniques with contemporary environmental concerns.

Community Engagement as Innovation Driver

The most successful culture precincts engage broader communities as active participants in innovation processes. Community gardens serve as testing grounds for urban agriculture technologies. Local maker spaces enable residents to prototype solutions to neighbourhood challenges. Cultural festivals provide platforms for showcasing and iterating new products and services.

Brisbane's Fortitude Valley demonstrates this community-driven approach. In addition to technology companies and creative studios, the precinct's innovation ecosystem includes community workshops where local residents collaborate with start-ups to develop solutions to urban challenges. This has produced innovations in everything from waste reduction, developed through community recycling initiatives, to aged care technology, which has emerged from intergenerational community programs.

The Creator Economy and Precinct Development

The emergence of the creator economy, valued globally at over US$100 billion, has changed how cultural precincts function as innovation environments. These spaces now serve as production hubs where creators develop content, products, and services for global audiences, evolving from venues for cultural consumption.

Hybrid Spaces for Hybrid Careers

Contemporary creators often work across multiple domains—a musician might also develop apps, a visual artist might run a social media consultancy, and a chef might create educational content. Culture precincts that support these hybrid careers, with co-working spaces accommodating recording studios, maker labs serving both artists and engineers, and event spaces that can host gallery openings as well as product launches, create environments where interdisciplinary innovation flourishes.

Digital-Physical Integration

Modern cultural precincts succeed by seamlessly integrating digital and physical experiences. Live music venues stream performances globally, galleries use AR to extend exhibitions into virtual spaces, and restaurants develop apps that connect diners with local food producers. This integration creates new possibilities for innovation by combining digital technologies with embodied cultural experiences.

Sustainability and Regenerative Innovation

The sustainability imperative has driven significant innovation in how cultural precincts operate and the innovations they foster. Rather than simply minimising environmental impact, leading precincts are becoming testing grounds for regenerative approaches that actively improve social and environmental conditions.

Circular Precinct Design

Contemporary cultural precincts increasingly operate as circular systems, where waste from one activity becomes input for another. Food waste from restaurants feeds community gardens, which supply ingredients back to cafés. Art installations are designed for disassembly and reuse. Event spaces are powered by renewable energy generated within the precinct.

Indigenous Knowledge Integration

Australia's cultural precincts increasingly recognise Indigenous knowledge systems as sources of sophisticated solutions to contemporary challenges. Traditional ecological knowledge informs sustainable design practices, Indigenous storytelling traditions inspire new approaches to data visualisation, and traditional governance systems provide models for more inclusive innovation processes.

Policy Implications for Precinct Development

The recognition of cultural precincts as innovation infrastructure has significant implications for planning, economic development, and innovation policy. Rather than treating cultural development and innovation support as separate policy domains, governments are increasingly recognising the need for integrated approaches.

Integrated Planning Approaches: Successful culture precincts require planning approaches that explicitly integrate cultural programming with innovation support. This means zoning that allows for mixed-use development, transport planning that prioritises accessibility for diverse communities, and public space design that facilitates both formal and informal interactions.

The New South Wales Government's approach to Tech Central suggests that this integrated thinking may be at work in combining infrastructure investment with cultural programming support and community engagement initiatives to create conditions for the sustainable development of an innovation ecosystem.

Supporting Cultural Entrepreneurship: Traditional business support programs often fail to address the specific needs of cultural entrepreneurs who work across creative and commercial domains. New approaches recognise that cultural innovators need access to both traditional business support (finance, legal advice, market development) and culturally specific resources (creative spaces, peer networks, cultural mentorship).

Measuring Precinct Success: Traditional economic development metrics, such as job creation, business formation, and revenue generation, fail to capture the value created by cultural precincts. Contemporary approaches require measurement frameworks that account for social cohesion, cultural vitality, community engagement, and the effects of innovation ecosystems.

Impact metrics for creative precincts should include cultural participation rates, the frequency of cross-sector collaboration, and community wellbeing measures, providing a more comprehensive understanding of how cultural investment drives broader innovation outcomes.

Future Directions: The Next Generation of Culture Precincts

Several trends will further strengthen the role of cultural precincts in driving innovation.

- *Climate Innovation Hubs*: Precincts designed to address sustainability challenges by combining climate tech companies with cultural programs that engage communities in environmental innovation.

- *Care Economy Integration*: Recognising that innovation in aged care, childcare, and community support often emerges from culturally rich environments that bring together different generations and communities.
- *First Nations Leadership:* Growing recognition that Indigenous-led precinct development offers unique approaches to innovation that combine cultural preservation with technological advancement.
- *Global-Local Connection:* Precincts that maintain a strong local cultural identity while connecting to global innovation networks enable distinctly Australian innovations to reach international markets.

Conclusion

The transformation of Australian cities through arts and culture precincts is one of the most significant innovations in creating conditions for economic and social prosperity. These spaces demonstrate that innovation involves creating environments where diverse communities can collaborate, experiment, and create new forms of value beyond technology transfer or R&D investment.

From Tech Central's integration of start-ups with street art to Melbourne's multicultural maker spaces, from Brisbane's community-engaged innovation labs to Perth's digital Indigenous cultural experiences, Australia's cultural precincts are proving that the future belongs to places that can effectively combine technological capability with cultural authenticity, global connectivity with local community engagement, and economic development with social and environmental regeneration.

The policy challenge is to support this organic development while avoiding the sterile "innovation districts" that fails to generate genuine cultural vitality. This requires integrated approaches that recognise culture and creativity as capabilities that enable innovation to flourish, well beyond the amenities designed to attract innovation.

As Australian cities continue to evolve, the regions that nurture arts and culture precincts, with their galleries and live music venues, multicultural restaurants, and community spaces, together with their seamless integration of work and creativity, will be the ones best positioned to compete in an innovation economy that increasingly rewards places where diverse minds can come together to imagine and create new possibilities.

6. Cultural Value, Social Meaning, and Economic Opportunity: The Hidden Wealth of Food

The analysis of Australia's food system appears constrained by a commodity paradigm that obscures where significant value is created.

While official accounts indicate that agricultural production, including both food and fibres (wool, cotton, hemp, etc.), contributes 2.4% of GDP, broader food system activities, including food and beverage manufacturing and related sectors, increase the contribution to significantly more, with estimates from various sources ranging from around 11% to 15%.

But these numbers do not include the role of food in the cultural, social, and broader sectors of the economy, which potentially contribute an even greater amount when viewed as integrated adaptive systems through the lens of complex value chains.

This apparent analytical myopia has structural foundations: Rural Research and Development Corporations, funded by producer levies and matched government payments, tend to focus on "farm gate" metrics, while support for institutions that promote consumption-oriented innovation, such as Food Innovation Australia Limited (FIAL), has been removed.

Meanwhile, food has become a primary medium for expression, identity formation, and social connection in Australia's diverse and multicultural society. In addition, the "food media revolution" and the material culture of cooking have transformed food from mere sustenance into a new phenomenon of consumption.

All of this necessitates the development of new conceptual and analytical frameworks. Policy thinking must shift from a production-centric approach to incorporate demand-side and consumer-led strategies that recognise food's role in community, creativity, and lifestyles that enable meaningful nourishment, enjoyment, and experiences, rather than merely efficient eating.

Context

Australia's food system analysis remains trapped in a commodity paradigm that fundamentally misrepresents where value is created and contested.

National Accounts show that agriculture contributes 2.4% of GDP, a figure that includes primary production, or "pre-farm gate activity". Broader definitions encompassing retail, food service, logistics, food media, and culinary tourism suggest contributions to GDP of much greater magnitudes. At this stage, however, consistent and standardised data is not readily available,

Food has become more than just sustenance. It serves a role in culture, identity, pleasure, and connection. Systemic analysis must evolve from being commodity-focused to include consumption-led frameworks.

Without this shift, policy and investment are likely to overlook the emergence of new forms of value creation.

The Structural Problem: Where Agricultural Research Misses the Mark

Australia's Rural Research and Development Corporations (RDCs) exemplify how institutional architecture shapes innovation priorities. Designed to co-invest producer levies with public funds, RDCs are structurally linked to farm gate metrics: yield, input efficiency, and export performance. This sectoral approach limits system-level innovation in areas such as multicultural food development, culinary tourism, artisanal manufacturing, and the overall food experience.

The 2023 closure of Food Innovation Australia Limited (FIAL) epitomises this commodity bias. As a rare cross-cutting intermediary focused on demand-driven innovation, FIAL's defunding signalled a policy preference for measurable production outputs over market signals and demand responsiveness.

Beyond Commodities: The Cultural and Economic Reality

The commodity paradigm captures measurable outputs but overlooks meaning, experience, and community. It reduces the social dimensions of food, such as hospitality, cohesion, and intergenerational knowledge transfer, to merely dietary outcomes. The role of food in pleasure, tourism, education, and aesthetic experience receives minimal consideration, despite representing growing economic sectors.

This narrow focus has cultural consequences. Unlike societies shaped by scarcity, Australia's abundance has led to a systematic undervaluation of food. This manifests in staggering food waste levels (7.6 million tonnes annually). This is not merely a logistical inefficiency, but symptomatic of treating food as a disposable commodity rather than as a cultural substance deserving of respect.

The assumption that efficient production automatically delivers positive consumption outcomes has become increasingly problematic. Current challenges relate less to scarcity than to meaning, connection, and cultural preservation. Despite an abundance of food, Australians face nutrition challenges, eroding food traditions, and declining opportunities for culinary learning and communal eating.

Cultural Transformation: Food as Identity and Performance

The "food media revolution" has fundamentally altered food's social role. Beginning with instructional cooking shows, it has expanded into celebrity chef entertainment, heritage documentaries, competitions, and digital platforms, making food content globally accessible.

Social media has given rise to food influencers, a photogenic meal culture, and viral culinary trends. These developments are rarely addressed in conventional food system analysis.

Food choices can also serve as identity statements. Veganism, keto, and low-FODMAP diets have become badges of belonging. Regional identities strengthen through "farm-to-table" movements, while generational preferences for artisanal coffee and sourdough reflect social distinctions.

Food photography restaurant reviews, and culinary tourism convey cultural capital. Cooking_classes and documentaries communicate skills, history, and values.

Food has become one of Australia's principal cultural languages. Analysis ignoring this transformation misses where future value will emerge.

Multicultural Innovation and Social Cohesion

Australia's multicultural society is a major force in reshaping its food system. Successive waves of migration have transformed what is grown, eaten, and ritualised. Italian, Greek, Vietnamese, Lebanese, Chinese, Indian, and other communities have influenced supply chains, farming, retailing, and definitions of "Australian food.". Asian greens in market gardens, halal meat processing, and ubiquitous Italian coffee culture demonstrate how adaptation drives innovation.

Food markets, shopping strips, and festivals operate as more than economic venues; they are sites of cultural exchange and mutual adaptation. Dishes like laksa, pho, sushi, souvlaki, and biryani have transitioned from the periphery to the mainstream. Fusion experimentation, often led by younger Australians, contributes to culinary creativity.

Suburban and regional farmers' markets exemplify the complexity of the food system, operating simultaneously as alternative distribution channels, craft venues, social gathering points, and educational spaces. Though seemingly marginal in official frameworks, they support resilience, cultural preservation, and social inclusion while fostering cross-cultural understanding in a highly diverse society.

Material Culture and Economic Opportunity

The material dimension of food has evolved dramatically. Australians invest heavily in specialty appliances, such as air fryers, sous vide machines, and fermentation kits, as well as in premium ingredients like single-origin flours and artisanal cheeses. This wave of gadgetry creates new economic activity in manufacturing and design while democratising professional expertise.

Policy and industry reporting lag behind these trends, focusing on large scale export processing while ignoring import dependence on specialty foods and opportunities for small-batch, design-driven manufacturing. Growth has been enabled by venture capital for start-ups, as well as by the emergence of incubators and food precincts near research organisations. However, limited access to expansion capital remains a barrier for small firms that lack economies of scale.

The Paradigm Shift: From Production to Consumption

A consumption-centred approach inverts traditional questions from "How do we produce efficiently?" to "How do Australians, and internationally, want to experience food, and how can systems support that?". This requires designing from desired experiences to necessary capabilities to production choices. Measurement must include cultural vitality, social cohesion, experiential richness, and equitable access.

However, consumption-oriented models face challenges. Market demand, shaped by marketing and convenience, has contributed to the increased consumption of ultra-processed foods. Gaps exist between stated preferences for health and connection and actual consumption patterns, which are influenced by commercial forces and practical constraints.

Policy cannot, therefore, simply respond to market demand but must support informed preferences through education and regulation, address access barriers, and enable choices that align with cultural values and community health. This recognises that choice is always influenced by information, opportunity, and social context.

Conclusion: Food as Cultural Commons

Food serves as a foundation for community, identity, learning, and wellbeing. Policies and analyses that fail to acknowledge these dimensions risk missing where real value and resilience are created. Consumers seek authenticity, health, sustainability, convenience, and new experiences. Innovation is now emerging at the intersection of production and consumption, with consumer values shaping the direction of the industry.

A forward-looking policy for Australia's food system should begin by asking how to create conditions that enable people to eat, live, and connect well. While efficient production remains important, designing systems that allow culture, creativity, and community to thrive will ultimately define Australia's food future.

Food is a cultural commons across innovation ecosystems that requires careful stewardship and investment in its full potential, and not just an economic opportunity.

**Comments and suggestions by Mirjana Prica, Chair & Managing Director of FIAL, on an earlier draft of this chapter are greatly appreciated.*

7. Hidden Innovation: The Connective Tissue of Innovation Districts

In innovation districts and precincts, much value creation occurs in ways that standard measures do not recognise. The usual focus on patents, research efforts, and the number of start-ups often overlooks organisational, service-related, and local factors that support successful collaboration.

Evidence suggests that these less obvious elements, shared platforms, translational infrastructures, governance systems, and coordinated alliances, are the foundations on which innovation is built and grows.

This understanding is especially important for innovation regions, districts, and precincts, where physical structures, such as buildings, laboratories, and anchor organisations, depend on an equally vital framework of coordination, service design, and governance.

This chapter examines the hidden innovations within Australian innovation districts, their implications for policy and evaluation, and how a diagnostic approach can guide future investments.

What Does Hidden Innovation Mean?

The idea of *hidden innovation* was introduced to describe these activities and acknowledge their economic and social value, despite being excluded from current measurement frameworks. As NESTA initially stated, innovation can still happen without patents or investments in business R&D, as long as it encourages adoption, solves challenges, or stimulates growth (Nesta, 2007).

NESTA identified four different types: creativity that is similar to measured activities but not included in statistical definitions; innovation without a strong science and technology foundation, such as organisational or business model changes; innovation through recombining existing technologies and processes; and small-scale, localised innovations that do not appear in aggregate data.

The 2018 Oslo Manual clarified this view by explicitly recognising business process innovation, extending beyond product development and R&D-heavy sectors (OECD/Eurostat, 2018). What may seem like administrative or organisational coordination is often the key innovation that enables knowledge transfer, accelerates adoption, and reduces the time required to bring new technologies to market.

Hidden Innovation in Practice

Service and Organisational Innovation

Innovation district operators act as facilitators and integrators rather than just property managers. Their innovation lies in developing programs, building communities, and fostering connections among research, industry, and government.

Tech Central in Sydney exemplifies this approach by establishing a unified framework that connects universities, healthcare providers, and local councils. The Macquarie Park Innovation District encourages collaboration at Macquarie Park through an incorporated not-for-profit organisation, aiming to transform simple transactions into cooperative problem-solving efforts.

These initiatives exemplify service innovation by introducing and implementing new methods of organising collaboration that boost overall efficiency.

Translational and Demand-Side Innovation

Many districts and precincts have incubator facilities and clinical translation resources to connect supply and demand. At Parkville, CSL's headquarters, an incubator space has been established that offers pathways for adoption, procurement opportunities, and regulatory insights for smaller companies. The Western Sydney Aerotropolis demonstrates how government processes for purchasing and planning infrastructure can create initial markets for innovative technologies.

Demand-side engagement offers a unique form of innovation. When major organisations or government agencies test and implement new solutions, they verify market viability and accelerate business development growth.

Platforms and Testbeds

Subtle innovation is also evident in the design of communal infrastructure. Maker spaces, prototyping facilities, and open-access labs are resources that offer technical support and exemplify new approaches to managing access within institutions. Many innovation districts combine prototyping facilities with co-working spaces and public engagement areas, such as galleries, to create a platform for interaction rather than merely a separate asset.

These examples demonstrate that platform innovation resides in the practices, service models, and integration efforts that transform physical assets into tools for collaboration.

The Intangible Economy and Precinct Dynamics

These "hidden" activities contribute to the development of intangible capital, which includes organisational processes, design, software, branding, and data. Corrado, Hulten, and Sichel (2006) demonstrated that investments in intangible assets are significant and enduring, but they are often underestimated in national economic statistics. Haskel and Westlake (2017) proposed that economies primarily driven by intangible assets display unique traits, including spillovers, scalability, and synergies.

Innovation districts and precincts exemplify small-scale instances of the intangible economy. Knowledge accumulates through networks and shared resources, with value emerging from interactions among individuals.

The collaboration of knowledge across different fields and industries is precisely what von Hippel (2005) referred to as user-driven problem-solving, in which professionals such as clinicians, engineers, and city managers work closely with researchers and suppliers to create solutions.

Cross-Cutting Observations

Three important insights arise from this discussion.

Enabling platforms are the innovation. Many innovation districts combine prototyping, co-working, and public engagement to create iterative processes for companies and researchers. Their significance lies more in the design of rules, access models, and services than in the physical infrastructure.

Governance Design as Innovation. Institutional frameworks can also be a source of innovation. Several districts demonstrate how a committed systems integrator can develop common agendas. Macquarie Park and the Gold Coast districts demonstrate how alliance governance operates across different regions. The Western Sydney City Deal, which focuses on Bradfield and the Aerotropolis, demonstrates how coordinating decisions about infrastructure, land use, and industry development under a single governance agreement is a novel approach to institutional design.

Connectivity and Programmatic Integration. While physical connections through transport or digital networks are vital, they are insufficient on their own. Programs that encourage collaboration between industry and universities, facilitate clinical translation, or provide shared testbeds transform resources into real outcomes. Macquarie Park transitioned from an office park to an innovation district by enhancing transportation, connecting hospitals to the campus, and fostering purposeful collaboration.

Policy and Measurement Implications

The discussion so far suggests some policy and measurement issues.

Measure what matters. If the value of districts lies in coordination, service models, and shared platforms, then the indicators should reflect these elements. The Oslo Manual includes process innovation, but applying this to district and precinct reporting requires new metrics. Nesta's strategy of using innovation stories tailored to specific sectors serves as a good example (Nesta, 2007; Miles & Green, 2008).

Invest in integrators and intermediaries. Orchestrators like Tech Central and Macquarie Park are valuable assets for innovation. Policy frameworks should recognise and fund these organisations as vital infrastructure.

Back platforms before projects. Shared facilities that serve multiple purposes offer "options value" to groups of companies. Many examples show that integrated platforms are more effective than stand-alone assets.

Use demand-side levers. Procurement, adoption programs, and regulatory flexibility can have transformative effects. Parkville's approaches to translating clinical research and securing resources are excellent examples.

Align place governance with innovation goals. The Western Sydney City Deal exemplifies how aligning governance frameworks with innovative results can yield significant benefits. Tech Central demonstrates how forming alliances can help navigate complex situations involving multiple jurisdictions.

These matters question the perspective that defines innovation activity solely in terms of tangible metrics, such as patents, research intensity, and the number of start-ups. Even then, start-ups succeed in settings that have carefully organised communities, interconnected pathways, open testing areas, and sustainable governance structures.

Hidden innovation is central, not marginal. It serves as the essential foundation for thriving innovation districts. It lies in the decisions that policymakers and leaders make regarding governance, access, coordination, and service models.

Conclusion

Innovation districts thrive when noticeable investments in infrastructure and research facilities are complemented by less obvious investments in collaboration, service design, and governance. Hidden innovation is fundamental element in innovation ecosystems, and acknowledging, assessing, and providing funding for it is crucial for districts to fulfil their commitment to promoting adoption, stimulating growth, and addressing societal challenges.

Policymakers face the challenge of broadening the discussion about innovation to include these often-overlooked aspects, ensuring that collaborative efforts are nurtured with as much intention as the physical infrastructure.

8. The Missing HASS: Why Innovation Ecosystems Need the Humanities, Arts and Social Sciences

Discussions about innovation policies worldwide are increasingly focused on innovation ecosystems. They are the preferred approach to drive economic growth, reform universities, encourage urban and regional development, boost productivity, and enhance national competitiveness. The frameworks we use to understand them are grounded in economics, urban design, management theory, and science and technology policy. These disciplines offer useful methods for analysing relationships and spotting gaps within the system.

The Humanities, Arts, and Social Sciences (HASS) are often overlooked in this effort. They sometimes appear as supplementary elements or afterthoughts, but are rarely regarded as essential for designing, managing, or assessing innovation ecosystems. Their absence limits our capacity to understand the complex factors that influence innovation practice, such as values, identities, histories, norms, ethics, and communication. These elements are not merely decorative additions; they are fundamental.

This chapter examines the extent to which *ecosystem frameworks that overlook or downplay the contributions of HASS may be fatally flawed.* Frameworks cannot always explain why innovation succeeds in some areas but not in others, why certain districts and precincts thrive while others remain stagnant, or why some initiatives are accepted while others face opposition. Policymakers and ecosystem developers should utilise the interpretive, critical, and creative insights provided by Humanities, Arts, and Social Sciences (HASS) disciplines.

What are We Overlooking?

Innovation is often described as either a process that moves from input to outcomes or as a network that connects institutions, individuals, and resources. These models can easily take on a mechanical nature. They prioritise what can be seen over what cannot be seen, what is physical over what is symbolic, and what can be measured over what has significance.

HASS disciplines provide deeper understanding and insight. They help in understanding how people and communities interpret, accept, oppose, or adapt to innovation. They pose various questions, such as:

- What stories motivate the desire to innovate?
- Whose stories get told?
- Who is being silenced?
- How does culture enable or block collaboration?
- What rules influence behaviour within districts and precincts, and how are they established?

Research and practice also indicate that innovation ecosystems are influenced by symbols, rituals, and common beliefs: language matters, identity is important, and design plays a significant role.

Innovation is likely to be more successful when it aligns with local aspirations, values, and personal experiences. Without this connection, even the most well-funded ecosystem might struggle.

Cultural Infrastructure and Placemaking

Innovation districts are usually assessed using economic or geographical metrics. How many start-ups are there? How much investment is attracted? How many square metres of commercial space are available? These indicators are significant, but they miss an essential aspect. Successful precincts over time are places where people want to gather. They are spaces for sharing ideas, shaped by events, cultural activities, community identity, and collective experiences.

The disciplines within HASS provide the essential theoretical frameworks to understand these dynamics. They explore how public art, heritage, festivals, music, and creative industries can boost a neighbourhood's appeal. They show what makes placemaking efforts succeed or fail. They highlight how design impacts human experiences by demonstrating how streets, signs, buildings, and communal areas can either promote or hinder casual interactions. Culture is often viewed as secondary, but in reality, it is fundamental.

There is growing recognition that a district's livability influences its ability to attract talent, retain community support, and maintain momentum. Cultural infrastructure is not an add-on; it is a vital part of the system.

Ethics, Equity, and the Social Contract

Innovation is never neutral. It creates both winners and losers. It changes the balance of power, job prospects, access, and engagement. Questions about who benefits, who makes decisions, and who bears the risks are not minor issues; they are core concerns.

Social scientists and humanists have long studied the ethical side of change. They examine the concepts of inclusion, justice, trust, and legitimacy. They explore how different communities react to technological and economic upheavals. They understand that innovation impacts groups—such as those based on gender, ethnicity, social class, age, and cognitive abilities—in diverse ways.

Without their input, designing ecosystems might miss or misunderstand patterns of exclusion. Strategies at the local level that overlook these dynamics may face opposition, resulting in community backlash, regulatory challenges, or damage to reputation. Social licence is just as vital for urban innovation as it is for the mining and energy industries. It must be earned and maintained.

Narrative, Communication, and Public Meaning

Adoption requires both convincing others and demonstrating effective actions. Ideas become more persuasive when they are clearly communicated, their meanings are understood, and they resonate with public feelings. HASS disciplines are vital here. Narrative frameworks help in understanding policy goals. They bring order to uncertainty, guide actions, and clarify differing interests. They also illustrate how stories change, how language evolves, and how legitimacy is formed.

Fields such as communication studies, design, and linguistics offer valuable tools for crafting effective messages and building audience engagement.

The absence of emotional connection is clear when innovation strategies are executed with technical accuracy but fail to engage people. Ineffective communication of meaning consistently limits the impact of policies.

Social Dynamics and Human Systems

Innovation is a social process that relies on connections between people, mutual trust, shared expectations, and unwritten guidelines.

HASS disciplines explore these dynamics, describing how social capital develops, leadership emerges, cooperation is fostered, and norms evolve over time. These elements are crucial to the functioning of ecosystems.

Insights from psychology and sociology can inform the development of incentives, governance frameworks, and collaborative processes. Institutional theory explains why some partnerships endure while others fail. Political science investigates how power and legitimacy operate within societies. Anthropology, by closely studying human behaviour, reveals how people experience innovation in their daily lives.

Together, these perspectives provide a more complete understanding of the human systems that underpin innovation.

The Consequences of Exclusion

Ignoring HASS perspectives carries significant risks. Diagnosis can become biased. Problems are often framed as technical or economic rather than cultural or social. Although solutions might seem logical, they often fail to gain support. Ecosystem interventions may feel imposed, leading to scepticism or resistance. Measurement frameworks tend to focus only on inputs and outputs, often missing important outcomes like civic pride, cultural value, and institutional trust.

There's also an opportunity cost. Precincts that do not invest in cultural infrastructure or neglect the creative sector may miss chances to stand out, inspire others, or achieve international recognition.

22@Barcelona demonstrates what can be achieved when cultural and economic strategies converge. Australian examples suggest the opposite: areas with effective planning and governance but lacking a strong identity are less attractive.

Rethinking Integration

The key to addressing this issue is to include HASS disciplines as essential contributors in strategy, governance, and assessment of ecosystems. This means more than consultation; it involves collaborative leadership, reciprocal respect, and equitable resource distribution.

Ecosystem design should prioritise cultural infrastructure and creative programming as fundamental elements rather than mere decorative features. Evaluation should broaden its focus from solely economic indicators to also assess cultural vitality, social cohesion, and public trust. Interdisciplinary careers bridging HASS and STEM should be recognised and supported, with funding mechanisms incentivising cross-disciplinary collaboration.

While there is no definitive model, pathways can be explored. Cities that prioritise cultural life, such as Barcelona, Kendall Square, Montréal, and Manchester, demonstrate that integrating Humanities, Arts, and Social Sciences (HASS) can create unique, adaptable, and resilient ecosystems.

Next Steps for Policy and Research

Australian innovation policy has begun to acknowledge the importance of inclusion, diversity, and local context. The integration of HASS remains inconsistent. Some initiatives, like Creative Australia, seek to empower and honour First Nations arts and culture. They highlight the creative industries and the ARC's support for interdisciplinary research, offering guidance for future efforts. Others are still trapped in limited paradigms.

There is a chance for a national dialogue about the future role of HASS in encouraging innovation. This would involve government officials, universities, cultural organisations, and community leaders. Such a conversation would explore which types of innovation we prioritise, the criteria we use to evaluate it, and how we ensure it remains inclusive, ethical, and meaningful.

The research agenda is also relevant. Research examining the cultural dynamics of neighbourhoods, the role of storytelling in innovation policy, the ethics of creating spaces, or the experiences of innovators would add valuable insights to the current literature. These are not fringe topics. They are part of the next stage of understanding ecosystems.

Conclusion

HASS demonstrate that innovation is not just a technical achievement; it is also a cultural act and a narrative, rather than simply a system. To succeed in economic, social, and political spheres, innovation ecosystems must be designed and evaluated from a comprehensive viewpoint.

Engaging HASS is not a concession. It is a crucial step in developing systems that reflect the diverse spectrum of human experiences and abilities. Policymakers and advisers who prioritise this will be better equipped to create effective strategies that work in real-world settings, not just in theory or on paper, within the communities they seek to serve.

9. Innovation Blind Spots: Can Biases Shape Who Gets to Innovate?

Innovation policy prides itself on being merit-based. It aims to fund the best ideas, support the most promising entrepreneurs, and celebrate breakthrough discoveries. The system appears to be designed to reward talent and creativity wherever they emerge. But what if this meritocratic vision obscures something more troubling?

Across Australia's innovation landscape, patterns are emerging that raise uncomfortable questions that should be at least acknowledges, if not addressed. Observations suggest that certain types of people seem to receive more opportunities, funding, and recognition. Others seem to be systematically excluded, despite their potentially valuable contributions. This chapter examines whether these patterns are coincidental or if they might reveal hidden biases that shape who participates in innovation.

This concern matters more than it may seem. If innovation systems favour particular groups while marginalising others, the consequences could extend far beyond concerns about fairness. We might be missing breakthrough ideas, which limits creative problem-solving and builds innovation ecosystems that serve narrow interests rather than broader social needs. These issues are explored under several headings below.

Is Innovation Really a Young Person's Game?

A walk through any start-up accelerator reveals a striking demographic pattern. Young faces dominate, reflecting what appears to be a deep-seated belief that innovation belongs to those under 30. But does this apparent age bias reflect a broader reality, and could it be limiting our potential for innovation? This issue is discussed in more depth in Chapter 4 of this part of the book.

While venture capital data shows overwhelming investment in younger entrepreneurs, the research in Chapter 4 indicates that successful start-up founders are much older. Older entrepreneurs often bring deeper domain knowledge, more developed professional networks, and tested management experience. They may have learned from previous business successes (and failures) and developed the patience needed for long-term innovation projects.

If innovation systems prioritise youth over experience, they may inadvertently create less resilient, more failure-prone ecosystems.

The irony appears particularly sharp in Australia. With an ageing population creating enormous opportunities in healthcare technology, aged care innovation, and lifestyle solutions, there is a risk of implying that the potential for innovation expires at age 35. The risk is missing the expertise needed to address some of our most pressing social challenges.

Perhaps more concerning is what this age bias might signal about an understanding of innovation itself. If innovation is seen as requiring risk-taking energy rather than accumulated wisdom, and rapid adaptation rather than patient development, it may be privileging certain solutions over others. In a worst-case scenario, policy and practice could be inadvertently favouring flashy disruption over sustainable improvement.

Does STEM Dominance Narrow the Vision for Innovation?

Science, technology, engineering, and mathematics drive important innovations. From breakthroughs in renewable energy to advances in medicine, STEM disciplines deserve significant policy attention. But this focus could create blind spots that limit innovation potential.

The patterns suggest that it might. HASS are consistently marginalised in innovation policy frameworks. They are often treated as secondary to "real" innovation occurring in laboratories and tech companies. This disciplinary hierarchy may result in missing crucial insights.

In artificial intelligence development, understanding human behaviour, ethics, and social systems may be as important as technical capability. In urban planning innovations, insights from sociology, anthropology, and design could determine whether smart city technologies actually improve residents' lives. The implication is that in developing tools that fail in practice, the disciplines that understand how people actually think, live, and work have been excluded.

When innovation policy treats technology as separate from its social context, it can produce a particular type of solution—technically impressive but socially irrelevant. It could be a systemic failure that favours innovations that work in laboratories over those that work in communities.

Multidisciplinary collaboration is often emphasised in policy documents, but institutional practice can tell a different story. Funding agencies often maintain strict disciplinary boundaries, universities reward narrow specialisation, and innovation metrics typically focus on patents rather than social impact. These structures may be preventing the cross-pollination that leads to breakthrough innovations.

Australia's innovation strengths lie at disciplinary intersections—sustainable agriculture that combines engineering with environmental science and Indigenous knowledge systems that integrate traditional practices with contemporary research. We must ensure that our policies are structured to recognise and support these hybrid approaches.

Do Class and Social Barriers Limit Access to Innovation?

Innovation ecosystems tend to cluster in inner city areas, where high property prices often exclude lower-income residents. Start-up accelerators can also operate in upscale districts, venture capital networks cluster in exclusive venues, and innovation events can take place in high-end locations. This spatial concentration could create class barriers to innovation participation.

The issue becomes more complex when funding patterns are examined. Access to venture capital may depend on factors such as family wealth, prestigious educational credentials, and established professional networks. Even government support programs might favour those with institutional knowledge and bureaucratic navigation skills. There is a risk that innovation systems could reward existing privilege rather than identify new talent.

A form of start-up mythology could exacerbate these barriers. Stories of entrepreneurs dropping out of university to build companies may obscure the family safety nets that make such risks possible.

When innovation policy celebrates risk-taking without acknowledging the resources required to take risks, it may inadvertently exclude those who cannot afford to fail.

Many practical obstacles confront working-class innovators. They might not be able to afford unpaid internships at tech companies, may lack mentors in professional networks, and could face cultural barriers in environments designed around middle-class assumptions. Innovation hubs may discuss accessibility, but their cultures may not be welcoming to people from diverse socioeconomic backgrounds.

These patterns could have measurable consequences. If innovation districts drive gentrification, they could displace existing communities while claiming to serve them. The problems that receive innovative attention may reflect affluent concerns rather than broader social needs. There is a potential risk in building innovation systems that solve first-world problems while ignoring more fundamental challenges.

If innovation reproduces class hierarchies, it may also limit creative potential. Different life experiences generate unique insights and problem-solving approaches. Socioeconomically homogeneous innovation teams may overlook solutions that would be obvious to individuals from diverse backgrounds.

Are Cultural Assumptions Limiting the Diversity of Innovation?

Despite celebrating multiculturalism, Australia's innovation policies may reflect narrow cultural assumptions. They appear to privilege Western models of entrepreneurship, individual achievement, and market-based solutions. This cultural framing may exclude alternative approaches to innovation.

Indigenous knowledge systems, which have sustained innovation in Australia for tens of thousands of years, are now beginning to be recognised in formal policy frameworks. Migrant entrepreneurs often face barriers due to credential recognition processes that do not acknowledge their international experience. Community-led innovation might be overlooked in favour of individual success stories. The result could be the systemic exclusion of innovations that do not fit Western entrepreneurial models.

Even subtle cultural assumptions can shape participation. Policy language, networking events, and institutional norms often seem to assume particular approaches to time, risk, and success that align with Anglo-Australian values. These barriers might be invisible to policymakers but can create real exclusion for people from different cultural backgrounds.

Cultural diversity can bring different approaches to problem-solving, alternative business models, and insights into global markets. When innovation systems remain culturally narrow, they may be missing opportunities to develop solutions that work across different contexts and communities

Australia's position in the Asia-Pacific region makes this issue important. Innovation partnerships with neighbouring countries generally require an understanding of different cultural approaches to technology, business relationships, and social organisation.

Domestic cultural diversity may represent an untapped strategic asset for international collaboration.

Does Cognitive Homogeneity Limit Creative Potential?

Innovation rhetoric celebrates creativity, breakthrough thinking, and unconventional approaches. However, innovation ecosystems might consistently favour narrow cognitive profiles that emphasise rapid processing, linear reasoning, extroversion, and confident self-promotion. This could create a paradox in which the very thinking styles that drive innovation are excluded.

Many breakthrough innovations may emerge from associative thinking, pattern recognition, and attention to detail. These cognitive strengths are often found among neurodiverse individuals. But recruitment processes, team dynamics, and performance evaluations in innovation ecosystems could systematically exclude individuals who think differently.

The bias might be largely invisible because it is coded as "cultural fit" or "communication skills", rather than as explicit discrimination. Open-plan offices may favour extraverts over those who need quiet spaces for concentration. Networking events can prioritise verbal confidence over substantive insights, while performance reviews can reward self-promotion over consistent contributions.

When innovation systems exclude cognitive diversity, they may lose access to various ways of processing information and approaching problems.

If teams become homogeneous in their thinking patterns, it can reduce creative potential and increase the risks of groupthink.

The irony is that many historical innovators would struggle in today's innovation ecosystems. The focused, detail-oriented thinking that produces breakthrough insights might not translate well to fast-paced, socially intensive innovation environments. The design of innovation systems must include the cognitive styles that are most likely to generate the innovations we claim to desire.

There are Innovation Performance Implications.

These potential biases may have implications beyond equity concerns. Innovation systems that exclude entire groups of people may not achieve optimal performance. They could be leaving talent unutilised, perspectives unexplored, and problems unsolved.

Complex challenges require diverse knowledge systems and interdisciplinary approaches. Climate change, urban sustainability, healthcare delivery, and social inequality all demand innovations that combine technical sophistication with a deep understanding of human behaviour and social systems. Narrow innovation pipelines risk producing limited solutions.

There may also be implications for legitimacy. Innovation ecosystems that primarily serve privileged communities could face public backlash and policy reversal. When benefits concentrate among existing elites while costs are distributed broadly, political support can become fragile. Inclusive innovation is more sustainable than exclusive excellence.

The global shift towards mission-driven innovation policies reflects a recognition of these realities. Countries are experimenting with participatory approaches that involve communities in setting innovation priorities.

They seem to be broadening success metrics beyond patents and profits to include social impact and systemic learning. They may be discovering that inclusive innovation produces better outcomes.

How Can Policy Design Address These Blind Spots?

If hidden biases shape innovation participation, what might policy responses look like? The challenge would be designing interventions that address systemic exclusion without compromising the quality of innovation.

As discussed in earlier chapters, definitions of innovation are being expanded beyond technological advancement to include social innovation, cultural preservation, and community problem-solving. This does not mean abandoning technological innovation but rather situating it within broader contexts of human need and social value. But it does mean:

- *Funding mechanisms* must be designed to eliminate any barriers faced by underrepresented groups. This could include simplified application processes, alternative assessment criteria, and targeted support for systematically disadvantaged groups.
- *Physical and virtual innovation spaces* should be designed to accommodate cognitive and cultural diversity. This means quiet spaces for concentration, flexible working arrangements, and the cultural adaptation of programs and services. Meaningful inclusion means moving beyond token representation to genuine power-sharing in governance and decision-making.
- *Measurement systems* must expand beyond narrow economic indicators to capture broader outcomes of innovation. They should track participation demographics, community impact measures, and long-term system changes rather than just short-term outputs.
- *Regional development* will be crucial for spatial inclusion. Investment in innovation concentrated in major cities can leave regional communities behind and limit access to various types of knowledge and problem-solving approaches. Distributed innovation networks can leverage local assets while connecting to global opportunities.

The Australian Context: Opportunity or Oversight?

Australia may have specific advantages in developing inclusive innovation ecosystems. Its cultural diversity can provide access to various traditions of innovation and global networks. Indigenous knowledge systems offer alternative approaches to sustainability and place-based innovation. Regional diversity creates opportunities for distributed innovation networks.

Current policies might not fully capitalise on these strengths. Commonwealth and State governments appear to favour narrow definitions of innovation and concentrate investment in established hubs rather than building broader ecosystems. There is a risk of importing innovation models designed for different contexts instead of building on our distinctive assets. We can learn from international practice; this does not mean we should try to replicate it.

The opportunity lies in reimagining innovation policy to reflect Australia's diversity. This could mean genuinely engaging with Indigenous knowledge systems, building on migrant entrepreneurship, and developing regional innovation capabilities. It may require some fundamental institutional changes:

- Innovation agencies require diverse leadership and staff
- Advisory bodies need voices from excluded communities
- Performance frameworks need to reward inclusion alongside economic outcomes.

Implications for Policy and Practice

The extent to which hidden biases shape participation in innovation has broader implications for how we understand innovation itself. If innovation ecosystems systematically exclude certain groups, they will fail to serve their purpose of expanding human capabilities and creating better futures. This chapter has sought to identify issues without offering concrete solutions.

Nonetheless, current approaches might be channelling resources towards familiar patterns while marginalising alternative approaches. They could also reproduce existing inequalities while claiming to reward merit and prioritising narrow definitions of success over broader social value.

Addressing these potential blind spots requires acknowledging that innovation policy is inherently political. It appears to make choices about whose knowledge counts, which problems matter, and what success means. Making these choices explicit opens up space for democratic debate about innovation priorities and approaches.

The task for policymakers is to create innovation systems that serve diverse communities while maintaining high performance standards. This means expanding expectations to include social impact alongside economic returns, rather than lowering standards.

Innovation's promise is in its potential to address complex challenges and improve human wellbeing. But realising this promise requires innovation systems that are as diverse, adaptive, and inclusive as the societies they aim to serve. The alternative is to waste human potential while claiming to maximise it.

Australia has an opportunity to lead in inclusive innovation approaches. The question remains whether policymakers will recognise and act on this potential. Can we build innovation systems that harness our full creative capacity, or will hidden biases continue to shape who gets to participate in creating our future?

PART IV: DESIGN LOGIC – BUILDING RESILIENT INNOVATION ECOSYSTEMS

Part IV turns to the design logic of innovation ecosystems, focusing on how they can be intentionally shaped and strengthened over time. While earlier sections examined what ecosystems are and how they create value, this part explores how to build their capability, resilience, and adaptive capacity.

These chapters examine practical frameworks for aligning systems, strategy, and structure. They highlight approaches such as smart specialisation, systems integration, and mission-oriented governance, drawing lessons from Australian and international examples. Topics include the role of enabling platforms, the importance of absorptive capacity, and the need for investment in skills, sustainability, and translational infrastructure.

The emphasis is on long-term stewardship and institutional learning—how to move beyond one-off interventions to create ecosystems that are robust, inclusive, and able to respond to changing technological and economic conditions.

1. Harmonising Systems, Strategy and Structure in Innovation Ecosystems

From a strategic management perspective, innovation ecosystem design should reflect the alignment of at least three interdependent elements: systems, strategy, and structure.

- *Systems thinking* provides the analytical framework to understand complex relationships, feedback loops, and emergent behaviours within innovation networks.
- *Strategy* supplies the directional clarity, priority setting, and resource allocation decisions that systems thinking alone cannot provide.
- *Structure* encompasses the institutional arrangements, governance mechanisms, and delivery capabilities that translate strategic intent into tangible outcomes.

When these three elements work in concert, they set up the conditions for sustained innovation performance. When they operate in isolation or imbalance, they generate the familiar patterns of well-intentioned initiatives that struggle to achieve their stated objectives.

Australia's innovation ambitions frequently falter at the intersection of enthusiasm and execution. Over the years, Australian policy documents have been filled with references to systems thinking, whole-of-government coordination, and collaborative networks. But the results often fall short of expectations. One reason for this is the persistent imbalance in the approach to innovation ecosystems. Australia tends to prioritise process over purpose, discussion over direction, and aspiration over architecture.

Systems thinking is rightly influential. It helps policymakers, researchers, and businesses step back from narrow, linear cause-and-effect thinking. It encourages recognition of feedback loops, dynamic interactions, emergent behaviours, and long-term consequences. Applied to innovation, it enables understanding of ecosystems not as static arrangements of actors but as adaptive, evolving networks.

However, there is a risk in overstating its capacity. Systems thinking does not tell policymakers *what* to aim for. It doesn't specify priorities, allocate resources, or weigh trade-offs. In this sense, it is best understood as a diagnostic lens rather than a direction-setting tool. It can illuminate leverage points, but it cannot decide which ones to pull or in what order.

Systems thinking, when adopted in isolation, often leads to intensive mapping exercises, consultation fatigue, and strategic drift. The risk is paralysis by complexity.

Without a clear strategic narrative or a delivery backbone, systems initiatives can become fragmented, unaccountable, and vulnerable to the very inertia they set out to challenge.

The Role of Strategy: Clarity, Priorities, Sequencing

Strategy is fundamentally about choice. It is the bridge between ambition and execution. In innovation ecosystems, strategy defines:

- Which challenges or missions are worth pursuing?
- What policy mix or market-shaping tools will help achieve impact?
- What are the trade-offs and what will be deprioritised?
- What is the optimal phasing of investments, reforms, and partnerships?

Australia's innovation ecosystems have struggled with strategy. Many policy initiatives conflate vision with strategy, mistaking inspirational rhetoric for executable plans. The absence of clear priorities and sequencing leads to diffusion of effort and dilution of resources. This is especially evident in the proliferation of pilot programs, demonstration projects, and short-lived task forces that seldom scale or persist.

Without strategy, systems thinking becomes a conceptual comfort zone. By contrast, well-articulated strategies give systems thinking its practical edge, targeting feedback loops worth amplifying, actors worth mobilising, and policies worth institutionalising.

Structure as the Engine of Delivery

Structure is the forgotten sibling of innovation policy. It encompasses the institutional arrangements, funding mechanisms, governance frameworks, and delivery capabilities that sustain innovation over time. It is structure that answers questions such as:

- Who is accountable for what?
- How are decisions made and coordinated?
- What incentives support long-term collaboration?
- What delivery vehicles turn intent into outcomes?

Too often, Australian innovation policies and associated announcements focus on designing programs without investing in the institutions, organisations, and platforms that deliver them. There is a reliance on traditional agencies, committees, and departmental silos to orchestrate ecosystems, even when their mandates and capabilities are misaligned with the complexity of the task.

Where structure is underdeveloped, even good strategies and well-mapped systems struggle.

Coordination costs rise. Momentum is lost. Promising initiatives are stranded without hosts or champions. Worse still, efforts become person-dependent rather than system-dependent, relying on charismatic leaders instead of institutional resilience.

What Happens When One Element Dominates

When only one element of the *systems, strategy, structure,* alignment dominates, the following consequences may follow.

- *System-centric without strategy or structure*: Endless diagnostics, stakeholder engagement, mapping exercises. Elegant models, but little movement. Such an approach is good at recognising complexity, poor at navigating it.
- *Strategy-centric without systems or structure*: clear targets, slogans, or missions ("become a global leader in X") without understanding how actors interact or what unintended consequences might arise. The approach can create linear implementation models that fail in complex environments.
- *Structure-centric without strategy or systems*: bureaucratic expansion, governance overdrive, institutional layering. The approach is typically reflected in new entities being formed without a unifying logic or adaptive capacity—activity mistaken for progress.

Balanced innovation ecosystems strike a balance among the three.

Conclusion: From Balance to Impact

The rhetoric of systems thinking is now well embedded in innovation policy circles. But its overuse, disconnected from strategy and structure, risks diminishing its value.

True innovation system performance depends on balance:

- Systems to understand complexity and dynamic change
- Strategy to choose direction and focus effort
- Structure to deliver, scale, and sustain results

We must stop treating these as alternative paradigms or optional layers. They are mutually reinforcing. Without structure, strategy is fantasy. Without strategy, systems become noise. Without systems thinking, both strategy and structure may be blind.

Australia's next wave of innovation policy will stand or fall on our ability to design ecosystems that are not just well-intentioned or well-informed, but also well-built.

2. Smart Specialisation: An Engagement Framework in the Hunter, New South Wales

John H Howard, with Todd Williams and Renu Agarwal[4]*, September 2016*

Australia's efforts to promote regional economic development are often disjointed, focusing primarily on infrastructure and investment pipelines while failing to adequately incorporate science, research, and innovation.

Regional Development Australia (RDA) Hunter's experiment with Smart Specialisation (RIS3) presents a promising new approach. This framework situates interactions within the innovation system using the Triple Helix model of connection between universities, industry, and government, and puts these interactions into practice in a manner that aligns with Australia's governance structure.

This chapter reflects on the findings from the Hunter RIS3 project undertaken in 2015-2016 and explores how Smart Specialisation, combined with strategic intermediation, brings clarity to a complex and crowded institutional environment. Ten years later, they are still just as relevant to discussions about regional innovation ecosystems.

Project Aims

The Hunter RIS3 project, which began in 2015, aimed to achieve five goals:

- Inform policy to ensure public and private innovation investment is better targeted
- Identify regional priorities based on strengths and emerging potential
- Recommend future opportunity areas for entrepreneurship and growth
- Encourage collaborative governance
- Support productive research and innovation activities.

The important strategic actions consisted of:

---

[4] This Chapter is an abridged version of a paper delivered at the XIV International Triple Helix Conference, 25-27 September Heidelberg Germany, 2016, *Triple Helix Models of Innovation: Addressing Ecosystem Challenges in the Era of Crises—Regional Dimensions of Triple Helix? Clusters, Cities and Geographic Boundaries.* The paper was awarded the Best Practitioner Case Award for the Conference.

- Establishing a Hunter Innovation Network as a cross-sector coordinating body
- Developing an inclusive leadership model with representation from education, business, and government
- Encouraging entrepreneurial education across schools and tertiary institutions
- Facilitating stronger university-business collaboration
- Defining the parameters for an Innovation Initiatives Fund to encourage investment.

From Theory to Operational Practice: Giving Life to The Triple Helix Effect

The Triple Helix (TH) model is commonly used as a tool for exploring discussions on innovation policy. It presents sustainable innovation ecosystems as requiring intentional and cooperative interactions among universities, businesses, and government agencies (Etzkowitz & Leydesdorff, 2000). The idea is solid: innovation does not come from a single sector acting alone but develops through collaboration, the exchange of ideas, and a shared purpose across different institutions.

The TH framework is helpful for diagnosing issues, but it is not very useful for guiding implementation. It can outline *the* types of interactions that are desirable but has difficulty explaining *how* these interactions can be supported, managed, or governed in practical situations.

The Triple Helix framework does not, by itself, provide a clear plan for organising interventions, any standards for prioritising investments, or any direction for creating accountability frameworks.

It particularly lacks in addressing *coordination and collaboration* among different levels of government, as well as between regions and central agencies.

These limitations are especially noticeable in regional areas like the Hunter, where multiple agencies and jurisdictions overlap with a complex network of public programs, private interests, and community expectations. Without a cohesive strategy or integrated governance, the TH framework may become merely a concept—often referenced but rarely put into practice.

The Regional Smart Specialisation (RIS3) framework of the European Union is effective in addressing these limitations. RIS3 does not replace the TH model; rather, it makes it actionable. It offers a structured and research-backed approach to aligning innovation objectives with local capabilities and global market trends. It can adopt a location-based strategy for economic change that relies on stakeholder engagement and seeks to ensure that research, education, and business investments are focused and effective.

While TH emphasises foundational principles, RIS3 moves forward into practical implementation. This statement applies the theory by emphasising the significance of establishing strategic priorities, collaborating in governance, and conducting ongoing evaluations. RIS3 provides practical strategies for effective collaborative innovation, particularly in regions that lack established coordination practices or governance structures, which the TH model does not address.

The use of RIS3 principles in the Hunter Region demonstrates how the TH framework can transition from a theoretical concept to a practical strategy. Effective regional innovation systems depend on the presence of institutions, the quality of relationships between them, and the mechanisms that align their goals and actions.

RIS3 as a Governance and Engagement Framework

The EU Smart Specialisation framework is a method for assessing funding eligibility and a planning tool. It is also a policy framework designed to change how regions engage with innovation, growth, and structural transformation. As a strategy focused on specific areas, it views regions as active participants rather than merely passive recipients of central policies. These regions play a crucial role in shaping their economic futures by aligning their resources, skills, and goals.

At its core, RIS3 necessitates shifting the emphasis of regional development from investing in infrastructure to orchestrating changes driven by innovation.

It urges regions to concentrate their public investment in specific areas where they can develop substantial strengths and long-lasting advantages, rather than allocating resources too broadly across various general or politically driven priorities. This issue is not just about selecting winners; it is about creating conditions where promising fields of specialisation, based on existing strengths and untapped potential, can thrive through targeted policy support and entrepreneurial innovation.

Crucially, RIS3 emphasises collaborative governance. It promotes the collaboration that the European Commission calls "entrepreneurial discovery processes", which includes universities, businesses (such as small and medium-sized enterprises), public agencies, and members of civil society. These processes identify strategic domains, test ideas, and build legitimacy around priority setting—not by fiat but through dialogue, experimentation, and evidence. This participatory element is essential. Without it, Smart Specialisation risks becoming just another technocratic planning exercise.

RDA Hunter's adaptation of RIS3 emphasised the challenges and opportunities presented by this framework in Australia. In a region lacking formal governance structures, coordinating efforts proves to be a difficult challenge. Many State and Commonwealth agencies share comparable responsibilities, while local governments lack the size and authority to lead substantial innovation initiatives. By acting as an innovation intermediary and organiser, RDA Hunter effectively linked distinct programs and individuals.

The RIS3 process allowed RDA Hunter to articulate a unified vision that linked local strengths, such as the research capabilities of the University of Newcastle, expertise in defence manufacturing, and the potential for energy transition, with broader market opportunities.

This also highlighted the flaws in how institutions collaborated, the inconsistencies in their policies, and the discrepancies in funding that had previously been obscured by their isolated methods.

Smart Specialisation functions as a new approach to *governance*. This is not just a method for selecting economic priorities; instead, it is a means of building the skills, relationships, and learning processes necessary for fostering a dynamic and adaptable regional innovation system.

Hunter as a Regional Case: Complex, Fragmented, but Capable

The Hunter Region, which has a population of more than 640,000 residents and a gross regional product of $38 billion, exemplifies the challenges faced by regional Australia. The economy is diverse, comprising sectors such as mining, healthcare, advanced manufacturing, and wine production. However, it encounters difficulties related to changes in the industry, the need for infrastructure, and inconsistent planning.

The regional innovation ecosystem is intricate, comprising a leading research university recognised for its robust STEM programs, a diverse business community that includes both global defence contractors and agile small to medium enterprises (SMEs), and three tiers of government that implement more than 130 programs and initiatives. This complexity has resulted in unnecessary repetition, inefficiency, and missed opportunities for collaborative growth.

The RIS3 process of RDA Hunter aimed to create a sense of coherence. Rather than establishing another rigid plan, it designed a strategic engagement platform. The development process was not guided by consultants or dictated by agencies; rather, it resulted from in-depth discussions, a careful analysis of regional strengths, and substantial participation from businesses, research institutions, and the public sector.

The Hunter RIS3 distinguishes itself by offering more than just a strategic plan. It led to specific commitments, such as $1 million in funding from the New South Wales Government for an Integrated Innovation Network. It also aligns with the University of Newcastle's NeW Futures Plan, which focuses on global impact clusters, engaged PhD programs, and practical innovation.

The Role of Intermediaries: Filling System Gaps

A key lesson from the Hunter experience is the crucial role of intermediaries—organisations that connect different participants, promote knowledge sharing, manage relationships, and secure resources. The RIS3 process revealed weaknesses in Australia's intermediary landscape, particularly the lack of organisations that can integrate all three components of the Triple Helix.

Based on previous research (Howard, 2007; Howells, 2006), intermediary functions are classified into four categories:

- *Consultants* provide expert advice
- *Brokers* facilitate transactions and collaborations
- *Mediators* build trust and convene stakeholders
- *Resource providers* unlock access to funding and facilities.

Hunter RDA played all four roles, supported by its independence, deep local knowledge, and networks across sectors. It did not act as an implementer but as a strategist, convener, and facilitator. It respected the boundaries of universities and businesses while enabling new collaborations between them.

This intermediary ability is uncommon in Australia, where only a few industry associations or regional organisations go beyond basic transactional roles.

The key takeaway is that for Smart Specialisation to be effectively applied in Australia, it is essential to invest in intermediaries dedicated to the mission, trusted by stakeholders, and adequately funded.

Aligning Parallel Strategies: The University Interface

The University of Newcastle's NeW Futures strategy enhances the TH ecosystem in the Hunter region. The goals of encouraging economic growth by supporting start-ups, promoting commercialisation, and developing skilled graduates are in line with the principles of RIS3. The methods it uses, global impact clusters, PhD programs linked with industries, and centres for business involvement, mirror the European models of engaged universities (European Commission, 2014).

Universities often struggle to balance global academic goals with local economic growth. They face challenges within their organisations that limit their regional participation and restrict their resources, making it difficult for them to respond effectively. The RIS3 framework creates a strategic foundation by linking a regional agenda to the strengths of universities. Universities contribute to RIS3 by offering research opportunities, training programs, and facilitating collaboration among individuals.

The Hunter case illustrates how this mutual reinforcement can manifest. The university helps support the regional strategy and gains advantages from it, even though it does not own the strategy. The difference is important. When universities assume responsibility for regional planning, their strategies can sometimes become impractical academic objectives. When they are absent, the depth of knowledge decreases. The RIS3 framework provides a method for maintaining balance.

Innovation Without Governance: The Australian Dilemma

Australia's federal structure complicates regional innovation policy. There is no framework for regional governance. Economic development responsibilities are divided among State and Commonwealth departments, which frequently pursue goals that conflict with one another. Local councils have limited capabilities. There are many regional strategies, but they are seldom coordinated. The delay in carrying out plans results in innovation being seen as a shared priority rather than an organisational obligation.

In such situations, RDA Committees, especially those driven and skilled, such as RDA Hunter, can play a significant role in connecting different elements. They are closely tied to local situations and also have strong connections to national policy processes. They can help maintain uniformity across various institutions and regions. However, this requires a shift in focus—from supporting infrastructure to incorporating systems.

The ongoing evaluation of the RDA program should consider this factor. To successfully promote Smart Specialisation, Regional Development Agencies (RDAs) must possess planning skills, strategic financial resources, and knowledgeable staff who understand innovation systems rather than merely handling grants. Their legitimacy should be based on their performance and collaboration rather than on political appointments.

Smart Specialisation as an Evolving Platform

The Hunter RIS3 is not merely a plan to be set aside; it is a platform that requires ongoing support. The process necessitates continuous enhancement, coordination of resources, and efficient political oversight. The important point is that it must be incorporated into the practices of government programs, university initiatives, and business development strategies.

Smart specialisation is not an easy solution to all challenges. The suggested solution will not resolve the fragmented STI system in Australia or reduce the differences between regions. However, it provides a framework for deliberate, evidence-based collaborative efforts. It focuses on what is most important: real skills, true partnerships, and measurable outcomes.

The Hunter case demonstrates the potential outcomes when TH theory is applied in practice within the RIS3 framework, supported by an effective intermediary.

It offers a framework that should not be replicated exactly but should be tailored to suit the unique conditions of different regions in Australia.

Conclusion

Australia has spent many years identifying the gap in innovation and lamenting the lack of connection between knowledge creation and economic growth. The Hunter RIS3 offers a distinct method: a practical, strategic, and adaptable framework for linking the various components. It accomplishes this by:

- Treating innovation as a regional and systemic challenge
- Engaging stakeholders through a structured, evidence-based process
- Leveraging intermediaries to coordinate and align
- Embedding universities as key players but not dominant actors
- Focusing on comparative advantages and strategic opportunities.

Innovation policy extends beyond Canberra. It is found in regions characterised by distinctive resources, organisations, and objectives. For Australia to transform into a knowledge economy, it must invest in systems that enable these sectors to function strategically. Effective Smart Specialisation provides a solid basis.

3. The Jazz Combo Leader: Why Innovation Districts Need Systems Integrators

Picture a jazz combo where each musician plays their own tune. Talented performers, quality instruments, intimate venue, but the saxophonist pursues a bebop riff while the drummer locks into a swing rhythm and the pianist explores modal harmonies. The result? Musical chaos instead of improvised brilliance.

This scenario plays out daily in innovation districts worldwide. Universities pursue research excellence, companies chase market opportunities, governments focus on policy outcomes, and start-ups scramble for survival. Each performs admirably within their domain, yet collectively they often miss the transformative potential that drew them together.

The missing element? A systems integrator: the leader who helps diverse players create something greater than the sum of their parts.

The Integration Imperative

Innovation districts represent substantial public and private investment across Australia and globally. Governments are establishing science parks, research precincts, and innovation quarters as platforms for long-term economic growth. However, many struggle to achieve their ambitious goals, despite having impressive infrastructure and institutional support.

The challenge is not the quality of individuals and organisations but how well they function as a coherent ecosystem.

Success depends on connections, the quality and depth of relationships that enable ideas, technologies, and talent to flow, recombine, and create value.

Most districts distribute activities across multiple institutions, each with distinct agendas, funding models, and organisational cultures. Without integration, this diversity leads to duplication, missed connections, and short-term thinking. Promising initiatives proceed on parallel tracks without intersection, while longer-term systemic issues remain unaddressed because they fall outside the mandate of any single organisation.

Systems integrators can counteract these tendencies. Rather than centralising control, they help stakeholders work more effectively together. They identify shared interests, spotlight emerging opportunities, and facilitate progression from informal coordination to structured collaboration.

Beyond Command-and-Control

Jazz music is an illuminating parallel. Jazz leaders do not dictate the performance but listen deeply, guide when necessary, and help musicians respond to each other in real-time. Similarly, effective systems integrators provide just enough structure to enable collaboration without imposing rigid frameworks that stifle creativity.

This function operates subtly but proves essential. It derives authority not from formal position but from earned credibility, established relationships, and a boundary-spanning perspective.

As districts tackle increasingly ambitious challenges, including sustainability, deep tech innovation, and digital transformation, integration becomes even more crucial.

Integration versus Governance

Systems integration and governance represent distinct functions, though they are often confused. Governance encompasses formal decision-making arrangements—who decides what, how accountability works, and how priorities get set. Integration operates in relational and adaptive modes, helping people connect their work productively across organisational boundaries.

Strong governance alone does not guarantee effective integration. A well-structured board may lack meaningful connection to ground-level activity. Conversely, a district with informal governance might achieve exceptional results through collaborative practice and adaptability.

Systems integrators bridge these realms. They work at the intersection of formal authority and informal collaboration, connecting different system components. They needn't hold official authority, but they must maintain presence and visibility—listening, connecting, and helping participants make sense of emergent patterns and opportunities.

Models That Work

Successful integration takes various organisational forms, each shaped by local context and institutional relationships.

Independent Non-Profit Platforms operate as mission-driven organisations created specifically to coordinate ecosystem activities. London's Knowledge Quarter connects over 100 academic, cultural, scientific, and media organisations through knowledge exchange forums and strategic initiatives. Brainport Development in Eindhoven serves as a strategic hub for regional innovation, convening stakeholders and coordinating investments while maintaining independence from any single institution.

University-Hosted Integration leverages major universities as anchor institutions. The University of Cambridge has historically functioned as an informal integrator for the Cambridge Cluster through commercialisation offices and translational institutes. This model uses institutional capacity and research leadership but risks excluding other stakeholders if the university appears to dominate decision-making.

Public Agency Models embed integration within government agencies focused on economic development or innovation policy. Barcelona's 22@ Innovation District was initially driven by the city government's planning department, which transformed an industrial zone through coordinated infrastructure investment and stakeholder engagement. Singapore's JTC Corporation leads both the physical and strategic development of the one-north district while serving as a convener for collaboration.

Hybrid Networks distribute responsibility across formalised networks of institutions operating under shared governance frameworks. Berlin Adlershof combines precinct management by a publicly owned company with research and business networks that shape strategy through formal agreements and working groups.

Managed Consortia involve precinct-wide partnerships including universities, government agencies, developers, and major tenants. Boston's Kendall Square has evolved through sustained collaboration among MIT, the Cambridge city government, venture investors, and pharmaceutical companies. While lacking centralised governance, MIT plays a crucial role in managing partnerships and maintaining long-term coherence.

Design Principles

Selecting a workable integration model requires an objective assessment of local conditions: trust levels between institutions, collaboration history, capacity for sustained commitment, and the availability of credible hosts. Several principles guide effective design, as outlined below.

The integrator must serve the ecosystem rather than compete within it. Trust-based neutrality works best when led by individuals or organisations perceived as knowledgeable yet neutral—parties that seek to enable rather than dominate.

Integration is a long-term commitment that requires investment in people. Effective integrators cultivate relationships through sustained development, iterative conversations, and a comfort with ambiguity. They balance patience, *moving at the speed of trust*, with the ability to catalyse action when opportunities arise.

Boundary-spanning capability enables integrators to move confidently across different domains. These cover policy, research, business, and community and understanding the priorities, constraints, and languages of each. This requires hybrid skills that combine systems thinking, negotiation, strategic communication, and program design.

Collaborative infrastructure creates environments conducive to cooperation through shared road mapping, collective agenda setting, working groups, and structured dialogues. These are the architecture for building shared understanding and practical alignment.

The Human Factor

The most significant investment is not in frameworks or structures, but in developing human capacity for thoughtful, connected leadership. Integration requires sustained investment but is too often treated as an afterthought or a volunteer activity.

Integration functions need explicit recognition and appropriate resources to succeed.

Effective integrators perform essential boundary-spanning functions: strategic alignment around shared challenges, knowledge translation across distinct communities, relationship brokerage between complementary capabilities, process coordination across different operational timelines, and conflict resolution through structured negotiation.

The Integration Advantage

Without effective integration, even well-resourced districts risk fragmentation and underperformance. Promising initiatives remain isolated, systemic challenges go unaddressed, and collective potential remains unrealised.

With thoughtful integration, innovation ecosystems achieve coherence and adaptability that enable them to tackle increasingly complex challenges. They become more than the sum of their parts—dynamic arrangements in which ideas, technologies, and talent flow, recombine, and create value.

The capacity for effective systems integration provides a critical indicator of an innovation district's potential. Assessment should examine whether dedicated integration functions exist, how comprehensively these functions span the ecosystem's dimensions, and whether integration capabilities receive sufficient recognition and resources.

As governments continue to invest in innovation infrastructure, the question is not whether to include systems integration, but how to design and resource it effectively.

The districts that master this challenge will create the collaborative advantage needed to address tomorrow's complex innovation challenges.

The jazz combo metaphor reminds us that technical excellence alone is not enough. Without the leader's guiding hand, even the most talented musicians produce discord. Innovation districts face the same choice: fragmentation or symphony. The presence of skilled systems integrators often makes the difference.

4. Productivity and Innovation Need Business Own Investment in Skills

Craig Fowler, 16 September 2025

Despite a modest increase in business R&D expenditure in Australia in the latest reported year (2023/24), Australia's performance on a GDP basis is still woefully worse than its international economic competitors. But at least business R&D is still being measured every two years by the ABS. There is no reliable, routine national data on business and industry spend on skills and training.

The last ABS survey 'Employer Training Expenditure and Practices, Australia, 2001-02' was released in 2003 – two decades ago. These data (e.g. training purpose, industry-specific sectors, total expenditure; spend per employee; training as % of gross wages/salaries etc) are routinely surveyed by business/industry and by governments in major competing economies e.g. US, UK, EU.

In Australia there have been a few limited ad hoc surveys (e.g. by CEDA, by Deloitte/RMIT, and others cited in the recent Jobs and Skills Australia report (Our Gen AI Transition, Implications for Work and Skills). The CEDA report (Learning Curve – Why Australia Needs a Training Boost) headlines the proportion of people undertaking work-related training has declined 14% since 2007.

Limited qualitative evidence indicates better resourced major corporates, and tech-savvy businesses are more nimble tech adopters and adapters (if not employers, then self-motivated employees). Businesses that do invest in skills, digital or otherwise, may use 'in-house' formal accredited training. But in the majority, they use informal and non-structured training, being disinclined to engage with tertiary education providers. Numbers of enterprise owned training providers have also declined.

There is no direct account of what kind of 'on-job' skills are learned 'in-house' e.g. technical, AI or 'enabling' skills, nor how much businesses invest.

The relevant Australian Accounting Standards (AASB 138) requires no distinction between types of training expenditure - whether accredited, non-accredited, in-company, or otherwise. It treats all training-related costs uniformly as expenses.

Worse, there is no national data on industry spend on training with tertiary institutions – it does not exist with either universities or TAFEs. Business's invest in 'earn while you learn' mode, apprenticeships and traineeships, but willingness to do this is majorly dependant on government wage subsidies. Numbers of commencing traineeships are especially vulnerable to cutbacks in government subsidies.

Regarding 'work integrated learning' the Australian Government has commenced new 'practical payments' addressing 'placement poverty' for nursing, care and teaching professions. The Higher Education Accord Review suggested financial support for unpaid work placements be funded by employers for all other fields – improbable dreaming.

Commendably in both the Higher Ed and VET sectors, multiple professional bodies invest resources in informing course accreditation in different professions, e.g. in medicine, dentistry, health, nursing, engineering, veterinary, teaching, accounting, law etc. But that's not tracked either.

It is VET/TAFE that creates training aligned to ecosystems is important for the talent pipeline. Not all employers want or need higher education graduates.

We measure productivity routinely, publish job ads in thousands, track house prices incessantly – yet do not measure or incentivise how businesses self-invest in skills. But business is big on skills advocacy.

This advocacy invariably extols governments to 'better align training with skill needs', that is, focus public funding/financing of post-school education and training courses on employer's 'on-job' skills needs. This has reinforced the politics of reframing university purpose and funding from being providers of quality education and research to deliverers of 'job-ready quals', TAFE-like plus research.

Such advocacy is based on an implicit and arguable bargain. Governments' key task is to fund and finance education, training and skills, and businesses provide jobs and (for profits) pay tax. The caustic view is that some businesses squeal about skills shortages, plan poorly, demand governments invest more in skills, leading to greater reliance, and growing controversy, on migrant labour.

All this has dwarfed any evidence of workforce benefits of those employers' that make direct or co-invest in training. But without routine evidence of business investment in skills and with governments under fiscal stress, this line of advocacy risks being increasingly impotent, despite imperatives for innovation, workplace upskilling and lifelong learning.

It is also increasingly corporately risky and irresponsible in an era where digital is the key economic driver and where AI, quantum and cybersecurity are potential master disrupters of future productivity, innovation and national security in the 21st century.

Australia has a patchy history of skills incentives and levies. The then government in its 2022-23 Budget announced a 'technology investment' and a 'skills and training boost for small businesses'. This was akin to the R&D Tax incentive. The training boost allowed businesses with an aggregate annual turnover of less than $50 million an additional 20% tax deduction for external training courses delivered to employees by registered training providers. It lasted 2 years, with no evidence of 'boost' outcomes.

Australia had a *Training Guarantee Act* and Levy 1990-94. It required all employers to spend a minimum % of payroll per year (1.5% in 1992) as a payroll threshold or risk a 'shortfall' payment to a Training Guarantee Fund. The scheme was of mixed effectiveness and problematic especially for SMEs.

Australia is leery of 'hypothecation' schemes, such as the TG Levy or the UK Apprenticeship Levy. There are however positive examples, like the producer levy on agriproducts, matched by government, funding the long running rural Research and Development Corporations. There are construction project levies in some States that create a funding pool for construction industry training for new industry entrants and safety etc. Governments may also include minimum training or 'earn-while-you-learn' obligations in its mega procurements e.g. within big tech contracts.

Limited evidence indicates that some Australian businesses invest significant resources in skills and training (including recently also greater R&D on AI), but this is overwhelmingly only in large corporates with resources and competitive need to invest. SMEs face resource and time barriers, and non-profits despite employing thousands have scant capacity.

The solution is never more detailed skills planning and lists of shortages. It requires greater entity self-investment in structured training and not just exhorting fiscally constrained governments. The issue is not simply about underfunding - it is systemic mis-design. Australia has no coherent fiscal mechanism that binds economic growth to the 'on-job' development of its human capital.

Coherence here goes well beyond individual SMEs getting a one-off government 'skills' subsidy. It's far more ambitious with e.g. AI/quantum/digital skilling across cooperating entities, ideally companies of fast growth potential in localised innovation districts and precincts.

This builds potential for bulking up both individual and aggregate absorptive capacity via local training networks, shared intellectual and physical infrastructure, and cost shared upskilling. Well executed, this builds performance and reputation for inbound investment by multinationals who expect local skills at global leading edge.

This is the expanded horizon needed in a national integrating architecture that unites the efforts of employers, educators and government. There is no such imaginative Government-initiated skilling incentive. When designed it needs to avoid subsidising what good businesses already do but subsidise networks of best up skilling practice, fiscally encouraging companies to participate.

The present malaise runs counter to the Prime Minister's economic outlook speech (Jul 2025) expressing the government was 'determined to create the right framework and make the right investments in skills and training to ensure Artificial Intelligence (AI) is a contributor, not a competitor'.

Maybe business feels the formal education and training system may not be delivering fast enough nor in a manner that meets the on-job urgency of business and employee uplift in AI/quantum, cyber and digital skills. But recognise, this is not like the COVID disruption where international bioscience ultimately delivered a universal vaccine salvation. Complacency does not AI upskill.

5. Absorptive Capacity: The Missing Link in Australia's R&D Collaboration Problem

Victor Pantano, 22 August 2025

For decades, Australian policy has been anchored on a familiar refrain: universities must do more to engage with the end users of research. The underlying assumption has been that if universities became better at outreach, collaboration, and commercialisation, the nation's innovation performance would follow.

Recent data paints a different picture. According to analysis reported in *Future Campus* ("Australian industry not playing ball on R&D collaboration" July 25, 2025), universities have significantly lifted their collaboration with industry, but almost all the growth has been with international companies, not Australian ones. That does not bode well for increasing the nation's productivity.

One implication of the growth in R&D investment from international companies is that universities may wish to focus more deliberately on marketing their capabilities to attract inward investment from these global players.

This trend validates that our research and the technologies emerging from it are indeed world-class. International companies are discerning investors; they have the resources and expertise to assess opportunities in the global research and technology market, and their willingness to invest here is a positive sign. However, this success also highlights a weakness. If Australian companies cannot engage with and benefit from these same research outputs, we risk boosting the innovation capacity and productivity of other nations while failing to strengthen our own.

The problem is not that universities are unwilling or incapable of working with industry. The problem is that much of Australian industry lacks the capacity to engage.

Absorptive capacity is a concept from innovation theory describing an organisation's ability to:

- Recognise the value of new knowledge.
- Assimilate it.
- Apply it to commercial ends.

This is not a passive process. It requires skilled people often with advanced education such as PhDs who can translate research into product development and operational improvement. It requires organisational processes to integrate external research with internal priorities. And it requires resources to sustain the effort from idea to implementation.

Large international companies tend to have high absorptive capacity because they employ more highly skilled, dedicated R&D and product development teams, maintain established processes for integrating new knowledge into products, services, or operations and can allocate budgets specifically for technology adoption and deployment, not just discovery.

The contrast with Australia is stark. Our industrial landscape is dominated by SMEs. These businesses often lack in-house R&D staff, have little experience engaging with universities (many do not know where to start) and struggle to commit resources beyond their immediate operational needs.

Even among the relatively few large Australian companies, absorptive capacity is often low compared to peers in Asia, Europe, and the United States.

This is reflected in the low levels of research engagement that the *Future Campus* article describes.

Too often, university-industry collaborations end at the delivery of research results. Universities complete a project, hand over the findings, and look for the next project. This is not enough.

From the industry side, even applied research results can be meaningless without people who understand the research and can translate it to their context, a plan for adoption that begins at project inception and resources to implement changes in processes, products, or services.

Larger companies overseas especially in sectors like pharmaceuticals integrate these considerations into their R&D strategy from day one. They involve implementers early, build cross-functional teams, and plan for scale-up alongside discovery. By contrast, many Australian projects reach Technology Readiness Level (TRL) 6 or 7, only for the results to gather dust. Millions of dollars can be spent without meaningful uptake.

Australia has had more success commercialising research by creating new companies rather than licensing technology into existing firms. Spin-outs can build absorptive capacity from scratch, bringing together the original researchers with commercial, product development, and manufacturing expertise. They are unencumbered by the cultural, political, and structural inertia that often exists in large organisations.

Another way large companies have addressed the absorptive capacity challenge is by investing directly in start-ups. This approach bypasses much of the internal inertia the cultural, social, and political complexities that can slow or derail adoption. Some large firms are better consumers of technology than developers or implementers.

Rio Tinto, for example, invests in start-ups and growth companies with the aim of becoming customers if those technologies succeed. This can be far more efficient than attempting in-house development, with all its attendant implementation hurdles. In practice, most large companies Rio included employ both strategies: building capability internally while also leveraging the agility of external innovators.

If we are serious about lifting R&D investment and impact, the focus must shift from blaming universities for poor commercialisation to growing absorptive capacity on both sides.

For industry, this means:

- Investing in specialist skills to interpret and integrate research.
- Embedding adoption planning into R&D investment from the outset.
- Bringing end users and implementers into the research process early.

For universities, this means:

- Going beyond delivering research outputs to actively helping partners absorb and apply findings.
- Providing expertise, training, and "technology deployment" support as part of the project.
- Embedding "transfer agents" individuals who bridge the research and industry worlds, clarifying expectations, solving problems, and driving implementation.

We have been grappling with Australia's absorptive capacity problem for decades, and it has long been a drag on our business R&D investment.

In the early 2000s, I helped build what became the largest industry research consortium in the country. At its peak it had 25 member companies, every one of them international. Not a single Australian company could be persuaded to join. They lacked the in-house expertise to engage on early-stage R&D, and they struggled to convince their management to invest. The mindset was too operationally focused, with little familiarity or comfort with partnering universities on research that might take years to bear fruit.

Fast forward to today, and I work with a large international company managing a diverse research portfolio. They are world-class at spotting research that could lead to breakthrough technologies, but they too have recognised a weakness in turning that promise into reality. Their solution was to create a dedicated "deployment" function within their innovation program, tasked with taking over once projects reach TRL 6/7. Importantly, this group will now get involved much earlier, working alongside research teams to prepare the relevant parts of the organisation for adoption. It's a pragmatic of the absorptive capacity challenge and one that Australia's businesses could learn a great deal from.

Without absorptive capacity, increased R&D investment will yield a poor return. Australia's industry makeup dominated by SMEs, with few large firms operating at scale makes this a particularly thorny problem.

Simply urging businesses to "invest more in R&D" will fail if they lack the skills, structures, and strategies to use what they develop.

If we want enduring, impactful research-industry partnerships, we must:

- Build absorptive capacity in Australian firms.
- Equip universities to support adoption, not just discovery.
- Foster roles and processes that bridge the gap between research and practice.

Otherwise, we risk continuing the cycle of promising research that never makes it to market a loss for industry, universities, and the nation's innovation future.

We should also acknowledge that absorptive capacity is also an ecosystem factor that goes beyond an organisation. As the case studies suggest we should think more about a system's absorptive capacity for graduates, new R&D, IP, and even VC funds. This is important for assessing ecosystem maturity and quality.

Victor Pantano is A technical generalist with an engineering and science background working across multiple industries: manufacturing, resources technology and minerals processing, clean energy, software development, IT, defence, space, medical products and health. His roles have included CEO, Director and General Manager and Associate Vice President covering functions as diverse as sales, operations, investment management, equity raising, R&D, industry and business development and executive management. He is a Member of the Acton Institute for Policy Research and Innovation

6. Innovation for Sustainability: Lessons from Australian and Californian Agrifood Systems

Rajesh Gopalakrishnan Nair[5] *November 2024*

Australia and California share many similarities concerning the diversity of geographical landscapes, multicultural societies and sustainability challenges.

Agriculture is crucial in both locations, and both are exploring opportunities and options for better integrating agrifood innovation systems to tackle the broadening and increasingly complex goals of the agriculture and food sectors.

Yet, these regions differ in their political governance structures, cultural influences and unique histories. This, in turn, shapes different strategies and approaches for deploying innovation to address the challenges of the agriculture and food sectors and to create new opportunities and pathways for prosperous and sustainable societies.

As part of the Science Summit at the 79th session of the United Nations General Assembly, a panel discussion took place on *Innovation for Sustainability: Collaborative Solutions from Australian and Californian Agrifood Systems.* The discussion centred on reimagining agricultural innovation in the context of the intertwined relationship between productivity, sustainability, and social inclusion.

Australian and Californian panel members were drawn from government, industry and research. This virtual event was held in September 2024 with a diverse audience of research, practice and policy stakeholders. The following is a synthesis of these discussions.

Redirecting Innovation for Sustainability.

Traditionally, agriculture, like all production-oriented sectors, has been driven by an agenda of mobilising innovation to improve productivity and international competitiveness. However, the sector's agenda is expanding. This is in response to global concerns related to climate

[5] Rajesh, an early-career researcher at CSIRO, Australia, was the rapporteur of the Australia-California joint panel at the Science Summit of the 79th UN-General Assembly

change, environmental sustainability and the challenges of creating just and equitable societies.

Innovation capabilities and processes must adapt if these broader agendas are to be tackled. What will these new innovation capabilities and processes look like?

A focal point of the discussion was the Circular Economy, highlighting two recent initiatives. BEAM Circular, a circular bioeconomy hub in North San Joaquin Valley (NJSV) in California's agricultural heartland, leads a portfolio of public-private projects for solutions that transform organic residues and by-products into renewable energy, industrial chemicals and consumer goods.

The Regional Circularity Cooperative Ltd (RCC), in the Bega Valley of New South Wales, Australia, is a regional development initiative aspiring to transform the Bega Valley into the most circular regional economy by 2030. Both BEAM and RCC demonstrate models for economic resilience, inclusion, and environmental stewardship. In partnership with the community, businesses, corporations, academia and government, they demonstrate CE transition pathways through pilots.

Six prompting propositions guided the panel discussion on this topic.

- As productivity and sustainability are inseparable and deeply interwoven, the forward-looking agenda should couple these objectives.
- The values of environmental sustainability, fairness and social inclusion underlying this agenda must be embedded in innovative action, practice and policy.
- An expanding repertoire of innovation practices and approaches is going to be needed to deal with the expanding and increasingly complex sector goals.
- Carefully choreographing a host of emerging partnerships with the community, industry, research, policy, and others is necessary for driving this agenda forward.
- It is important to recognise and build new innovation system functions that enable intermediation between actors in the innovation systems and help set the new direction for innovation-action and practice.
- There is a need to expand the set of policy options and policy levers beyond R&D and integrate policy responses across different cross-cutting sectors and portfolios.

Unique Strategies and Shared Commitments

California, having taken pro-active steps in climate change policy and agricultural sustainability, has embraced strategies such as supporting *pre-competitive collaboration* and *nature-based solutions*. This means public policy supporting cooperation in addressing common challenges before competition becomes a factor while also increasing reliance on nature-based solutions.

These approaches help to enhance inclusiveness and minimise environmental impacts through the participation of diverse communities, including First Nations people. Policies like the Sustainable Groundwater Management Act and cap-and-trade mechanisms drive California's focus on inclusiveness. These policy innovations have prepared a strong ground for a sustainable transition that leaves no one behind.

In parallel, Australia's agricultural sector faces its own set of challenges, primarily climate change and transitioning to a circular economy. Australian efforts have increasingly focused on inclusivity, particularly through the involvement of First Nations communities, women, and people with disabilities.

Significant R&D investments have led to technological advancements and economic growth. Collaborative action and sustainability have always been prioritised. The National Statement on First Nations Agriculture is a major move for improving inclusiveness. The national statement envisions the integration of grassroots knowledge and practices with the national agricultural policy.

The narratives from both Australia and California share a commitment to sustainability but also highlight unique strategies. For example, while California focuses on market-based solutions and strategic funding for de-risking green transition, Australia's focal areas include biosecurity and environmental stewardship.

Both regions aim for net zero emissions and have developed policies employing "Nature Positive" approaches and carbon markets as tools for meeting sustainability goals.

Key takeaways

Four key reflections emerged from the discussion

1. While technology is necessary, it is not sufficient for addressing the complex and expanding agenda of the agriculture and food sectors.
2. Innovation should be inclusive, deeply engaging and bringing together different communities, voices, perspectives, resources, ideas and knowledge. Further, the inclusiveness should also be in terms of its outcomes, making it relevant to different communities and stakeholders. This essentially requires building innovation ecosystems that embed knowledge and action, connecting communities and empowering people to operate within them.
3. Market-driven solutions alone will not resolve sustainability challenges. Pro-active government intervention is necessary in convening pre-competitive collaboration. Public policy should foster intentionality and directionality in the pathways of development.
4. Scaling sustainability efforts is not merely deploying more technology and diffusing it widely. On the contrary, scaling involves creating and diffusing new industries, economic models and socially and environmentally relevant value propositions. This will need far-reaching changes in society and in political arenas so that planning and action are driven by explicit values of sustainability, inclusion and circularity. Deep scaling of this sort is a key priority.

Together, these four ideas reinforce the importance of the ongoing dialogue between Australia and California in a collaborative learning process. The shared challenges and necessity-driven quest for solutions position them as potential leaders in addressing the global agrifood and environmental issues through sustainable and inclusive innovation[6].

---

[6] Note: This article is based on a conversation held as part of the 79th session of the United Nations General Assembly organised by CSIRO (VSFSP)-DAFF (ABARES) Innovation Policy Project in collaboration with the California Department of Food and Agriculture (CDFA) on 18 September 2024. A Memorandum of Understanding between the Australian Government and the Government of the State of California, signed in 2023, was the impetus for this event. The MoU provides a forum for cooperation and information sharing on issues including sustainability in agriculture. The panel included Karen Ross (Secretary, CDFA, California), Jared Greenville (Acting Deputy Secretary, DAFF, Australia), Joshua Viers (Professor, School of Engineering, University of California, Merced), Cindy Cassidy (Executive Director, SNSW Innovation Hub, Australia), Karen Warner (CEO, BEAM Circular, California) and Andrew Taylor (CEO, Regional Circularity Cooperative in Bega, New South Wales). The panel was moderated by Josette Lewis (Vice President and Chief Scientific Officer, Almond Board of California) and Peat Leith (Director VSFSP, CSIRO) and was chaired by Andy Hall (Senior Principal Research Scientist, CSIRO, Australia).
The author thanks Andy Hall for valuable insights in refining the report.

7. Large Scale Industrial Legacy Adaptation: Transforming the World's Biggest Coal Port

Roy Green, 25 January 2025[7]

The Port of Newcastle's evolution from a traditional coal export hub to a diversified maritime gateway exemplifies how legacy infrastructure can adapt to emerging economic and environmental imperatives. This transformation, occurring against a backdrop of global decarbonisation efforts and shifting trade dynamics, offers valuable insights into large scale industrial adaptation.

When I assumed the chair position in December 2017, the world's biggest coal port faced a critical strategic inflection point. The facility's overwhelming dependence on coal exports, which represented 85 per cent of revenue, created significant vulnerability as global markets began to show structural shifts away from fossil fuels. This realisation catalysed an ambitious diversification strategy that would fundamentally reshape the port's operational model.

Three distinct but interconnected challenges emerged. First, flattening global coal markets threatened to strand valuable assets. Second, controversial privatisation decisions in 2013-14 led to regulatory constraints, particularly regarding container terminal development. Third, the acceleration of climate action policies worldwide necessitated rapid adaptation to new energy paradigms.

The port's response demonstrates how industrial assets can pivot towards emerging opportunities.

A new entrepreneurial management team redesigned the port's strategy and vision in consultation with the workforce as part of a structured process of business model innovation. This followed management theorist Peter Drucker's observation that, "Innovation is the specific instrument of entrepreneurship... the act that endows resources with a new capacity to create wealth".

There was an urgent need at the outset for the port to adjust the access charge for coal producers, which, remarkably, had not been increased in the previous 19 years of public ownership. In effect, New South Wales

[7] Emeritus Professor Roy Green is the outgoing chair of the Port of Newcastle.

taxpayers were cross-subsidising coal exports. After a lengthy court battle, this was achieved, providing security for future initiatives.

Simultaneously, the port's successful campaign to remove an anti-competitive container terminal restriction imposed at the time of privatisation opened new pathways for significant capacity expansion. This development will address critical inefficiencies in Australia's east coast container handling operations, where existing ports rank in the bottom 20 per cent globally for productivity.

Strategic planning has also positioned Newcastle as an emerging clean energy hub, supported by federal government investment. The development of hydrogen infrastructure presents dual opportunities: serving domestic industrial requirements while establishing export prospects focused on North Asian and European markets. This initiative aligns with research indicating a growing demand for green hydrogen among key trading partners.

Historically, the port's prosperity has been intertwined with that of the Hunter region, and now its transformation will promote regional industrial renewal. The potential for green iron and steel production, utilising zero-carbon technologies, exemplifies how traditional capabilities can be reimagined for a low-carbon economy. This is an opportunity to take advantage of global trends in sustainable manufacturing and circular economy principles.

However, the port's privatised ownership structure presents unique challenges for implementing transformative projects. While private ownership can drive innovation and challenge operational assumptions, it also introduces complex stakeholder dynamics. Current understanding suggests successful transformation will require careful balancing of shareholder returns with broader community benefits.

Several additional promising developments are already underway. The port's emergence as a preferred RO-RO destination for vehicle imports demonstrates effective market diversification. Additionally, the handling of wind turbine components, coupled with potential domestic manufacturing under the Future Made in Australia program, builds capacity in renewable energy infrastructure.

The ongoing transformation of the Port of Newcastle illuminates several key insights for infrastructure adaptation in an era of rapid change.

First, legacy assets can be successfully repositioned when transformation strategies align with emerging market trends. Second, regulatory frameworks significantly influence adaptation capabilities. Third, private ownership structures introduce unique dynamics in managing a long-term program of innovation and change.

Successful infrastructure transformation requires careful attention by boards to multiple stakeholder interests, particularly in privatised assets with a significant role in regional economic development. The Port of Newcastle's experience demonstrates how industrial assets can navigate complex transitions while maintaining operational viability and building future growth potential.

This transformation model offers valuable lessons for other infrastructure assets facing similar challenges globally. Evidence indicates that success factors include clear strategic vision, stakeholder engagement and alignment with broader economic and environmental trends. The Port of Newcastle's continuing journey from coal dependency to diversified maritime gateway demonstrates how industrial assets can adapt to changing market conditions while contributing to regional economic resilience.

8. RMIT's Enabling Impact Platforms: Building System Integration for Impact

RMIT University has taken an ambitious step in its research and innovation strategy through the *Enabling Impact Platforms (EIPs)* initiative, which has evolved over almost a decade, beginning in 2016 with what were then known as Enabling Capability Platforms (ECPs). The original ECPs were formed to bring together a critical mass of research capabilities for which RMIT has demonstrated excellence, relevance, and differentiation.

As RMIT's strategy evolved to focus more strongly on research and innovation for impact, the ECPs transitioned into EIPs in 2021. This marked a trajectory which began with building the foundations of interdisciplinary research excellence towards deployment of these capabilities to achieve impactful outcomes

These eight platforms, covering advanced manufacturing, biomedical innovation, sustainable technologies, urban futures, social change, and other fields, are designed to bring people together across disciplinary, sectoral, and cultural divides. Their purpose is to accelerate the journey from discovery to impact by creating the conditions for collaboration and enabling diverse pathways for research translation and impact.

The EIPs achieve this through their five key strategic roles:

- Catalysing interdisciplinary and cross-sectoral collaborations
- Leading the development of impactful initiatives at scale
- Fostering impact-focused culture and capabilities
- Enabling diverse pathways to research translation
- Providing strategic thought leadership.

The idea resonates strongly with international experience. Innovation systems thrive when they include entities that do more than generate knowledge: they integrate it and enable ways to create value from research. System integrators connect capabilities, orchestrate partnerships, and align efforts with societal needs. By creating the EIPs, RMIT has established an internal mechanism to fulfil this integrator role both within the university and in the wider ecosystem.

The Integrator Opportunity

System integration has long been recognised as the missing ingredient in complex innovation systems. While excellence in disciplines is essential, impact requires much more: it requires mechanisms that cut across disciplinary and structural silos and enable collective action. Interdisciplinary approaches also enable more effective problem finding and produce more innovative solutions, which are especially important when addressing the complex problems faced by society today. The EIPs represent exactly this type of mechanism, and over the last decade has served as an important driver of cultural change within RMIT.

The EIPs are structurally very lean, with researchers and professional staff being affiliated to one or more Platforms, each led by highly accomplished, service-oriented EIP Directors with modest operational staff support.

Their design allows researchers from different faculties to quickly form interdisciplinary teams. They provide small but catalytic funding to seed ideas and test new collaborations. They convene events and networks that involve government, industry, and community partners. In doing so, they create opportunities for academics to see their work adopted and applied in new contexts and to access networks that would otherwise be out of reach.

Enabling diverse pathways to create impact from research is critically important in maximising the potential benefits of translating research. This is because impact comes in many forms and is therefore realisable through different pathways.

While less immediately obvious but equally important, RMIT's decade-long commitment to the EIPs and its activities has also created an environment for continual experimentation and rapid learning to support its enablement model.

Just as importantly, the EIPs position RMIT to contribute to national and international missions. In particular, the EIPs have been instrumental in helping identify 'white spaces' for the deployment of research capability, exemplified by the High Impact Potential Initiatives (HIPIs) program, which aims to develop impact-focused, industry-engaged activities at scale.

By aligning the platforms with pressing priorities, including health, sustainability, advanced materials, and digital futures, they give the university a visible role in responding to challenges that are important to society. This is much more than internal collaboration; it is about building the capacity to partner externally on a scale that can make a difference.

Learning from International Exemplars

The value of such integrator platforms is confirmed by global experience. The examples below highlight the role of integration and the conditions that allow it to flourish.

MIT's Deshpande Center – integration through entrepreneurship

The Massachusetts Institute of Technology's Deshpande Centre for Technological Innovation illustrates the integrator role at the scale of individual projects. The Centre provides small "ignition grants" to faculty, enabling them to test ideas with commercial potential. Crucially, the funding is coupled with mentoring from entrepreneurs and industry experts. This combination has helped many projects transition from laboratory prototypes to venture-backed companies.

The Deshpande story demonstrates the power of seed-to-scale pathways. Small grants alone rarely transform research, but when coupled with mentoring and clear routes to application, adoption, and use, they can unlock extraordinary impact.

For RMIT, the opportunity is to frame EIP seed funding as the first step in a larger journey, potentially leading to ARC and NHMRC grants, Cooperative Research Centres, and commercial ventures. Positioning the platforms as gateways rather than endpoints makes participation more attractive to researchers.

University College London – integration through Grand Challenges

University College London's (UCL) Grand Challenges initiative is also an instructive model. Launched in 2009, it brings academics together around themes such as global health, sustainable cities, and cultural understanding. The Grand Challenges program has supported hundreds of interdisciplinary projects, many of which have led to policy influence, public engagement, and new external funding.

UCL's experience highlights the importance of linking integration to institutional performance. Outputs from the Grand Challenges are recognised in the UK's Research Excellence Framework (REF), ensuring that interdisciplinary work contributes to the university's overall standing. For RMIT, this suggests that embedding EIP outcomes into institutional performance metrics, whether in research assessments, rankings, or industry engagement measures, will be crucial for long-term sustainability.

Fraunhofer Institutes – integration through applied research

Germany's Fraunhofer Institutes are perhaps the most widely cited examples of integration in practice. Established after the Second World War to bridge universities and industry, they now number more than 70 institutes across the country. Each is focused on a domain of applied research, such as laser technology, materials science, or industrial automation. Their collective role is to provide the connective infrastructure for the national innovation system.

The Fraunhofer model shows that integrators succeed when trusted by multiple partners. They are funded partly by government block grants, partly by competitive project income, and largely by contracts with industry.

This blended model means they must always remain relevant to industry needs while still contributing to long-term national capabilities. The lesson for EIPs is that legitimacy as an integrator increases when platforms create value for both academic and external partners. Even small-scale seed projects can build credibility if they demonstrate responsiveness and trust.

Catapult Centres – integration as national infrastructure

In the United Kingdom, the Catapult Centres play a similar integrative role but with a more explicit focus on the national industrial strategy. Launched in 2011, they now cover areas such as high-value manufacturing, digital technologies, offshore renewable energy, and cell and gene therapy. Catapults provide shared facilities, technical expertise, and networks that allow businesses, particularly small and medium enterprises, to access capabilities they could not build on their own.

The catapult experience underscores the importance of long-term institutional support. Each centre receives multi-year core funding from the government, allowing it to plan strategically and provide continuity for its partners. By implication, the impact of its EIPs will be amplified if they can be linked to state or national initiatives in ways that provide continuity beyond annual project and funding cycles. By aligning with emerging missions, such as those supported by Australia's National Reconstruction Fund, the platforms can position themselves as indispensable connectors.

Other lessons from global practice

Examples from other sources reinforce these points. Arizona State University has embedded integration by restructuring faculty into transdisciplinary schools, ensuring that collaboration is an integral part of the academic fabric. ETH Zurich has created competence centres that pool resources across faculties and external partners, while TU Delft has interfaculty institutes that share doctoral programs and infrastructure. These cases remind us that integrators thrive when they are *structurally embedded* rather than treated as peripheral add-ons.

Taken together, these international examples suggest that the EIPs are part of a broader global movement.

Continuing Opportunities

For researchers, these platforms open new avenues to apply their expertise, build interdisciplinary teams, and pursue funding opportunities that require cross-cutting collaboration. EIP activities and programs help guide RMIT researchers and their collaborators towards a better understanding of societal needs.

By connecting with industry and government partners, they can see their work adopted in ways that expand its relevance and reach. In doing so, the EIPs help enhance the capabilities of researchers, enabling them not only to conduct excellent research but also to deliver potentially impactful findings that are relevant to society, the economy, and the environment.

For the university, EIPs provide a way to showcase impact that goes beyond publications or grant income. RMIT's initiative is focused on convening partnerships, generating solutions, and contributing to national and global missions. This can only strengthen the university's reputation as an institution committed to applied, relevant, and impactful research.

For policymakers, EIPs offer a model of how universities can function as system integrators within regional and national innovation systems. They illustrate the potential for higher education institutions to take a pro-active role in connecting fragmented capacities and aligning them with societal priorities. In a policy environment that increasingly emphasises collaboration, translation, impact, and mission orientation, this is an extremely valuable contribution.

Building on Momentum

The platforms have already demonstrated momentum. The Strategic Impact Fund has supported dozens of projects that span multiple disciplines, with a particular focus on enabling impact through unconventional pathways. The Platform Activity Fund has seeded workshops and pilots that brought new teams together and built new capabilities for interdisciplinary collaboration.

Enabling Impact Networks formed under the EIPs have engaged hundreds of researchers and professional staff. There are currently over 40 such networks sponsored through the EIPs, all inherently interdisciplinary or transdisciplinary, all self-driven by researchers with minimal enterprise-level control, many involving industry participants from inception and increasingly being established across RMIT's global footprint, which includes Southeast Asia and Europe.

These early outcomes demonstrate the catalytic potential of even modest investments.

Additionally, the EIPs have evolved an internal ecosystem that fosters leadership development beyond traditional research leadership. The myriads of seed-funded projects and network activities has particularly benefited emerging researchers by providing leadership opportunities that are impact-focused, purpose-led, and values-based. The EIP Directors themselves are exemplary role models of collegiate, cross-disciplinary collaboration.

The next stage is to build upon this foundation. By linking EIP activities more explicitly to external funding pathways, these activities can become launchpads for major initiatives such as the HIPIs. By embedding participation in academic performance systems, it can become integral to careers rather than an optional extra. By cultivating external partnerships, they can strengthen their role as boundary organisations trusted by the government, industry, and community alike.

From Potential to Practice

RMIT's Enabling Impact Platforms represent a positive step in the evolution of Australian university research and innovation. They demonstrate that it is possible to design mechanisms that act as system integrators, connecting disciplines, convening partnerships, and aligning knowledge with societal needs.

The opportunities are considerable. For individual researchers, these provide pathways for collaboration and visibility. For RMIT as an institution, it signals a commitment to impact and relevance. For the national innovation system, they demonstrate that universities can do more than generate knowledge; they can integrate it to maximise value from knowledge creation and exchange.

The lesson from international experience is that integration works best when it is embedded, supported, and sustained.

With its very lean structure and relatively modest resources, it is also an exemplar of organisational commitment to long-term strategy and direction, with courageous leadership to learn as it evolves and to make course corrections and strategic re-alignment as required.

RMIT's EIPs are still young, but they have the potential to evolve into platforms of enduring significance. By embracing the integrator role, they can help transform fragmented research capacity into a coordinated impact. In doing so, they not only advance RMIT's mission but also contribute to shaping innovation ecosystems in Australia and beyond.

The assistance of Professor Swee Mak, Associate Deputy Vice-Chancellor, Research and Innovation Capability at RMIT, is greatly appreciated in preparing this chapter is greatly appreciated.

9. Why Australia Needs Fraunhofer Institutes

John Howard, Peter Moar, 29 July 2025

Australia's innovation system is constrained by a persistent failure to translate world-class research into industrial capability and commercial outcomes. Drawing lessons from the German city of Kaiserslautern and its transformation through Fraunhofer Institutes, this chapter argues that Australia must have applied research institutes dedicated to bridging the gap between universities and industry.

The Fraunhofer model, built on public-private partnership, applied research excellence, and ecosystem catalysis, provides a proven pathway to sovereign capability, economic growth, and regional transformation. Strategic establishment of Fraunhofer-style institutes, leveraging existing assets and the CSIRO, would address Australia's innovation paradox, grow industrial competitiveness, and ensure economic and strategic benefits for future generations.

Context

Kaiserslautern, a German city of just 100,000 residents, demonstrates that geographic scale is not a prerequisite for innovation leadership. Despite its modest size, this southwestern German city has created one of Europe's most successful technology ecosystems, anchored by two Fraunhofer Institutes that have helped to translate world-class research into commercial success.

The city's transformation from a traditional industrial centre to Germany's premier hub for applied mathematics and software research offers compelling lessons for Australia's innovation policy. At the heart of this success lies a proven model: *applied research institutes that systematically bridge the gap between university research and industrial application.*

For Australia, grappling with its own innovation challenges while seeking to build sovereign capability and drive regional economic development, the Kaiserslautern experience provides a roadmap for success that is both scalable and replicable.

Australia's Innovation Paradox

Australia faces a well-known and frustrating innovation paradox. Our universities consistently rank among the world's best in research output and quality: they rank as fifth globally in research citations per capita and eighth in the Global Innovation Index for innovation inputs. Yet we consistently underperform in research commercialisation and industry collaboration, ranking 25th overall in the Global Innovation Index and struggling to translate research excellence into economic outcomes.

This "valley of death" between brilliant research and commercial application costs Australia billions in unrealised economic potential. The OECD consistently ranks Australia poorly on business expenditure on R&D as a percentage of GDP, and our venture capital investment levels lag significantly behind comparable economies. Meanwhile, our best and brightest researchers often see their innovations commercialised overseas, contributing to the economic growth of other nations rather than our own.

The situation is particularly acute in critical technology areas where Australia desperately needs sovereign capability.

From advanced manufacturing and critical minerals processing to defence technologies and renewable energy systems, we often excel at the fundamental research but struggle to translate that knowledge into domestic industrial capability and export opportunities.

This is precisely the challenge that Fraunhofer Institutes were designed to solve, and Kaiserslautern's experience demonstrates they can deliver transformational results even in modest-sized cities.

The Fraunhofer Formula: A Proven Success Model

Kaiserslautern's remarkable transformation rests on four critical elements that make Fraunhofer Institutes uniquely effective innovation catalysts.

The Fraunhofer-Gesellschaft Benchmark

Fraunhofer has been the backbone of Germany's advanced manufacturing export-based economy since it was established after World War II. There are now 76 institutes and close to 32,000 employees in Germany. Internationally, they have been invited by numerous national and state governments to set up operations in 30 locations. The UK and US Governments identified the Fraunhofer as the benchmark model to emulate for bridging the gap between research and industrial application.

The UK's Catapult Network was explicitly modelled on Fraunhofer-Gesellschaft, both in structure and funding design. The government also supported the establishment of several Fraunhofer Centres.

The US Obama Administration, after reviewing the decline in the traditional manufacturing sector, invited and financially supported the establishment of an additional five Fraunhofer Centres to the two that were already in existence. MIT invited Fraunhofer to set up on campus in recognition that the model was better at translating their applied research into industrialised outcomes.

Canada has jointly funded the establishment of a Fraunhofer Project Centre for Composites Research at Western University in London, Ontario.

Strategic Public-Private Partnership

The Fraunhofer model is one of the world's successful examples of public-private research collaboration. Unlike traditional university research funded primarily by government, or private R&D confined to individual companies, Fraunhofer Institutes operate through a sophisticated funding model that aligns public investment with industry needs.

In Kaiserslautern, this approach has created research institutions that maintain long-term research capability, while remaining laser-focused on industry relevance. The institutes receive base funding from government for fundamental capabilities but earn approximately 70 per cent of their revenue through industry contracts and collaborative projects.

For Australia, this model could revolutionise how we approach research commercialisation and technology transfer.

Rather than hoping universities will become more industry-focused or expecting private companies to invest in longer-term research, Fraunhofer Institutes create dedicated institutions optimised for the crucial interface between fundamental research and practical application.

In Western Australia, for example ECU and Fraunhofer Institute for Ceramic Technologies and Systems (IKTS) have partnered to establish the region's first research and innovation lab. The Lab will concentrate on developing advanced ceramic solutions for industrial water treatment and hydrogen generation, implementing efficient testing methods for industry-standard and next-generation materials, and exploring ceramic technologies essential for sustainable energy solutions.

Applied Research Excellence

Two Fraunhofer Institutes, the Institute for Industrial Mathematics (ITWM) and the Institute for Experimental Software Engineering (IESE) are in Kaiserslautern. They have become global leaders by focusing on applied research excellence. ITWM is positioned as one of the world's largest mathematical research institutes, with its research methodology firmly grounded in industrial applications: modelling, simulation, and optimisation for real-world challenges.

This focus on applied research creates a unique value proposition. Universities typically excel at fundamental research but often lack the industry connections and application focus necessary for commercialisation. Private companies excel at near-market product development but rarely have the resources or risk appetite for the longer-term, higher-risk research necessary for breakthrough innovations.

Fraunhofer Institutes occupy a crucial middle ground—conducting research that's too applied for universities but too fundamental for private companies. In Kaiserslautern, this has resulted in institutes that serve industries ranging from automotive and mechanical engineering to finance and telecommunications, while maintaining world-class research reputations.

Ecosystem Catalysis and Regional Transformation

Perhaps most important for Australian policy makers, Fraunhofer Institutes serve as powerful catalysts for broader innovation ecosystems. Kaiserslautern's experience demonstrates how these institutions can drive comprehensive regional economic transformation.

In the early 1990s, the city realised it was on the path to becoming a "rust belt" with unemployment at 15 per cent and the loss of 10,000 traditional manufacturing jobs. The turning point came when the city acknowledged the problem and committed to investing in institutional and regional infrastructure, recognising that the future would be increasingly digital.

The focus shifted to digital transformation through software engineering, artificial intelligence, information technology, robotics, and Industry 4.0.

Fast forward to today: unemployment is projected to fall to three per cent, and the city has created over 10,000 high-value, future-focused tech jobs. Sustained and continuous growth has attracted global companies, including John Deere, which established its EU R&D centre there after assessing 28 cities across Europe.

More recently, companies such as Amazon (2,000 jobs), battery manufacturer ACC (a joint venture of Stellantis, Mercedes, and SAFT with over 2,000 jobs), and Corning have also begun establishing operations in the city.

The city now hosts one of Germany's concentrations of research institutions relative to its population, creating what researchers call "innovation density", the critical mass of expertise necessary to attract international companies, support start-up formation, and create high-value employment opportunities.

Companies like ThinkParQ (creators of the BeeGFS parallel file system) and LUBIS EDA (semiconductor design software) have emerged from this ecosystem, demonstrating its ability to generate globally competitive technology ventures.

The institutes have also been instrumental in Kaiserslautern's recognition as a member of Germany's Software-Cluster alongside technical universities in Darmstadt, Karlsruhe, and Saarland.

This network effect amplifies the impact of individual institutes, creating research and innovation capabilities that far exceed what any single institution could achieve independently.

The Australian Opportunity: Strategic Imperatives Converging

Australia's current strategic context creates an unprecedented opportunity for Fraunhofer Institute success. Multiple policy imperatives are converging in ways that make applied research institutes not just beneficial, but essential:

Sovereign Capability Development: Recent global disruptions have highlighted Australia's vulnerability in critical technologies and supply chains. The government's Critical Technologies List identifies areas where Australia needs enhanced domestic capability, from quantum computing and artificial intelligence to advanced manufacturing and critical minerals processing. Fraunhofer Institutes offer a proven pathway to building these capabilities while maintaining integration with global innovation networks.

National Development: The $15 billion National Reconstruction Fund is Australia's most significant manufacturing policy intervention in decades. The $22.7 billion Future Made in Australia package recognises immense opportunities for Australia and its people lie at the intersection of industry, energy, resources and skills. However, success will require more than financial investment—it demands the applied research capabilities to support industry transformation and technology adoption.

Fraunhofer Institutes could provide the technical backbone for NRF and Future Made in Australia success across priority areas, including medical devices, renewable energy, defence capabilities, and advanced manufacturing.

Regional Economic Development: Australia's innovation ecosystem remains heavily concentrated in Sydney and Melbourne, creating challenges for regional economic development and contributing to skills drain from regional areas. The Kaiserslautern model demonstrates how applied research institutes can create innovation hubs in smaller cities, supporting regional economic diversification while building national innovation capacity.

Industry 4.0 and Digital Transformation: German industry's leadership in Industry 4.0 and digital manufacturing owes much to the Fraunhofer network's applied research capabilities. As Australian manufacturers and resource companies pursue digital transformation, they need access to world-class applied research support, exactly what Fraunhofer Institutes provide.

Australia has substantial capability in manufacturing research excellence, which is thinly spread across the country in many innovation districts and precincts, and has been led by CSIRO over many years in multiple locations. To realise this potential through a Fraunhofer connection, there are potentially two ways to go:

Two Possible Approaches

Approach 1: Leverage Current and Potential Capacities and Capabilities in Well-Performing Innovation Districts

Drawing from Kaiserslautern's success in university research partnerships, industry cluster alignment, and regional development potential, several locations emerge as prime candidates for Australian Fraunhofer Institutes.

Australian innovation districts are rapidly strengthening their translational/ applied research and industry engagement models. This has covered a greater translational focus, co-location, shared infrastructure, industry-shaped research, and workforce integration.

Gaps remain in systemic industry partnerships at scale, late-stage scale-up, and full multidisciplinary integration, presenting clear opportunities for Fraunhofer-style applied research institutes to deliver impact through structured, long-term, industry-led research programs

Below, for indicative purposes, is a high-level mapping of what would appear to be the most relevant districts to host a Fraunhofer capability.

City/ Region	District/Precinct	Specialisation Relevant to Fraunhofer Focus	Institutional Anchors/Notes
Adelaide	Lot Fourteen	AI, software engineering, defence, space, industrial mathematics	AIML, SmartSat CRC, Defence/Space Landing Pad
	Tonsley	Advanced manufacturing, automation, applied software in industry	Flinders, TAFE SA, Siemens Energy, Factory of the Future
Melbourne	Parkville	Biomedical research, translational data science, some AI	University of Melbourne, Monash
	Fishermans Bend	Advanced manufacturing, mobility, automation, simulation	Several Melbourne Metropolitan Universities, including University of Melbourne, RMIT and Swinburne Universities; Industry partners

City/ Region	District/Precinct	Specialisation Relevant to Fraunhofer Focus	Institutional Anchors/Notes
	Clayton Technology Precinct	AI, advanced manufacturing, software engineering	Monash University,
Sydney	Tech Central	Software, digital, AI, start-up/scale-up, urban technology	Atlassian HQ, Canva, USyd, UTS, UNSW
	Macquarie Park	Corporate innovation, digital health, software integration	Macquarie University, multinational firms
	Bradfield/ Aerotropolis	Advanced manufacturing, automation, aerospace	New development, Western Sydney University
Newcastle	Port of Newcastle Clean Energy Precinct	Renewable energy, hydrogen, digital health, advanced manufacturing	University of Newcastle (NIER), CSIRO Energy Centre
Geelong	ManuFutures/Pivot City	Advanced manufacturing, digital fabrication, industry-university integration	Deakin University, SAGE Automation
Bendigo	Bendigo Regional Manufacturing Group	Advanced manufacturing, digital health, AI, agricultural technology	La Trobe University, Bendigo TAFE Regional Manufacturing Group, City of Greater Bendigo, ANU
Perth	Automation & Robotics Precinct	Mining tech, automation, AI (industrial focus)	Curtin, University of WA, Pawsey Supercomputing Centre
	Hendeson Defence Precinct	Naval shipbuilding and defence industry cluster leveraging strategic infrastructure investment and proximity to Australian Marine Complex facilities.	University of Western Australia, Edith Cowan University
Brisbane	CBD & distributed precincts, Gold Coast	Advanced Manufacturing, Biotech	University of Queensland, Griffith University, QUT, CSIRO, TRI, ARM Hub

Approach 2. A Central Role for CSIRO and CRCs

Ideally, CSIRO would consolidate its assets across Australia and become the R&D engine for specialised stand-alone facilities in conjunction with universities, CRCs, and key industry partners, including knowledge-based FDI companies. This will both require and engender significant cultural change among all stakeholders.

Other models derived from Fraunhofer Institutes could be adapted, including UK HVM Catapults, ManufacturingUSA Institutes and NSF Regional Innovation Engines.

CSIRO already operates as Australia's closest equivalent to the Fraunhofer model, providing immediate advantages for institute development. With major facilities in every proposed Fraunhofer location, CSIRO offers ready-made research infrastructure, eliminating the need for costly duplication.

The organisation's 100-year heritage of applied and basic research, extensive industry networks conducting over 1,000 projects annually, and proven technology transfer capabilities through ON Prime and Main Sequence Ventures align perfectly with Fraunhofer objectives.

This integrated approach would create a powerful three-tier research ecosystem: universities providing fundamental research and talent development, Fraunhofer Institutes conducting industry-focused applied research, and CSIRO delivering technology transfer and commercialisation pathways.

The result would be uniquely Australian applied research institutes that combine our research excellence with proven industry engagement methodologies, creating the missing link in Australia's innovation ecosystem.

Universities would contribute world-class research capabilities and talent pipelines, while CRCs would provide proven collaboration methodologies and industry networks. Each institute would operate as a joint venture between CSIRO, university partners, and industry, with governance reflecting this partnership.

This model offers compelling advantages: accelerated implementation by building on existing capabilities rather than greenfield development, reduced risk through leveraging proven institutions, enhanced credibility with industry partners, and cost effectiveness through shared infrastructure. Most importantly, it amplifies rather than competes with Australia's research excellence, creating synergies that enhance the effectiveness of all participants.

Economic and Strategic Returns: The Investment Case

Kaiserslautern's experience demonstrates that Fraunhofer Institutes deliver economic returns far exceeding their establishment and operational costs. The city's transformation from a traditional industrial centre to a globally recognised technology hub has created thousands of high-value jobs, attracted billions in investment, and generated export revenues that continue to grow.

For Australia, similar institutes could address multiple strategic priorities simultaneously. Economic modelling suggests that each institute could generate economic returns of 3-5 times their operational costs within a decade, while building sovereign capabilities for long-term economic competitiveness.

The strategic benefits extend beyond direct economic returns. Fraunhofer Institutes would enhance Australia's research commercialisation performance, support industry transformation under the National Reconstruction Fund and Future Made in Australia, create high-value regional employment opportunities, and build the applied research capabilities for technological sovereignty.

Perhaps most importantly, they would help solve Australia's innovation paradox—transforming our world-class research capabilities into economic and strategic advantages that benefit all Australians.

The Competitive Imperative: Why Australia Can't Afford to Wait

As global competition for innovation capacity intensifies, countries that translate research excellence into economic advantage will prosper, while those that do not will fall behind. The Kaiserslautern experience shows this translation requires more than hope: it requires institutional innovation and sustained strategic commitment.

Australia has the research excellence, industry foundations, and policy frameworks to replicate Kaiserslautern's success. What we need is the strategic vision to bridge the gap between research and application through proven institutional models.

The Fraunhofer Institute model offers Australia a pathway to transform its innovation ecosystem, building on existing strengths while addressing persistent weaknesses. The question is not whether Australia can afford to establish Fraunhofer Institutes: it is whether we can afford not to.

If it is good enough for the US, UK and Canadian Governments, why not Australia? There is a greater imperative for Australia to establish Fraunhofer in the country due to our remote location from major international population-based markets and small population spread over such a large land mass.

Australia must be more committed to overcoming these additional hurdles. Introducing Fraunhofer into Australia's ecosystem will ensure the world's best benchmark is operating in the country and instantly connected into the global operations of the Fraunhofer network. Australia will be seen as a more attractive country to invest in technological innovation

The Kaiserslautern story tells us that in the knowledge economy, applied research excellence is the foundation of competitive advantage. Australia has the opportunity to build that foundation. The Commonwealth and the States require the strategic vision and courage to act.

PART V: FURTHER INSIGHTS FROM THE CASE STUDIES

Part V synthesises insights from more than 90 international case studies, offering a comparative perspective on the drivers of ecosystem performance. It examines the critical importance of infrastructure, connectivity, visionary leadership, and anchor institutions, as well as the risks of property-led approaches and fragmented governance.

This part highlights how global exemplars such as Kendall Square, 22@Barcelona, and MaRS Discovery District have succeeded through patient investment, system integration, and clear mission focus. It also addresses vulnerabilities, blind spots, and future trends, providing a practical evidence base for policymakers and ecosystem leaders seeking to strengthen Australia's innovation capacity.

1. Introduction

This analysis examines innovation districts and precincts, focusing on their development, integration into local economies, and comparative performance. The study of innovation ecosystems gains particular power when anchored in actual places, institutions, and histories, spanning from Australia to international examples.

Australia's innovation districts present an intriguing paradox. Despite geographical distance from major global innovation hubs, Australia has developed one of the most methodical approaches to creating knowledge-intensive urban ecosystems. From government-academia collaboration in Canberra to Shenzhen-style manufacturing integration at Tonsley, Australian innovation districts represent a distinctive approach between Silicon Valley's market-driven spontaneity and the centrally planned industrial parks of authoritarian states.

The Australian Innovation Archetype

Australian innovation districts share core characteristics with international counterparts while demonstrating unique features. They blend strong institutional governance, world-class research capabilities, and strategic government intervention at unmatched scale.

Unlike organically grown districts such as Kendall Square, Australian districts emerged through conscious policy orchestration designed to leverage existing strengths and address market failures. Government involvement extends beyond traditional infrastructure provision. In Tonsley, the South Australian Government purchased Mitsubishi's entire automobile plant and managed its restructuring. The New South Wales Government's commitment to Tech Central in 2025-26 demonstrates strategic leadership in ecosystem development.

This government leadership model offers both advantages and disadvantages. Benefits include comprehensive planning, coordinated infrastructure, and long-term investment bridging political cycles. Fishermans Bend's transformation from automotive manufacturing into aerospace and artificial intelligence demonstrates how strategic intervention can create new industry clusters. However, government intervention raises questions about measuring public return on investment and risks stifling entrepreneurial dynamism through excessive management.

The University-Anchor Institution Strategy

No characteristic distinguishes Australian innovation districts more clearly than systematic leverage of universities as primary anchors. While international districts often feature universities as important components, Australian districts place them at the strategic centre of ecosystem development, reflecting Australia's exceptional university sector and understanding of how research institutions drive innovation-led economic development.

Monash-Clayton Innovation District exemplifies this approach. With over 50,000 students and world-class research facilities including the Australian Synchrotron, the district creates integrated ecosystems where academic research, corporate partnerships, and start-up formation reinforce each other. Companies like Vaxxas, emerging from University of Queensland research and now operating state-of-the-art manufacturing at Boggo Road, demonstrate how university-anchored districts generate globally competitive commercial outcomes.

This university-centric model creates distinctive advantages in talent development and research commercialisation. Unlike corporate-anchored districts depending on individual company decisions, university-anchored districts benefit from educational institutions' inherent stability and long-term commitment. Universities provide continuous talent flows, research breakthroughs, and international connections while serving as neutral convening platforms.

However, the university-anchor model presents challenges in developing entrepreneurial culture and commercial focus. Academic institutions excel at research and education but often struggle with speed, risk tolerance, and market focus required for innovation commercialisation. Successful districts like Tech Central and Macquarie Park address this by combining university anchors with major corporate presences, creating hybrid ecosystems balancing academic excellence with commercial dynamism.

Sectoral Specialisation and Competitive Advantage

Australian innovation districts demonstrate strategic focus in developing sectors where Australia possesses genuine competitive advantages. Rather than replicating Silicon Valley's broad technology focus, Australian districts carve out specialised niches leveraging existing institutional strengths, geographic advantages, and market opportunities.

Biomedical concentration in Boggo Road and Parkville Biomedical Precinct reflects Australia's world-class medical research institutions, Asia-Pacific market proximity, and biotechnology regulatory expertise. Companies like Vaxxas, developing needle-free vaccine technology crucial during COVID-19, demonstrate how sector-focused districts generate innovations with global impact. Defence and aerospace focus in Fishermans Bend and Canberra leverages Australia's Asia-Pacific strategic position and major defence spending commitments including AUKUS.

This specialisation strategy creates advantages over broader technology districts through deeper expertise, specialised infrastructure, and stronger industry networks. Parkville-Carlton's concentration of medical research institutions, clinical facilities, and biotechnology companies enables collaboration and knowledge transfer impossible in diversified environments. Companies access specialised talent, regulatory expertise, and industry-specific infrastructure reducing development costs and accelerating commercialisation timelines.

The sectoral approach aligns with Australia's broader economic strategy of competing on knowledge intensity rather than scale. While Australia cannot match Silicon Valley's venture capital availability or Shenzhen's manufacturing scale, it can develop world-leading capabilities in specialised domains. Quantum computing research at ANU, advanced manufacturing at Tonsley, and precision medicine in Parkville-Carlton represent areas where Australian districts can achieve global leadership through focused investment and strategic coordination.

Scale Ambitions and Coordination Challenges

Australian innovation districts embrace geographically ambitious scales by international standards. Tech Central's six-square-kilometre span across multiple Sydney neighbourhoods is nearly 40 times Kendall Square's geographic footprint, while Fishermans Bend's 480 hectares rivals the largest global urban renewal projects. This scale ambition reflects both Australian cities' space availability and understanding of how innovation ecosystems require diverse components to achieve critical mass.

Large scale creates opportunities for comprehensive ecosystem development smaller districts cannot achieve. Tech Central accommodates diverse sectoral clusters, different company types at various growth stages, and supporting infrastructure from start-up incubators to corporate headquarters. Integration of residential development, cultural amenities, and recreational facilities creates authentic urban communities rather than sterile business parks.

However, scale creates coordination complexity representing perhaps the greatest challenge facing Australian innovation districts. Managing coherent development across multiple local government areas, state agencies, and private stakeholders requires governance mechanisms few cities have successfully implemented. The Sydney Start-up Hub's relocation to Tech Central demonstrates how complex stakeholder management becomes when multiple organisations with different missions must coordinate activities.

International Comparisons and Geographic Isolation

Australian innovation districts reveal striking similarities and important differences with international peers. Convergence on mixed-use urban development, transport connectivity, and corporate anchor strategies suggests certain fundamentals apply regardless of national context. Postindustrial area transformation into knowledge-intensive districts, seen in Barcelona's 22@, London's King's Cross, Fishermans Bend and Tonsley, are global trends towards urban regeneration through innovation-led development.

However, differences are equally revealing. Kendall Square's venture capital ecosystem maturity, with over 100 venture capital firms within walking distance, has no Australian equivalent. This reflects both Australia's smaller capital markets and different financing patterns in government-coordinated versus market-driven innovation ecosystems.

Australia's geographic isolation creates both constraints and opportunities shaping innovation district development uniquely. Unlike European districts benefiting from multi-country proximity or Chinese districts integrating into vast domestic supply chains, Australian districts must overcome distance to achieve global connectivity. This isolation forces development of distinctive capabilities in international partnership building and digital connectivity.

The response to isolation often drives innovation rather than hindering it. Australian districts excel at developing international partnerships, attracting global talent, and leveraging digital technologies to overcome distance constraints. Companies like Atlassian anchoring Tech Central or Cochlear emerging from Macquarie Park demonstrate that geographic isolation need not prevent global commercial success.

Future Trajectories and Strategic Implications

Australian innovation district maturation over the next decade will likely determine whether Australia maintains its position as a developed knowledge economy. The districts represent Australia's most significant institutional innovation in economic development, but ultimate success depends on addressing current limitations while building on established strengths.

Venture capital ecosystem development is perhaps the most critical challenge. While government funding and university partnerships excel at supporting early-stage R&D, transitioning to growth-stage companies requires private capital Australia's financial markets have historically struggled to provide.

International connectivity becomes increasingly important as global innovation networks evolve. Australian districts must balance systematic development approaches with greater integration into global innovation flows, requiring superior transport links, digital infrastructure, and cultural adaptation towards more open, experimental business development approaches.

The governance innovations developed in Australian districts may prove their most significant global contribution. The combination of government coordination, research institution anchoring, and private sector engagement is a distinctive model other countries with strong public institutions might adapt.

Australian innovation districts represent a remarkable experiment in systematic innovation ecosystem development. While differing significantly from international peers in governance, financing, and cultural approaches, they achieve impressive outcomes in research commercialisation, urban development, and economic diversification. Their continued evolution provides valuable insights into alternative pathways for innovation-led economic development and government coordination's potential to complement market forces in building knowledge-intensive economies.

2. Infrastructure as the Foundation of Innovation Districts

Across more over 80 case studies of innovation districts, precincts and hubs, the term *infrastructure* appears almost one thousand times. This frequency is not an accident. It reflects the central role that infrastructure plays in determining whether an innovation ecosystem flourishes or stalls.

Infrastructure provides the physical and organisational foundations that enable innovation to take place. It is not confined to roads and pipes. It includes the digital networks that allow data to flow, the research facilities that anchor knowledge production, the housing and amenities that sustain talent, and the governance arrangements that make delivery possible.

This Insight explores the significance of infrastructure as revealed in the case studies, showing how different categories of infrastructure shape outcomes, and why a systemic view of infrastructure is essential for policymakers and district leaders.

Why Infrastructure Matters for Innovation

Innovation districts, precincts, and hubs depend on concentration, connectivity and collaboration. Infrastructure underpins each of these dynamics.

- *Concentration* is enabled by research infrastructure: laboratories, incubators, testbeds and shared platforms that bring actors together.
- *Connectivity* relies on transport and digital infrastructure that move people, goods, ideas and data quickly and reliably.
- *Collaboration* is sustained by social infrastructure—housing, public space, cultural venues, cafés—where networks and relationships form.

Without infrastructure, "innovation district" becomes a branding exercise rather than a functioning ecosystem. The case studies confirm that infrastructure is not incidental but constitutive of innovation.

Placemaking as Innovation Infrastructure

The case studies consistently demonstrate that physical design decisions fundamentally shape innovation outcomes. This extends beyond providing adequate office space to creating environments that foster the "creative collisions" essential for knowledge transfer and collaborative innovation.

Physical Design for Collaboration

Kendall Square's evolution illustrates this principle clearly. The transformation from industrial wasteland to "the most innovative square mile on the planet" required deliberate attention to physical connectivity, mixed-use development, and transport infrastructure. MIT's strategic real estate investments, valued at $4.9 billion by 2024, created purpose-built environments where academic research could transition to commercial applications through proximity and shared facilities.

Similarly, Melbourne's Parkville-Carlton Innovation District demonstrates how architectural choices can embody institutional missions. The Translational Research Institute's eight-level design integrates laboratories, clinical facilities, and manufacturing capabilities, physically manifesting the "bench to bedside" research model. This architectural integration enables rapid progression from research discovery to prototype manufacturing and clinical testing within a single facility.

Adaptive Reuse and Industrial Heritage

Several case studies highlight the strategic value of adaptive reuse in creating innovation environments. Adelaide's Tonsley Innovation District is perhaps the most dramatic example, transforming the former Mitsubishi automotive plant into a world-class innovation hub. The strategic retention of the Main Assembly Building's 5-hectare outer shell, adapted into collaborative space, demonstrates how industrial heritage can provide both symbolic continuity and functional infrastructure for innovation activities.

Barcelona's 22@ district similarly leveraged industrial heritage through model restorations like Palo Alto and Ca l'Aranyó textile mill. The Industrial Heritage Protection Plan, consolidating protective measures for 114 elements, shows how innovation districts can respect historical identity while fostering technological advancement. This approach creates authentic environments that attract knowledge workers seeking meaningful connections between past and future.

Scale and Connectivity Considerations

The case studies reveal significant variation in geographic scale, from London's Knowledge Quarter at one square mile to Sydney's Tech Central spanning six square kilometres and Fisherman's Bend covering 480 hectares. However, successful placemaking appears less dependent on absolute size than on achieving optimal density and connectivity.

Shenzhen's success demonstrates how manufacturing density can create innovation advantages. Huaqiangbei's concentration of suppliers and technical expertise has created what many describe as the world's most efficient prototyping environment. This physical density enables rapid iteration cycles and collaborative problem-solving that would be impossible in dispersed environments.

Transport Infrastructure: Enabling Access and Density

Transport appears repeatedly as a decisive factor. Tech Central in Sydney is literally built around Central Station, providing unrivalled access to heavy rail, light rail, metro and buses. Macquarie Park was transformed from a car-dependent office park into an accessible innovation precinct only once the Sydney Metro was extended, connecting three stations directly into the district.

By contrast, Monash Technology Precinct and Fishermans Bend in Melbourne illustrate the costs of neglect. Monash has long lacked a direct rail link, limiting its density and integration, although the Suburban Rail Loop promises eventual relief. Fishermans Bend was rezoned for transformation without adequate transport in place, forcing reliance on limited tram lines and congested roads. Paris-Saclay faces similar constraints, with growth tied to the long-delayed Metro Line 18.

The lesson is that transport infrastructure is a platform for success, not an add-on. Districts that embed high-frequency, multi-modal transport early build credibility and attract talent. Those that defer transport struggle with legitimacy and vibrancy.

Digital Infrastructure: The New Backbone

Digital infrastructure, broadband, fibre, data centres, 5G networks, smart city systems, is now as essential as roads and rail.

Amsterdam Science Park thrives because it hosts the AMS-IX internet exchange, multiple data centres and national research networks, enabling low-latency connections to most of Europe. Dublin Docklands has created a live testbed for IoT and 5G applications, supported by metropolitan-scale data centres. Barcelona's 22@ district embedded fibre into its streetscape, while Konza Technopolis launched with a National Cloud Data Centre linked to submarine cables.

Where digital infrastructure lags behind demand, innovation ecosystems struggle to reach their potential. Lagos illustrates these constraints starkly. In Yaba, unreliable grid electricity forces technology firms to rely on expensive diesel generators, with some businesses devoting nearly half their operating budgets to power. This dependency raises costs, prevents round-the-clock operations, and diverts capital away from hiring or product development.

An infrastructure deficit becomes a direct tax on innovation, limiting the ability of promising ventures to scale. Even in well-resourced contexts, policymakers cannot assume infrastructure will automatically keep pace with demand. Proactive investment and coordination between authorities, utilities and developers remains essential to support knowledge-intensive activities.

The case material underscores that building digital infrastructure and capability is not optional. It is a precondition for data-driven research, advanced manufacturing, creative industries and platform-based business models.

Research Infrastructure: Anchors and Shared Platforms

Universities, hospitals and laboratories provide the hard and soft infrastructure for discovery and translation. The case studies highlight multiple layers:

University campuses with research facilities, teaching programs, and student flows create demand that attracts complementary firms and services. Success depends on coordinated public investment in transport and other enabling infrastructure, financial sustainability, and aligned timing across multiple stakeholders. Even well-intentioned institutional commitments can stall when these conditions are not met.

Shared platforms, including incubators, accelerators and testbeds, lower the cost of entry for firms and entrepreneurs by providing access to expensive equipment, specialised facilities and professional networks that individual start-ups could not afford independently. Shared laboratory space, prototyping equipment and mentorship reduce early-stage risk and accelerate development timelines. Macquarie University's Incubator and deep tech facilities, Dublin's startup spaces, and Boston's LabCentral illustrate this role.

Specialist facilities, such as biomedical labs, pilot plants, and clean rooms, enable advanced research translation. Cambridge's science parks, Seestadt's smart energy labs and Singapore's Biopolis are examples. RMIT has recently established a specialised medical device pilot manufacturing facility (with heavy support and endorsement from industry and peak bodies) to catalyse growth around what is emerging as a significant precinct.

Research infrastructure is expensive, long-lived and lumpy, but it creates the conditions for sustained knowledge production. It also shapes the identity of a district—whether it is seen as biomedical, digital, manufacturing or creative.

Social Infrastructure: Making Places Liveable

Innovation districts are not just sites of research and business. They are urban environments where people live, work and interact. Social infrastructure—housing, childcare, schools, cultural venues, green space—determines whether districts can attract and retain talent.

The Tonsley Innovation District integrates residential areas, greenways and public transport, creating a balanced environment. Dublin Docklands invested in amenities that make it attractive not only for global firms but also for residents. Medellín's innovation strategy included public libraries, cultural centres and education facilities as infrastructure for inclusion.

Districts that neglect social infrastructure risk becoming sterile "office parks." Research Triangle Park exemplifies this: without housing, retail or public space, it has long been criticised as a "ghost town" after hours. Social infrastructure builds legitimacy, vibrancy and resilience.

Governance and Institutional Infrastructure

Less visible, but equally vital, is governance infrastructure. Innovation districts require coordination across multiple actors and sectors. Governance infrastructure provides the institutional capacity, decision rights and delivery mechanisms that make complex projects possible.

The Bradfield Aerotropolis in Sydney demonstrates this with its dedicated Development Authority, empowered to align transport, land use, utilities and innovation initiatives. Barcelona's 22@Alliance provided a platform for firms, universities, and city officials to coordinate investments. Dublin's Smart Docklands set up an "honest broker" model to align technology vendors, researchers and city agencies.

Without governance infrastructure, investments fragment and momentum stalls. With it, districts can align multiple categories of infrastructure into coherent, strategic platforms.

Comparative Insights from the Case Studies

Reading across the case material, several comparative insights emerge.

- *Strong integration of multiple infrastructure types* is characteristic of leading districts (e.g. Amsterdam Science Park, 22@Barcelona, Dublin Docklands).
- *Transport or digital gaps* are the most common drag on performance (e.g. Monash, Fishermans Bend, Paris-Saclay, RTP, Lagos).
- *Research anchors* consistently drive identity, but their impact depends on complementary infrastructure. Fishermans Bend's university investment will not achieve full potential without better transport and housing.
- *Social infrastructure* is critical for legitimacy. Medellín demonstrates that investment in cultural and social facilities can underpin inclusive innovation.
- *Governance capacity* is decisive for sequencing and integration. Where governance is weak, infrastructure investments are piecemeal and disconnected.

These aspects are captured in the following examples drawn from the case studies and reflected in the Innovation District Infrastructure Framework.

Case Study	Transport	Digital	Research	Social	Governance
Tech Central (Sydney)	Strong - Central Station, metro, light rail	Emerging - high-speed networks planned	Strong - UTS, Sydney Uni, CSIRO nearby	Emerging - Mixed-use under development	Strong - NSW Govt, precinct authority
Macquarie Park (Sydney)	Improved - Sydney Metro stations	Emerging - Smart precinct connectivity	Strong - Macquarie Uni incubators, labs	Adequate - Some housing, limited vibrancy	Strong - CMPID nonprofit governance
Monash Tech Precinct (Melbourne)	Lagging - Awaiting Suburban Rail Loop	Lagging -No major fibre/compute assets	Strong - Monash Uni, CSIRO anchors	Lagging - Limited housing/ amenities	Adequate - University, local govt partnerships
Tonsley (Adelaide)	Strong - On-site rail station, greenways	Adequate - Digital health, university IT	Strong - Flinders Uni, TAFE, labs	Strong - Housing, retail, green space	Strong - Renewal SA as coordinator
Dublin Docklands	Strong - LUAS, DART, MetroLink	Strong - IoT/5G testbed + data centres	Strong - Trinity College Dublin, research centres	Strong - Mixed-use, housing, waterfront	Strong -City + Smart Docklands broker
22@ Barcelona	Strong - Metro, tram, bus integration	Strong - Dark fibre, Cisco IoT testbed	Strong - Universitat Pompeu Fabra, research hubs	Strong - Mixed-use, housing, cultural assets	Strong - 22@Alliance multi-actor model
Amsterdam Science Park	Adequate - Strong urban rail links	Strong - AMS-IX, data centres, SURFnet	Strong - Nikhef, universities, quantum facilities	Adequate - Housing, limited retail	Strong - Science Park Foundation
Konza Technopolis (Kenya)	Developing -Planned BRT, road links	Strong - National Cloud Data Centre	Developing -Kenyan universities + labs	Developing -Housing planned in city vision	Strong - Konza Development Authority
Cyberjaya (Malaysia)	Adequate - Expressways, local links	Strong - Power continuity, ICT hub	Strong - Malaysian ICT and R&D anchors	Adequate - Housing, ICT parks, retail	Strong - Malaysian govt ICT policy

Case Study	Transport	Digital	Research	Social	Governance
Aspern Seestadt (Vienna)	Strong - Public transport + cycling	Strong - Digital twin + smart grid	Strong - ASCR labs, smart energy research	Strong - Housing, schools, green space	Strong - ASCR public-private partnership
Research Triangle Park (USA)	Weak -Car-dependent, drive-in/out	Lagging - Legacy broadband, limited ICT	Adequate - University anchors, biotech firms	Weak -No housing, A Ghost town	Weak - Fragmented governance
Medellín Innovation District	Improved - MetroCable and buses	Adequate - Civic digital programs	Strong - Local universities, Ruta N labs	Strong - Libraries, cultural centres, housing	Strong - Ruta N public agency

Policy and Strategy Implications

Three implications stand out for policymakers and district leaders.

- *Infrastructure must be treated systemically.* Transport, digital, research, social and governance infrastructure are interdependent. Investment in one without the others creates bottlenecks. Successful districts align multiple categories.
- *Sequencing matters.* Delivering transport or digital early creates credibility. Embedding research anchors and social amenities at the right stage sustains momentum. Governance infrastructure must be in place from the outset.
- *Infrastructure is long-lived.* Districts succeed where infrastructure is planned for decades, not election cycles. Seestadt's twenty-year program in Vienna and Konza's staged delivery in Kenya illustrate the importance of patience and persistence.

Conclusion

The near-thousand references to infrastructure across the case studies are a clear signal: infrastructure is the foundation of innovation ecosystems. It is not background, but foreground. It shapes access, density, legitimacy and collaboration.

Innovation districts are much more than clusters of firms and universities.

They are complex systems built on interdependent infrastructures: transport to connect, digital to enable, research to discover, social to sustain, and governance to integrate. Where these infrastructures are planned and aligned, ecosystems thrive. Where they are weak or fragmented, potential is lost.

Specific aspects if infrastructure delivery in transport and digital are addressed separately below.

3. Public Transport as Platform Infrastructure

Innovation districts are celebrated for concentrating talent, research assets, and entrepreneurial energy into proximate spaces. Policymakers, planners, and university leaders frequently emphasise governance arrangements, anchor tenants, real estate strategies, and placemaking. Yet, across the international case record, one factor repeatedly distinguishes vibrant, connected districts from those struggling to realise their potential: efficient public transport.

Transport does more than move people: it defines catchment areas for talent, enables the density that underpins knowledge spillovers, and legitimises districts as accessible and liveable.

This chapter examines the role of transit through examples from the case studies of innovation. It reveals that transport must be treated as platform infrastructure, equal in importance to broadband, laboratories, and incubators. Where well-designed, multi-modal transit systems are embedded, ecosystems thrive. Where absent, delayed, or fragmented, the promise of collaboration and serendipity is diluted.

Why Transport Matters

Innovation ecosystems depend on concentration and circulation. Researchers, entrepreneurs, investors, and students must move quickly and easily between institutions, shared facilities, and informal gathering spaces. Transit provides three core functions:

- *Labour market access*: Expanding the pool of workers, interns, and collaborators within a 30–40-minute travel schedule.
- *Time–distance compression*: Reducing frictions that shape the frequency of meetings, chance encounters, and collaboration.
- *Urban density*: Supporting mixed-use precincts where cafés, housing, and cultural amenities are viable, sustaining vitality beyond office hours.

Transit is an enabler of the dense, mixed-use, and serendipitous environments that innovation literature highlights as critical (Katz & Wagner, 2014; Clark, 2020). Without it, districts risk becoming isolated office parks or "drive-in, drive-out" science parks.

Cases of Success: Transit as Catalyst

Several districts illustrate how strong public transport can underpin growth, legitimacy, and attractiveness.

Tech Central, Sydney. Tech Central is literally wrapped around Sydney's Central Station, one of the country's largest rail interchanges. Heavy rail, light rail, buses, and the Sydney Metro converge here, placing the district at the intersection of national and metropolitan flows. The case notes that this connectivity is "fundamental to Tech Central's success," enabling global firms, start-ups, and students to access the precinct seamlessly.

Macquarie Park Innovation District, Sydney. Originally conceived as a suburban office park, Macquarie Park was limited by car dependence. The opening of the Sydney Metro Northwest in 2019, with three stations serving the district, was transformational. The metro placed the precinct within a 20-minute ride of the CBD, broadening its labour market catchment and strengthening its competitiveness. The case highlights this as the single most important infrastructure shift in its evolution.

Tonsley Innovation District, Adelaide. Tonsley sits directly on a rail line, complemented by pedestrian-friendly design and greenways linking to surrounding suburbs. This connectivity supports mixed-use development and makes it viable for students, researchers, and local residents to circulate through the precinct daily.

Dublin Docklands (Silicon Docks). The Docklands' rise as "Europe's tech capital" coincided with expanded light rail (LUAS), commuter rail (DART), and the MetroLink project. Case materials describe transit investment as essential in managing growth pressures, diffusing benefits, and legitimising the district in the eyes of the wider community.

Kendall Square, Cambridge, Massachusetts. Anchored by MIT, Kendall Square is inseparable from the MBTA Red Line, which links Cambridge directly to Boston. Transit allows thousands of workers and students to flow through daily, sustaining the cafés and bars where much of the district's informal networking takes place.

Barcelona's 22@ District. Barcelona has embedded transport connectivity optimisation as a formal lesson in its replication strategy. Tram, metro, and bus services were synchronised with land use changes, making the district accessible to the wider metropolitan population.

Bellevue Innovation District, Washington. A suburban city transformation into Pacific Northwest technology hub through transit-oriented development of BelRed corridor and Spring District, anchored by Amazon, T-Mobile, and Meta headquarters alongside emerging startup ecosystem.

Houston's Ion District. Unusual in a car-dependent city, the Ion is located adjacent to a major transit hub. The case notes that this positioning is crucial for connecting distributed assets across Houston, though it also highlights the wider challenge of poor metropolitan transit integration.

When Transit Fails: Constraints and Missed Potential

The dataset also reveals multiple cases where weak, delayed, or non-existent transit has constrained outcomes. These districts often show strong anchors, investment, and placemaking ambitions, but their growth is capped by accessibility.

Fishermans Bend, Melbourne. Australia's largest urban renewal project was rezoned without adequate transport in place. With no heavy rail and limited tram links, accessibility has constrained growth. A 2018 Framework prioritised active and public transport, but delays continue to destabilise investor confidence and vibrancy despite strong university and industry anchor commitments.

Monash Technology Precinct, Melbourne. The Suburban Rail Loop will finally bring direct rail to Clayton, home of Monash University and CSIRO. Case evidence acknowledges that limited transit access has been a long-standing barrier to density and cross-institutional circulation.

Paris-Saclay. Despite extraordinary research assets, Paris-Saclay remains partly isolated. The case stresses that completion of Metro Line 18 is "essential" for growth and integration. Until then, reliance on cars and buses undermines the district's cohesion.

Research Triangle Park, North Carolina. Once the archetype of a research park, RTP has long struggled with vibrancy. The case points to "improving transportation infrastructure" as a priority, with the district described as a "ghost town" after hours. The absence of integrated transit has entrenched a drive-in, drive-out commuting model that stifles serendipity.

Edinburgh BioQuarter. Future growth is tied explicitly to tram extensions and improved transport. Without these, the precinct risks remaining peripheral to the city centre's talent and cultural flows.

Comparative Insights

Across cases, a clear comparative picture emerges:

- Districts with integrated transit (Tech Central, Kendall Square, Dublin Docklands, Tonsley) show vibrancy, strong catchment areas, and legitimacy with communities.
- Districts awaiting or lacking transit (Paris-Saclay, Monash, Fishermans Bend, RTP, Edinburgh BioQuarter) suffer from limited density, weaker attractiveness for global firms, and reduced afterhours activity.
- Intermediate cases (Ion District, South Lake Union in Seattle) highlight how partial improvements can help but are insufficient without metropolitan-scale systems.

Transit is thus not peripheral but a key variable in ecosystem performance, explaining variation even where governance and research assets are strong.

Policy and Strategic Implications

The findings suggest three main implications for policymakers and ecosystem managers.

- *Treat transit as platform infrastructure.* Transit investment must be sequenced alongside, not after, real estate development. Delivering metro, rail, or bus-rapid services early ensures that innovation districts are connected from inception. Tech Central, Macquarie Park, and Boggo Road in Brisbane illustrate how front-loading transport creates credibility and attractiveness.
- *Design for walkable first kilometres.* Even the best metro stations fail if district design does not place anchors, labs, and cafés within short walking distance. Tonsley's rail station and greenways provide a model. Walkability magnifies the value of transit by sustaining face-to-face interactions.
- *Use transport to anchor legitimacy and inclusion.* Community legitimacy comes when districts are seen as accessible to the public, not enclaves for elites. Dublin Docklands demonstrates how investment in commuter and light rail systems can diffuse benefits and sustain public support.

Conclusion

The evidence from global case studies confirms the centrality of public transport for innovation districts. Transit does not guarantee success, but without it, the chances of reaching potential are greatly reduced. Transport expands talent access, compresses distances, sustains vibrancy, and legitimises projects in the eyes of citizens.

Districts such as Tech Central, Kendall Square, and Dublin Docklands illustrate the dividends of integrated, early, and multi-modal transit. Paris-Saclay, Monash, and RTP caution against neglect. The strategic lesson is straightforward: if innovation districts are to thrive, transport must be conceived as platform infrastructure, not as an afterthought.

4. Digital Infrastructure, Data Centres and the Performance of Innovation Districts

Digital infrastructure has moved from enabling function to core platform for innovation districts. Bandwidth, latency, power quality, resilience and trusted platforms for data exchange shape firm location choices, collaboration patterns and the feasibility of data-intensive research and commercialisation.

The case studies show clear relationships between strong digital foundations and ecosystem growth, and they also reveal where lags in connectivity or power quality hold places back.

What "Strong" Digital Foundations Means

Amsterdam Science Park illustrates how world-class networks anchor a district's global role. Three co-locations of the Amsterdam Internet Exchange within the campus and more than 600 network hubs have made the park one of Europe's densest, fastest digital nodes, with reach to 80 per cent of Europe in 50 milliseconds. Recent decisions to host a European quantum computer integrated with the Dutch national supercomputer lift the park's compute profile and strengthen AI- and data-heavy research and business formation.

Dublin Docklands has treated the city as a testbed. Smart Docklands operates a two-square-kilometre urban lab with corporate partners, trialling use cases for LoRaWAN, Sigfox, NB-IoT, and 5 G. At the same time, the Dublin area continues to expand data centre capacity, with Microsoft constructing a large facility at Clondalkin that signals the region's role in cloud and enterprise compute.

The Sunshine Coast shows how international links can change regional trajectories. A direct international submarine internet cable now connects to key business precincts, including the University of the Sunshine Coast, improving bandwidth and reliability for digital content, immersive media and other compute-intensive sectors.

Cyberjaya (Malaysia). Early planning guidelines mandated high continuity of electricity supply (99.99 percent availability), dual power connections and on-site backup. This de-risked co-located ICT operations and underwrote data centre and cloud workloads.

Konza Technopolis (Kenya). The district's first phase "National Cloud Data Centre" is backed by resilient power and fibre links to submarine cables, with a policy stance centred on secure, locally hosted cloud services for government and industry.

Aspern Seestadt (Vienna). A 3D geospatial digital twin supports planning, operations and energy-system coordination district-wide. The ASCR consortium applies AI and machine learning to building and grid management, turning the precinct into a live digital-operations platform.

What Leading Districts are Doing to Keep Up with Demand

Several common moves recur across the case studies.

- Districts are building or securing dedicated backbone digital capacity. Amsterdam leverages co-location of an internet exchange (AMS-IX), a national research network node and proximate data centres to keep latency low and throughput high for universities, labs and scale-ups. The configuration is purposeful rather than incidental.
- Some places hard-wire fibre and edge capability into urban works. Barcelona's 22@ treated dark fibre as public realm infrastructure, then layered a city-scale IoT testbed so pilots can move to deployment without re-digging streets.
- "Living-lab" arrangements shorten the path from trial to service. Dublin's Smart Docklands uses a civic "honest broker" model to line up spectrum, assets and street furniture for LoRaWAN, NB-IoT and 5G trials with corporate anchors and researchers. Fishermans Bend's smart-pole program similarly pre-approves sites and power for radios and sensors.
- Regions with cable landings are localising high-bandwidth activity. The Sunshine Coast's submarine cable shifts content and compute closer to firms and the university, enabling data-heavy sectors such as immersive media and cybersecurity to grow away from capital city cores.
- Some new build districts pair fibre with power quality guarantees. Cyberjaya's design rules for uptime and redundant supply de-risk data centre loads and enterprise ICT. Konza's national cloud facility combines resilient power with terrestrial fibre links to undersea cables, aimed at sovereignty, latency and cost.
- Governance matters for timely delivery. Bradfield's development authority has coordinated "roads, utilities and digital connectivity" with a private fibre partner, setting requirements for high-capacity

fibre and broad mobile coverage into the master plan, so the digital grid is built as streets and buildings go in.

- Districts are digitising their own operations. Seestadt's live digital twin and the ASCR program illustrate how precinct managers can use real-time data to schedule loads, optimise energy, and support building automation. This tightens the link between infrastructure provision and day to day performance.

Policy Implications

The case studies suggest three practical conclusions for precinct leaders.

Treat digital like core utility. Where districts made backbone and edge assets part of the capital works program, the ecosystem scaled faster and with fewer bottlenecks. The contrast between Amsterdam/22@ and the Knowledge Quarter's stress on legacy broadband and network congestion is instructive.

Co-design with operators and anchors. Smart Docklands' "honest broker" role and Bradfield's procurement of high-capacity fibre with an operator partner show that early joint design reduces transaction costs for pilots, upgrades and co-location.

Power quality and resilience are non-negotiable. Lagos's experience shows unreliable power and higher-cost connectivity can cap growth despite market momentum, while Cyberjaya and Konza demonstrate how explicit reliability standards and sovereign cloud can anchor digital investment.

5. Vision and Leadership: It Matters–A Lot

Across dozens of cases, a consistent pattern appears. Where a specific person or organisation articulated a long-term direction, set an investable narrative, and coordinated actors, districts accumulated capabilities and weathered political cycles. Where vision was absent or contested, projects drifted into land speculation, stalled infrastructure, or episodic programming.

How Vision and Leadership Show Up in Successful Districts

The case studies reveal several distinct categories of vision and leadership.

University-anchored Vision

Kendall Square, Cambridge (USA). MIT's leadership explicitly framed the Kendall Square Initiative as a multi-year, mixed-use strategy to knit academic, corporate and civic activity into one place. The Institute shepherded an inclusive planning process, secured special permits, and delivered new research, housing and museum assets — a deliberate act of placemaking by a single anchor with staying power.

Parkville Biomedical Precinct, Melbourne. Parkville's evolution from co-located institutions to an integrated biomedical innovation ecosystem came through coordinated, decades-long investment and institutional leadership that oriented research, clinical care and commercialisation towards a shared precinct strategy.

Macquarie Park Innovation District, Sydney. Macquarie University and corporate partners intentionally launched the district in 2016 as an industry-led collaboration, then merged efforts into the not-for-profit CMPID to give the vision durable organisational form. The governance model formalised coordination and advocacy rather than relying on proximity alone.

MaRS Discovery District, Toronto. MaRS was conceived and championed by founding chair Dr John Evans, then scaled under CEO Ilse Treurnicht. That combination of personal and institutional vision created Canada's flagship urban innovation hub.

The Ion District, Houston. Rice University articulated the district's purpose, converted a historic Sears building into the anchor, and signed a community benefits agreement with the city. The university's role provides a focal vision and an implementation vehicle.

Government-mandated or Authority-led Vision

22@Barcelona, Spain. Barcelona City Council launched the 22@ Plan in 2000 to transform Poblenou's industrial area into a knowledge economy district, with a council-run development entity coordinating land use and cluster development. Strong mayoral and municipal leadership sustained the project over a decade.

One-North, Singapore. The state development agency JTC led an integrated precinct strategy at one-north, including Biopolis and Fusionopolis, to grow knowledge clusters in biomedicine, information and communications technology, and media. The agency model aligns planning, assets and tenant mix to a national industry strategy.

Bradfield City Centre and the Western Sydney Aerotropolis. The Western Sydney City Deal established the institutional framework, and, in July 2024, the Bradfield Development Authority received a focused mandate for master planning, infrastructure coordination and investment attraction — an authority designed to hold and execute the vision through cycles.

Brainport Eindhoven, Netherlands. Long-standing corporate–public leadership, including Philips and ASML, has been consolidated into regional vehicles that invest in housing, skills and infrastructure to sustain the technology base. Recent corporate funds reinforce that shared regional vision.

Cortex, St Louis (USA). Five anchor institutions created a 501(c)(3) master developer to drive a 200-acre innovation district. The founding coalition narrowed governance and gave the district a patient, executing entity.

Entrepreneur or Philanthropy-catalysed Vision

Tech Central, Sydney. Atlassian's decision to locate its global HQ at Central signalled a credible tech vision and accelerated the precinct narrative. The government followed through with staged public investment, and universities aligned their research commercialisation efforts in that direction.

South Lake Union, Seattle. Paul Allen and Vulcan pursued a multi-decade redevelopment thesis to create a biotech and technology hub. That private vision, later matched by Amazon's urban campus, shifted a former light-industrial area into a dense innovation neighbourhood.

Station F, Paris. Xavier Niel's private investment converted a historic freight hall into what is widely described as the world's largest start-up campus, with curated programs and corporate partners. The single-owner vision simplified delivery and brand position.

City-orchestrated Ecosystem Vision

Canberra innovation ecosystem. The creation of the Canberra Innovation Network (CBRIN) in 2014 consolidated a city-level vision across the ACT Government, ANU, UC, UNSW Canberra and CSIRO, catalysing programs such as Griffin Accelerator and First Wednesday Connect and lifting spin-outs across defence, quantum and space.

Medellín's Ruta N. The mayor's office and city partners established Ruta N to articulate and coordinate a citywide innovation agenda, using the Medellín Innovation District concept to channel investment into science and entrepreneurship.

Industrial Transition with a Clear Repurposing Vision

Tonsley Innovation District, Adelaide. After Mitsubishi's closure, the South Australian Government set a clear repurposing strategy and recruited Flinders University and TAFE SA as anchors. The district now hosts thousands of workers and students, with notable exits such as SAGE Group.

Dublin Docklands. Dermot Tuohy of IDA Ireland played a pivotal role by targeting Silicon Valley firms after the dotcom crash, securing Google's European headquarters in 2004. This decision catalysed the "Silicon Docks" identity, attracting Facebook, Twitter and others.

Boston Seaport. Mayor Thomas Menino was central, launching the "Innovation District" strategy that repositioned the Seaport as a hub for life sciences and technology. The Fallon Company's acquisition of the Fan Pier site in 2005 provided a private anchor, showing investor confidence. The $14.6 billion Big Dig transformed connectivity, making large scale redevelopment viable.

Where Vision has Been Weak or Fragmented

Fishermans Bend, Melbourne. Early rezoning preceded strategic planning, creating a contested planning context and low community confidence. Subsequent reviews and the Victorian Auditor-General have pointed to delays and missing enabling infrastructure, indicating that vision and governance were not aligned in the crucial early years.

Adelaide's Multifunction Polis (MFP). The 1987–98 new city proposal illustrates how a large-scale concept without a coherent, locally anchored vision struggled to survive political cycles and governance disputes. The project was ultimately abandoned.

The Australian Technology Park. ATP's early vision was championed by Tom Forgan, its founding CEO, who drew on his expertise in technology commercialisation to establish the park at the old Eveleigh Railway Workshops. He worked to embed collaboration between government, universities, and industry tenants. After his tenure, the absence of sustained champions left governance adrift, and priorities shifted to property management. ATP's later sale to Mirvac marked the loss of its innovation purpose, turning a bold vision into a missed opportunity

Songdo International Business District, Incheon. While impressive as an urban project, multiple assessments have noted top-down governance, a property-led delivery model, and underperformance against early innovation and livability rhetoric. The critique centres on a gap between an infrastructure vision and an ecosystem vision.

What the Case Examples Collectively Show

Vision needs a durable holder. This can be an anchor university (MIT; Melbourne's Parkville), a dedicated authority (Bradfield Development Authority; JTC), a coalition of anchors (Cortex), or a committed founder or philanthropist (Xavier Niel; Paul Allen). Without an entity that can maintain strategic intent through cycles, projects revert to parcel-by-parcel development.

Vision is translated through governance. The act of creating a precinct-level organisation or authority is a concrete test of seriousness. CMPID, CBRIN, and Cortex each embody this.

Vision needs an investable narrative. Atlassian at Tech Central, Rice at the Ion, and Philips–ASML in Brainport provided catalytic commitments that align public infrastructure, private development and talent.

Vision must span land use and industry development. The standout cases explicitly couple land, transport and real economy strategies. Barcelona's 22@ and one-north are textbook examples of planning tightly linked to cluster development.

Absent or unclear vision raises risk of drift. Fishermans Bend's early rezoning, followed by years of rework, shows how difficult it is to retrofit coherent infrastructure and governance after the fact.

Briefly, the same pattern appears in Research Triangle Park, where the Research Triangle Foundation carried a multi-decade vision linking three universities to industry; in Zhongguancun, where municipal policy instruments and proximity to national academies created a platform for successive private waves; and in Tel Aviv's CityZone, where the municipality codified a living-lab vision with university and park partners.

Practical Implications

Several implications follow from these observations.

- Set a single, testable vision statement for the district that defines purpose, sector focus and the role of key actors.
- Give that vision a legal and financial home through a dedicated authority or a not-for-profit with board-level participation from anchors.
- Lock in early catalytic commitments — an anchor R&D facility, an HQ relocation, or a programmatic facility that the market can price.
- Align planning controls, transport and capital budgets to the vision, and publish a transparent delivery schedule with accountabilities that survive election cycles.

6. Anchors Matter Too: Why Universities and Hospitals Drive Health-Biotech Clusters, but Manufacturing and Technology Struggle

Innovation districts thrive when they are able to draw together anchor institutions, firms, talent, and infrastructure into concentrated ecosystems. Across more than 90 case studies of innovation precincts, districts, and corridors internationally, a striking divergence is evident.

In health, medical, and biotech ecosystems, clustering is consistently and deliberately built around universities and public hospitals, which act as anchor institutions and business organisations in their own right. In manufacturing and technology districts, universities matter, but they are rarely sufficient to generate dense clustering without the presence of strong corporate anchors, global supply chain connections, or large-scale public investment in testbeds and prototyping infrastructure.

This chapter examines the evidence, compares the two models, and considers the policy implications.

Anchors in Health, Medical, and Biotech Precincts

The health-biotech sector repeatedly demonstrates the centrality of universities and public hospitals to clustering. This is visible across Australia and internationally.

Australia's Experience

Monash-Clayton integrates Monash University, Monash Health hospitals, the Australian Synchrotron, and CSIRO research. The co-location of research, clinical facilities, and biomanufacturing has attracted partnerships with Moderna, embedding a complete pipeline from discovery to commercialisation.

Parkville Biomedical Precinct, Melbourne, coalesces the University of Melbourne, Royal Melbourne Hospital, the Walter and Eliza Hall Institute, and CSL's commercial presence. This integration has generated one of the world's leading biomedical clusters.

Adelaide BioMed City brought together the $2.4 billion Royal Adelaide Hospital, SAHMRI, and two universities in a deliberately co-located $3.8 billion precinct, creating scale and concentration that repurposed Adelaide's economy after automotive decline.

International Evidence

Edinburgh BioQuarter aligns the University of Edinburgh with NHS Lothian's Royal Infirmary, creating an integrated campus purpose-built for translational medicine.

South Lake Union, Seattle is anchored by Fred Hutchinson Cancer Center, University of Washington Medicine, and Seattle Children's Hospital. Co-location of research, hospitals, and biotech companies has created a dense translational environment.

Cortex, St Louis, was established by Washington University School of Medicine and BJC HealthCare. It deliberately combines academic research, hospitals, and commercial space in a single governance framework.

Medicon Valley, spanning Denmark and Sweden, illustrates the scale of clustering possible: 17 universities and 32 hospitals are explicitly part of the ecosystem.

MaRS Discovery District, Toronto markets its location as "adjacent to the University of Toronto and its affiliated research hospitals", reinforcing the hospital-university nexus as a defining feature.

These cases confirm that the *structural logic of health innovation* demands proximity to hospitals. Clinical trials, patient cohorts, and translational pathways are all predicated on the hospital-university partnership.

Anchors in Manufacturing and Technology Precincts

Manufacturing and technology precincts present a different picture. Universities are important but rarely sufficient as clustering anchors. Instead, these ecosystems often pivot around corporate R&D laboratories, industrial supply chains, and engineering universities.

Australian Examples

- *The Australian Technology Park (ATP)* began as a university-led consortium, but without strong industrial anchors, it evolved into a commercial real estate development, highlighting the limitations of university clustering alone.
- *Fishermans Bend, Melbourne,* builds on a manufacturing legacy (Holden, Boeing) and the RMIT has established engineering presences, but corporates remain central to its viability. A University of Melbourne commitment to the district is on hold.

- *Bradfield Aerotropolis, Western Sydney,* revolves around advanced manufacturing, aerospace, and defence tenants. Universities are partners, but the anchor is a government built Advanced Manufacturing Readiness Facility.

International Examples

- *Brainport Eindhoven, Netherlands,* revolves around Philips, ASML, and the High-Tech Campus. Eindhoven University of Technology is a key partner, but the system is corporate-led rather than hospital-led.
- *Shenzhen, China,* has become the "Silicon Valley of hardware" anchored by Huawei, ZTE, DJI, and Tencent. Universities play supporting roles, but supply chains and entrepreneurial density drive clustering.
- *Zhongguancun, Beijing,* is anchored by Tsinghua University and Peking University, but its dynamism comes from the agglomeration of ICT and venture-financed firms, rather than hospital infrastructure.
- *Munich Innovation Corridor* connects TUM, LMU, and Fraunhofer Institutes with Siemens, BMW, and global corporations. Hospitals play only a marginal role except in the dedicated life sciences node at Martinsried.
- *Zurich Innovation Park* builds on ETH Zurich and University of Zurich, but anchors are global corporates like Google and IBM, establishing R&D centres.

In these precincts, clustering relies less on the automatic gravitational pull of hospitals and more on attracting corporates, securing investment in research infrastructure, and embedding into global production networks.

A summary of these observations is provided below.

Dimension	Health-Medical-Biotech Precincts	Manufacturing & Technology Precincts
Primary Anchors	Universities and teaching hospitals	Corporates, engineering universities, testbeds
Clustering Drivers	Clinical trials, translational research, patient access	Supply chains, prototyping, global markets
Role of Universities	Central to research and training	Important but rarely sufficient alone
Role of Hospitals	Core anchors, act as business organisations	Minimal role (except in healthtech crossovers)
Role of Corporates	Supplementary (pharma/biotech firms join later)	Core anchors (Philips, Huawei, Siemens, Atlassian)

Dimension	Health-Medical-Biotech Precincts	Manufacturing & Technology Precincts
Spatial Logic	Proximity to patients and clinical settings	Proximity to suppliers, logistics hubs, transport
Governance Models	Formal alliances (e.g., Camperdown-Ultimo Alliance, Cortex 501(c)3)	Often fragmented; risk of property-led development
Policy Levers	Invest in translational infrastructure, clinical trial capacity, biomanufacturing facilities	Support corporate anchors, create shared testbeds, mission-oriented research centres

Mixed and Hybrid Models

Some precincts integrate both logics.

- *Cambridge Biomedical Campus* combines clinical care and research with pharma investment, notably AstraZeneca's headquarters. Cambridge's broader cluster also covers computing and electronics, but the biomedical expansion has been fuelled by co-location with hospitals.
- *Eindhoven Brainport* has developed a HealthTechXL stream, but its dominant identity is semiconductors and photonics.
- *Munich's Martinsried* shows that when the sector is life sciences, teaching hospitals are deliberately built into the cluster, even within a technology corridor.

These hybrid examples underscore the broader point: the anchor differs by sector.

Why the Difference? Structural Explanations

Several reasons explain why clustering around universities and hospitals is stronger in health-biotech than in manufacturing and technology.

Clinical Translation Needs: Biomedical research must link seamlessly to clinical trials and patient care, requiring hospital co-location. Manufacturing and technology research can be commercialised through supply chains and distributed networks.

Hospitals as Business Organisations: Hospitals are not just service providers; they are large scale organisations with budgets, staff, procurement capacity, and embedded research infrastructure. Their scale and resources make them de facto industry anchors.

Corporate Anchors in Manufacturing: By contrast, in advanced manufacturing and ICT, corporates (Philips, Huawei, Siemens, Atlassian) are the equivalent anchor institutions. Universities are partners, but not the gravitational centre.

Spatial Logic: Biomedical research requires patient proximity. Manufacturing requires proximity to suppliers, testing, logistics, and global markets—these do not necessarily cluster around universities.

Policy Implications

The divergence between sectors highlights important issues for policymakers to consider.

Health-biotech strategy should double down on strengthening the hospital-university nexus. Policy should recognise hospitals as business organisations with significant innovation potential and ensure governance frameworks allow them to act as central ecosystem anchors. Investment in translational research infrastructure, clinical trial capacity, and biomanufacturing will magnify clustering benefits.

Manufacturing and technology strategy must place equal emphasis on corporate anchors, shared testbeds, and integration with global supply chains. Simply building precincts around universities is insufficient. Dedicated institutional mechanisms, such as advanced manufacturing facilities, Fraunhofer-style applied research centres, and mission-oriented platforms, are required to create clustering effects.

Governance models matter. Health-biotech districts have benefited from alliances (e.g., Camperdown–Ultimo Alliance in Tech Central; Cortex St Louis' 501(c)3 model) that give hospitals, universities, and industry a formal voice. Manufacturing districts often falter when governance is left to fragmented property development interests.

Avoiding false equivalence: Policymakers should resist assuming that all innovation districts work the same way. The logic of clustering differs fundamentally by sector. What works for biomedical precincts (hospital-university integration) does not translate into manufacturing or ICT ecosystems.

Conclusion

The comparative evidence is clear. Health, medical, and biotech innovation ecosystems cluster tightly and predictably around universities and hospitals, which function as both research anchors and business organisations. Manufacturing and technology ecosystems, by contrast, require corporate anchors, global supply chain integration, and sector-specific infrastructure to achieve similar clustering.

Recognising these structural differences is essential for effective policy. Australia's health-biotech success is not accidental—it reflects the gravitational pull of hospitals and universities in tandem. By contrast, the recurring difficulties in manufacturing and technology precincts reflect the absence of equivalent anchors. For policymakers, the implication is simple but profound: anchors matter, but they differ by sector. Strategies must be tailored to the structural logics of each domain, rather than assuming one model of clustering applies universally.

7. Innovation Districts and the Experience Economy

Over the past two decades, the global spread of innovation districts, precincts and hubs has highlighted the convergence of placemaking, economics, business, and governance.

Much of the analytical focus has been on anchors, connectivity, governance structures and investment flows. But in walking through these districts, participating in their events, and observing how people use the space, it becomes evident that there is another critical dimension: the lived experience.

This chapter explores whether "the experience" is itself a driver of success in innovation districts, how it influences investment, legitimacy and reputation, and what policymakers and managers can do to design for experience in the era of the experience economy.

Experience as a Market Signal

The experience economy refers to creating memorable, engaging encounters that go beyond goods and services. It is important because distinctive experiences attract talent, investment, and visitors, strengthen legitimacy, and enhance the cultural and economic vitality of innovation districts

The design and curation of experiences within innovation districts send strong signals to investors and partners. A district that feels alive with people, activity, and opportunity communicates a reduced coordination risk and a heightened potential for deal flow. Brookings' work on innovation districts highlights that successful places are "living, breathing communities" rather than sterile research parks.

This vibrancy reassures investors that innovation is not only happening but is also visible, legible, and social. The experience becomes part of the due diligence: if a district cannot be felt or seen to be active, its market proposition is weakened.

Experience and Talent Attraction

The experience economy literature makes clear that people are drawn to memorable experiences. This is not confined to leisure but extends to where knowledge workers choose to live and work. Districts that combine high-quality workplaces with cultural amenities, learning opportunities, cafés, green space, and informal third places are more likely to attract and retain talent.

The Global Institute on Innovation Districts stresses that quality of place and programming are as vital as laboratories and venture capital. Talent is mobile; the feel of a district, its festivals, exhibitions, music, food and serendipitous encounters, plays a decisive role in anchoring talent locally.

Experience as Legitimacy

Experience also plays a political and social function. Where innovation districts are perceived as welcoming, accessible, and valuable to residents, they generate legitimacy. This legitimacy provides a licence to operate and expand. In Monterrey, the DistritoTec initiative built legitimacy through co-designed projects, collaborative governance, and community-facing experiences such as the INCmty festival. This demonstrated that the district was not an enclave for elites but a shared urban common. Legitimacy is reinforced when experiences are inclusive, affordable, and integrated with the life of the city.

The Visitor Economy Dimension

Innovation districts also feed into the visitor economy. While the scale of spend from tourists, school groups, and conference attendees is modest compared to traded-sector activity, the cumulative effect strengthens brand and outreach. Cities such as Barcelona actively promote guided tours of 22@, presenting innovation as part of the cultural visitor offer.

In Kendall Square, the MIT Museum and the Kendall Innovation Trail draw visitors into a living campus of science and entrepreneurship. Station F in Paris and MaRS in Toronto utilise curated tours, public atria, and food halls to showcase innovation to the public. These experiences generate mindshare, support international positioning, and create channels for soft power.

Who Does It Well?

Several districts have integrated experience as a strategic layer:

- *22@Barcelona* has fused industrial heritage with contemporary architecture, supported by a strong programming calendar and routinised community building events. Guided tours and a clear narrative of transformation turn the district into a living case study of urban regeneration and innovation.

- *Kendall Square, Boston–Cambridge,* has deliberately made science and entrepreneurship legible through the Innovation Trail and the relocation of the MIT Museum into the heart of the district. This creates a walkable story of innovation that resonates with visitors and locals alike.
- *One-North, Singapore,* was designed by JTC as a series of themed precincts with meeting venues, programming capacity, and cultural amenities. This deliberate layering of experience has helped attract global talent and investment.
- *DistritoTec, Monterrey* has embedded experiences such as INCmty, which make the district outward-facing and community-oriented. It is as much a civic and cultural project as an economic one.
- *Station F, Paris* has built a signature experience around its monumental interior street and food hall, drawing in non-tenants and sustaining a brand as a hub for entrepreneurship and creativity.
- *MaRS Discovery District, Toronto* has created an iconic atrium and invested heavily in public events and programming, making innovation visible and attractive beyond the professional community.
- *South Lake Union, Seattle* demonstrates how urban design frameworks, walkability, and architecture tours can make a tech-led district accessible, intelligible, and desirable.
- *Tech Central, Sydney* has embedded global events such as SXSW into its calendar, generating a global footprint of attention and making the district visible to investors, partners and talent pipelines.

Can Experience Drive Investment?

Experience by itself does not attract capital flows. Anchors, research capability, and proximity to markets remain essential. However, experience amplifies these fundamentals. Investors seek signals of activity, community, and opportunity. A lively district reassures them of future deal flow. In Barcelona, strong leasing activity is coupled with a visible, legible innovation story.

In Sydney, international events provide a platform for investment attraction and brand-building. In Monterrey, festivals and public-facing initiatives support investor pipelines. Experience therefore works as a multiplier.

Can Experience Drive Legitimacy?

Yes. Legitimacy stems from openness and inclusivity. Experience ensures that a district is not an isolated corporate campus but part of the city. DistritoTec's co-governance and events with local residents illustrate how experience underpins legitimacy. Creative placemaking research shows that cultural and public experiences build pride, safety and trust.

A district perceived as secretive or exclusionary risks losing its social licence. Experience is therefore a governance tool as much as a cultural one.

Does Experience Attract Tourists?

Experience-driven districts can attract tourists and support visitor spend. Barcelona and Paris actively market innovation precincts as attractions. In Boston and Seattle, innovation trails and architecture tours are positioned as part of urban discovery. These visitor flows may not be transformative economically but do reinforce global reputation and provide a stream of small-ticket revenues for local cafés, bookshops and gift outlets. They also educate citizens and students, embedding the district into wider social narratives.

Policy and Management Implications

If experience is a credible dimension of success, policymakers and district managers should reflect on the following:

- *Make the story walkable.* Develop interpretive trails, signage and visitor centres. Create simple narratives that make the district's innovation visible and understandable.
- *Program the public realm.* Fund a regular calendar of events: talks, maker fairs, showcases, festivals. Ensure programming serves residents, students, entrepreneurs, and investors alike.
- *Blend heritage with new build.* Retain and repurpose distinctive industrial architecture. Authenticity strengthens brand and makes for memorable experiences.
- *Design third places.* Invest in cafés, parks, and informal learning and working spaces. Encourage activity beyond working hours and on weekends.
- *Stage marquee events.* Anchor the district's reputation through one or two globally visible festivals or exhibitions. Use these as platforms for investment and international positioning.

- *Invest in accessibility.* Walkability, transit, and intuitive wayfinding are as much a part of the experience as architecture and events.
- *Govern the experience.* Establish teams responsible for place activation, marketing, and measurement. Include community representatives in governance to ensure inclusivity and legitimacy.
- *Measure experience.* Add KPIs such as footfall, repeat visitation, tour participation, and conversion of visits into partnerships or investment enquiries.

Conclusion

The experience of innovation districts is neither a peripheral concern nor a substitute for hard fundamentals. It is an amplifier. Experience makes innovation visible, legible, and social. It builds trust, attracts talent, reassures investors, and strengthens legitimacy. It can support tourism and visitor spend, adding another dimension to economic impact.

The districts that succeed in the experience economy are those that deliberately curate culture, learning, openness, and storytelling into their fabric. Policymakers and managers who overlook the experience dimension risk building sterile campuses. Those who embrace it have the opportunity to turn their districts into living laboratories of innovation that resonate not only in economic terms but also in cultural and civic life.

8. Policy Overlays: The Architecture of Innovation District Success

Innovation districts have emerged as critical instruments of economic development, yet their success depends on far more than clustering technology companies within designated boundaries. The most effective districts operate through intersecting policy overlays that extend well beyond science and technology to encompass commercial property development, investment attraction, housing policy, transport integration, urban renewal, placemaking, and economic development strategy.

These overlays reveal a fundamental tension in contemporary urban policy. While jurisdictions often approach innovation districts through narrow sectoral lenses, the evidence from leading national and international examples demonstrates that success requires systematic integration across multiple policy domains. The districts that achieve genuine economic transformation are those where overlays reinforce rather than compete with one another around a coherent strategic purpose.

The Property Development Challenge

Commercial property development is perhaps the most visible overlay, yet it frequently operates according to logics that conflict with innovation objectives. In Australia, this tension is particularly evident in Tech Central and Macquarie Park, where the dominance of private ownership structures has led to ad hoc innovation integration. Development patterns largely follow traditional commercial office yield models, with innovation remaining peripheral to core investment decisions.

Macquarie Park exemplifies this challenge. Originally conceived as a corporate park, it is being incrementally repositioned through densification and co-location initiatives, yet the underlying commercial logic constrains innovation outcomes. The result is a district where innovation businesses occupy space rather than shape it.

Bradfield presents a contrasting model. Master planned by the Western Parkland City Authority, land assembly, zoning, and infrastructure sequencing remain under government control. This enables innovation logic to lead rather than follow property market forces, creating the potential for genuine alignment between development patterns and economic objectives.

International experience reinforces this lesson. Barcelona's 22@ district remains the strongest example of public leverage over property development, using zoning gains to finance affordable housing and public realm improvements. The public authority mandated that 30 per cent of development be allocated to innovation uses, with financial gains recycled into social infrastructure. This allowed greater public leverage over real estate outcomes while ensuring that commercial development served broader strategic goals.

Kendall Square in Cambridge offers a different trajectory. While now a global innovation hub, it emerged through aggressive private sector redevelopment in partnership with MIT. The result is a high-rent environment with strong venture capital presence but limited long-term affordability. Shenzhen's state-led approach ensured synchronisation of infrastructure and development, albeit with limited flexibility and minimal attention to inclusion.

The fundamental insight from these experiences is clear: public land ownership or strong zoning authority is essential for aligning commercial development with innovation goals. Without such mechanisms, real estate cycles dominate, and innovation becomes subordinate to property investment returns.

Investment Attraction and Sectoral Focus

Investment attraction strategies reveal similar patterns of integration versus fragmentation. Australian districts typically adopt passive approaches based on institutional presence and urban branding. Tech Central emphasises digital innovation and anchor tenants like Atlassian but lacks a defined sectoral proposition. Macquarie Park benefits from Macquarie University and multinational tenants, but its attraction strategy remains uncoordinated across government levels and institutional partners.

Bradfield is attempting to challenge this pattern by positioning itself for investment through targeted industry engagement in aerospace, advanced manufacturing, and logistics, supported by Commonwealth and State incentives. This is a more active approach to investment attraction, but its success will depend on execution capacity and institutional coordination.

International examples demonstrate the power of place-based strategies embedded within broader economic frameworks. Toronto's MaRS Discovery District leveraged strong public-private partnerships and philanthropic capital to attract health tech and social innovation ventures. Barcelona's 22@ investment strategy was tightly linked to broader city economic plans, aligning the district with Catalonia's digital and design sectors. Toronto's MaRS positioned itself as a health innovation hub, anchored by university hospitals and provincial health policy.

Kendall Square succeeded without a formal attraction strategy, relying instead on the gravitational pull of MIT, Harvard, and Boston's innovation ecosystem to draw firms and venture capital. Shenzhen attracted foreign investment through special economic zone status, preferential taxation, and rapid infrastructure build-out.

These experiences suggest that attraction strategies are strongest when sectoral focus, institutional density, and physical infrastructure align with national or regional economic goals. Australia lacks many of the institutional arrangements and investment vehicles that underpin targeted sectoral attraction overseas, making Bradfield's more deliberate approach particularly significant.

Affordable Housing Policy

Housing access and affordability have been the most neglected overlay across both Australian and international case studies. Until now, the absence of housing policy in Australian innovation districts has been particularly notable. High property prices in Tech Central and Parkville-Carlton have contributed to displacement and undermined diversity. Macquarie Park offers limited residential integration.

Bradfield's greenfield status presents a rare opportunity to embed inclusionary housing policy from the outset; however, there has been little public articulation of how housing supply, typology, and pricing will be managed. This is a significant policy gap, particularly given the district's ambitious economic objectives.

International experience provides both cautionary tales and potential models. Kendall Square's rapid gentrification has displaced lower-income residents, and no substantive affordability policies have been implemented. The city has struggled to impose inclusionary zoning retroactively, demonstrating the difficulty of addressing housing pressures after districts have achieved commercial success.

Barcelona's 22@ district offers a more integrated approach, including affordable housing through a mandatory 30 per cent inclusion rate for new developments. This ensured that urban transformation did not entirely displace existing communities while maintaining commercial viability.

The fundamental insight is that innovation districts that do not integrate housing risk excluding the workforce they seek to attract. Housing should be treated as economic infrastructure rather than a peripheral consideration.

Transport Integration and Urban Connectivity

Transport integration is a comparative strength in the Australian context. Tech Central operates within Sydney's rail and metro hub. Macquarie Park is connected via the Metro Northwest. Bradfield is being developed around the new Western Sydney Airport and a dedicated metro line, with freight, road, and active transport linkages incorporated into planning.

This infrastructure-first approach reflects lessons from international experience. All successful innovation districts emphasise the crucial role of transportation in district performance. Kendall Square is served by the MBTA Red Line, linking it to Boston's universities, airports, and tech centres. Barcelona's 22@ integration with metro, tram, and cycling networks has made it a model of walkable urbanism. London's Knowledge Quarter benefits from King's Cross-St Pancras as a multimodal and international hub.

The common success factors include multi-modal connectivity timed with development phases and integrated with land use planning. High walkability and access to airports and education hubs consistently emerge as critical features.

Economic Development as the Integrating Overlay

Among all policy overlays, economic development and industrial strategy exert the greatest systemic influence. Where it is present and well-articulated, other overlays align around it. Where it is absent, overlays operate in parallel with limited coherence or strategic impact.

In most Australian precincts, industrial strategy is poorly articulated. Tech Central lacks a coherent sectoral agenda. Parkville-Carlton aligns with health and biomedical policy, but other overlays lag. Bradfield is

Australia's strongest example of spatial-industrial integration, explicitly aligned with space, defence, freight, clean energy, and advanced manufacturing.

International examples consistently demonstrate that economic development determines the coherence and impact of all other overlays. China's industrial modernisation agenda underpins Shenzhen's entire urbanisation trajectory. Kendall Square reflects a tight coupling between academia, capital, and strategic sectors, particularly biotech. Barcelona's 22@ aligns with Catalonia's priorities in the creative economy and green technology.

Without strong economic development overlays, innovation districts risk becoming enclaves of commercial real estate, disconnected from national priorities. The result can be underperformance in innovation impact, economic transformation, and public legitimacy.

Implications for Australian Policy

These findings carry significant implications for Australian policymakers. Purpose-led design must precede spatial designation. A district must be planned around a sectoral and strategic purpose rather than retrofitted with one. Planning authorities need cross-portfolio capabilities to manage complex overlay interactions. Real estate frameworks should prioritise long-term economic goals over short-term development returns.

Transport, housing, and placemaking require governance alignment to achieve their full potential as supporting infrastructure for innovation ecosystems. Capital flows, both public and private, should be mobilised around long-term strategic sectors rather than dispersed across competing priorities.

Conclusion

Australia's innovation districts demonstrate growing policy maturity, but they still fall short of full system integration. The challenge is structural, involving overlapping levels of government, unclear institutional roles, and reliance on market-driven development models.

Bradfield will be an important test case. Its governance arrangements, planning authority, and industrial alignment provide a template for designing policy overlays that reinforce rather than compete with one another. International comparisons demonstrate that genuine innovation ecosystems are designed rather than simply designated.

Without deliberate industrial strategy, precincts risk becoming real estate developments with little enduring economic or social value. Policymakers must now decide whether to treat innovation precincts as strategic assets to be cultivated or marketing zones to be branded. This choice will shape Australia's innovation economy for decades to come.

The most competitive innovation districts globally are those with the greatest alignment between economic purpose, infrastructure delivery, investment mechanisms, and civic amenity. Success cannot be retrofitted; industrial logic, governance authority, and placemaking must be designed together from the outset. Where strategy leads, systems follow.

9. Other Insights Emerging from the Case Studies

Risk Factors and Ecosystem Vulnerability

The case studies reveal that innovation districts are particularly vulnerable during economic downturns and market disruptions. The COVID-19 pandemic provided a natural experiment in ecosystem resilience, highlighting both strengths and weaknesses across different models.

London's Knowledge Quarter experienced significant challenges when the pandemic disrupted international travel and event-based activities. TechHub, a shared office space provider that housed hundreds of tech start-ups, filed for administration in August 2020 after revenues fell by 75%. This demonstrates how districts relying heavily on co-working models face particular vulnerability to disruptions that affect physical presence and networking activities.

Conversely, Toronto's MaRS Discovery District demonstrated greater resilience, with many portfolio companies successfully pivoting to address pandemic-related challenges. Companies like Maple led Canada's telemedicine revolution by securing $75 million in investment while doubling in size. This suggests that districts with diverse sectoral representation and strong digital capabilities may prove more resilient to external shocks.

Development Pressure and Gentrification

Several case studies highlight the challenge of managing success without destroying the conditions that enabled innovation in the first place. Rapid appreciation in property values can price out the start-ups and creative workers who initially drove district development.

Sydney's Tech Central faces this challenge acutely, with the 2015 closure of the Australian Technology Park providing a cautionary example. When the park was sold to property developer Mirvac, many smaller, innovative firms were displaced, revealing tensions between the imperatives of commercial property development and the needs of the innovation ecosystem.

Barcelona's 22@ district has managed development pressures more successfully through zoning requirements that mandate mixed-use development and affordable housing components. The requirement for developers to dedicate specific percentages to productive activities, subsidised housing, and green areas has helped maintain ecosystem diversity while accommodating growth.

Talent Competition and Retention

As innovation districts mature and achieve success, they face increasing competition for talent from global technology companies establishing major operations locally. This creates both opportunities and challenges for existing ecosystem participants.

Toronto's experience illustrates this dynamic clearly. The establishment of major engineering centres by Google, Microsoft, and Amazon has created substantial employment opportunities but also intensified competition for skilled workers. While this validates the ecosystem's success in attracting global investment, it creates pressure on smaller companies that may struggle to compete on compensation packages.

Kendall Square faces similar challenges, with high real estate costs making it difficult for young professionals and researchers to live near their work. The lack of affordable housing could limit talent attraction and retention over time, potentially undermining the ecosystem's long-term competitiveness.

Technology Transfer and Intellectual Property Dynamics

Commercialisation Pathway Innovation

The case studies reveal diverse approaches to technology transfer, ranging from traditional licensing models to integrated research and commercial facilities. The most successful districts have developed multiple pathways for moving research discoveries to market applications.

Brisbane's Boggo Road Innovation Precinct demonstrates particularly innovative approaches through its integration of research, clinical testing, and manufacturing capabilities within the Translational Research Institute. The co-location of Patheon by Thermo Fisher Scientific's biopharmaceutical manufacturing facility adjacent to research laboratories enables rapid progression from discovery to prototype development and clinical testing.

This integrated model contrasts with traditional approaches that rely on licensing agreements and separate commercialisation processes. The proximity of manufacturing capabilities reduces transaction costs and accelerates development timelines while providing researchers with direct exposure to commercial considerations.

Intellectual Property Management

Different innovation districts have adopted varying approaches to intellectual property management, reflecting both institutional cultures and strategic objectives. These approaches significantly influence the ease of collaboration and the distribution of commercial benefits.

Shenzhen's "shanzhai" culture is a unique approach to intellectual property that emphasises rapid iteration and improvement over strict protection of original innovations. While initially viewed as copying, this culture has evolved into a distinctive form of iterative innovation that has proven particularly effective in hardware development.

Personal relationships, the free flow of information, localised modularisation of tasks, and supplier-driven sourcing of components, combined with an open approach to IP, have made Shenzhen an important innovation hub. This approach creates conditions for rapid prototyping and collaborative development that would be difficult under more restrictive IP regimes.

Research-Industry Interface Models

The most successful innovation districts appear to have developed institutional mechanisms that facilitate ongoing interaction between researchers and industry practitioners. These mechanisms go beyond formal partnerships to create environments where knowledge transfer occurs through regular professional interaction.

Melbourne's Parkville-Carlton Innovation District exemplifies this through the co-location of university research facilities, teaching hospitals, and corporate research centres. The proximity enables informal collaboration and knowledge sharing that complements formal research partnerships. The district's success in generating spin-out companies reflects the effectiveness of these integrated environments.

Similarly, Cambridge's Kendall Square benefits from dense networks of researchers, entrepreneurs, and investors who interact regularly through both formal and informal channels. The concentration of venture capital firms within walking distance of research institutions creates opportunities for rapid connection between research discoveries and commercialisation resources.

Talent Development and Human Capital Dynamics

Successful innovation districts create systematic connections between educational institutions and industry needs, ensuring that talent development aligns with ecosystem requirements. These connections span from secondary education through university and professional development programs.

Canberra's innovation ecosystem demonstrates effective pipeline development through programs that connect university students with industry partners. The Innovation ACT student-led start-up competition encourages entrepreneurship development while providing practical experience with technology commercialisation. These early exposures to innovation and entrepreneurship create cultural foundations that support ecosystem development.

London's Knowledge Quarter has developed community engagement programs that create pathways for local residents to participate in the knowledge economy. The Francis Crick Institute's outreach program engages over 20,000 young people in Camden's state schools, while Camden STEAM connects students with local employers through 200 STEAM Ambassadors from 60 local organisations.

International Talent Attraction

Innovation districts compete in global markets for skilled and talented knowledge workers, requiring strategic approaches to international talent attraction and retention. The case studies reveal diverse strategies adapted to different regulatory environments and competitive positions.

Shenzhen's success reflects particularly effective international talent attraction, with over 65% of the city's 17.8 million permanent residents originating from outside the city. The government's 30-point talent incentive program provides employment subsidies, innovation funding, housing assistance, and household registration incentives, all designed to create an attractive environment for skilled professionals worldwide.

Australian districts face distinct challenges in attracting international talent, with geographic isolation and visa requirements creating barriers that European and North American districts do not typically encounter. However, Australia's positioning in the Asia-Pacific region provides unique advantages for attracting talent from growing Asian markets.

Knowledge Worker Mobility

The circulation of talent between institutions within innovation districts creates knowledge spillovers that strengthen the entire ecosystem. This mobility enables rapid diffusion of new ideas and practices while creating professional networks that facilitate collaboration.

The case studies suggest that districts benefit from creating conditions that encourage talent circulation while maintaining sufficient stability to support long-term projects. This requires balancing flexibility in employment arrangements with institutional continuity that enables sustained R&D efforts.

Toronto's MaRS Discovery District exemplifies effective talent circulation through its role as a hub where individuals move between start-ups, established companies, and research institutions throughout their careers. This circulation creates cross-pollination between different types of organisations while building professional networks that strengthen the entire ecosystem.

International Benchmarking and Competitive Positioning

Comparative Performance Analysis

The case studies enable comparison of different approaches to innovation district development, revealing the strengths and limitations of various models. These comparisons provide insights for policy development and strategic planning.

When compared to other international innovation districts, the Knowledge Quarter consistently ranks highly across various indicators of research collaboration and impact, particularly in co-publication rates with local organisations. This suggests that the district has successfully created conditions for institutional collaboration despite the complexity of coordinating multiple independent organisations.

Similarly, Kendall Square's economic performance provides benchmarks that other districts aspire to achieve. The ecosystem's $5 billion in annual economic activity and 50,000+ knowledge workers demonstrate the potential scale of impact that innovation districts can generate over decades of development.

Model Adaptation and Local Context

While successful districts share common characteristics, the case studies demonstrate that effective approaches must be adapted to local institutional contexts and competitive environments. Simple replication of successful models from other contexts is unlikely to succeed.

Australia's systematic approach to innovation district development reflects the country's federal system and strong tradition of research excellence. The emphasis on government coordination and university anchoring builds on existing institutional strengths while addressing market failures in commercialisation and ecosystem coordination.

This contrasts with Shenzhen's approach, which leverages China's manufacturing capabilities and proximity to Hong Kong's financial markets. The "shanzhai" innovation culture and integrated manufacturing innovation model reflect specific advantages that would be difficult to replicate in other contexts.

Measurement and Evaluation Frameworks

Economic Impact Assessment

Effective management of innovation districts requires measurement frameworks that capture both direct economic impacts and broader ecosystem effects. The case studies reveal various approaches to performance measurement, each reflecting different objectives and stakeholder priorities.

MaRS Discovery District has developed a comprehensive tracking of portfolio company performance, measuring job creation, revenue generation, capital raised, and GDP contribution. The organisation's ability to demonstrate $29.6 billion contribution to GDP and $7.5 billion in tax revenue provides compelling evidence of a return on public investment. While these self-reported figures suggest substantial economic returns, they represent an important approach to demonstrating the impact of innovation districts.

The methodology and attribution of economic outcomes always warrant careful consideration and review by policymakers and practitioners. Moreover, quantitative measures may not capture the full value of innovation districts. The spillover effects, knowledge transfer, and cultural changes that drive long-term competitiveness are difficult to measure but may represent some of the most important outcomes of ecosystem development.

Innovation Metrics and Indicators

Beyond economic measures, successful districts require indicators that capture innovation activity and ecosystem health. These might include patent applications, publication rates, start-up formation, collaboration networks, and talent attraction metrics.

The case studies suggest that the most useful measurement frameworks combine quantitative indicators with qualitative assessment of ecosystem dynamics. Regular stakeholder feedback, case studies of successful collaborations, and assessment of institutional relationships provide insights that complement numerical data.

Future Challenges and Emerging Trends

Technological Disruption and Adaptation

Innovation districts must continuously adapt to technological changes that alter competitive landscapes and create new opportunities. The rise of artificial intelligence, quantum computing, and biotechnology creates both opportunities and challenges for existing ecosystems.

Several case studies demonstrate districts positioning themselves for emerging technology opportunities. Sydney's Tech Central is developing quantum computing capabilities through the Quantum Terminal, while Fisherman's Bend is investing heavily in AI infrastructure through NEXTDC's $2 billion facility.

However, technological disruption also creates risks for districts that fail to adapt quickly enough. The transition from traditional manufacturing to advanced manufacturing, the shift towards digital platforms, and the growing importance of sustainability all require districts to evolve their capabilities and strategic focus.

Climate Change and Sustainability

Environmental sustainability is becoming increasingly important for innovation district development, both as a driver of innovation opportunities and as a requirement for long-term viability. Districts that position themselves as leaders in clean technology and sustainable development may achieve competitive advantages in attracting investment and talent.

Barcelona's 22@ district has integrated sustainability objectives with innovation development through requirements for green buildings, renewable energy systems, and sustainable transport options. This approach demonstrates how environmental considerations can enhance rather than constrain innovation district development.

Implications for Policy and Practice

The case studies offer several important insights for policymakers to consider in developing innovation district strategies.

- Successful districts require sustained commitment over decades, with returns often taking 10-15 years to fully materialise. This challenges traditional policy cycles and requires governance mechanisms that can maintain strategic direction across political changes.
- The most effective approaches combine public sector strategic vision with private sector operational efficiency. This requires developing partnership models that align incentives while maintaining appropriate accountability for public investment.
- Physical planning decisions have fundamental impacts on innovation outcomes. Creating environments that foster collaboration and knowledge transfer requires deliberate attention to connectivity, density, and mixed-use development rather than simply providing adequate commercial space.
- Financial ecosystem development requires attention to capital availability across all company development stages. Early-stage funding mechanisms alone are insufficient if companies cannot access growth capital to scale their operations.
- Effective governance requires striking a balance between coordination needs and institutional autonomy. The most successful models create shared strategic direction while allowing individual organisations to maintain their distinct missions and operational approaches.

Towards Convergent Innovation Ecosystems

The transformation of innovation districts worldwide reveals a fundamental shift in how we understand economic development. Rather than pursuing traditional sectoral approaches, successful districts integrate placemaking, economics, business strategy, and governance as interconnected systems.

The case studies demonstrate that successful innovation districts emerge when placemaking, economics, business strategy, and governance converge around shared objectives rather than operate as separate optimisation domains. This convergence creates conditions for breakthrough innovation that cannot be achieved through any single approach.

The Australian experience, with its systematic government coordination and research institution anchoring, provides valuable lessons for other regions seeking to build innovation ecosystems. However, success requires adaptation to local contexts, sustained commitment over decades, and governance mechanisms that balance coordination with institutional autonomy.

As global competition for knowledge-intensive economic activity intensifies, the ability to achieve this convergence will increasingly determine which regions succeed in building sustainable innovation ecosystems. The case studies suggest that success requires mastery of the integration challenges that enable placemaking, economics, business strategy, and governance to work together in creating dynamic, adaptive innovation environments.

The future of innovation districts lies in developing the governance capabilities and institutional mechanisms that enable these four domains to reinforce each other in creating environments where breakthrough innovation can flourish. This is both a significant challenge and an opportunity for policymakers committed to building knowledge-intensive economies that can compete effectively in global markets while generating broad-based prosperity and social progress.

PART VI: CONCLUDING COMMENTS

Part VI draws together the key arguments and insights presented across the book and reflects on their implications for policy and practice. It considers what must change in Australia's approach to innovation if ecosystems are to achieve their full potential.

The discussion emphasises the need for deliberate system design, long-term commitment, and integration across the four domains of placemaking, economics, business, and governance. It calls for policy settings that survive electoral cycles, invest in system integrators and translational infrastructure, and build legitimacy through inclusion and shared public value.

This final section challenges policymakers, university leaders, and business strategists to translate analysis into action and focus on building coherent, resilient, and globally connected ecosystems.

The State of Play: Achievements and Limitations

The review of over 90 case studies confirms that Australia has made substantial progress in developing innovation districts, research precincts, and collaborative programs. However, the analysis also reveals that, from international comparisons, the rhetoric may have outpaced institutional practice.

Despite significant investment in infrastructure and programs, many initiatives remain fragmented, driven by short-term funding cycles, narrow metrics and public administration failures.

The "input fallacy" – the assumption that simply increasing funding will deliver innovation outcomes - is a mindset that must be changed.

The Case for Systemic Integration

The evidence points to a consistent conclusion: enduring innovation outcomes emerge from the *orchestration of systemic interactions* between research organisations, businesses, governments, and communities.

Success stories such as Kendall Square, MaRS, and 22@Barcelona demonstrate that the convergence of placemaking, economic strategy, business development, and governance create self-reinforcing dynamics that cannot be achieved through siloed interventions.

These stories underscore the importance of adaptive governance, institutional depth, and long-term stewardship.

Areas for Policy Consideration

The background work and case studies reinforce the case for continuing policy attention in at least six areas:

- *Building Translational Infrastructure.* Policy must prioritise facilities and capabilities that move ideas from research to market—prototyping labs, clinical trial platforms, pilot manufacturing facilities, and digital testbeds. These assets are as critical as basic research infrastructure for closing commercialisation gaps.
- *Strengthening Governance and System Integrators.* Australia needs well-resourced intermediary organisations that coordinate stakeholders, align priorities, and provide continuity across electoral cycles. Governance structures must balance shared strategic direction with institutional autonomy, enabling responsiveness to local contexts.
- *Financing Growth and Scaling.* Early-stage capital is not sufficient if companies cannot access follow-on funding to scale. Policy frameworks should expand patient capital, blended finance models, and procurement pathways that create demand for innovative products and services.
- *Investing in Human Capital.* Talent remains the most critical input. Future strategies must connect education, skills development, and migration settings with the evolving needs of ecosystems. This includes supporting interdisciplinary training and embedding entrepreneurship and design thinking in curriculums.
- *Embedding Inclusion and Legitimacy.* Innovation ecosystems must benefit the wider community to be politically and socially sustainable. Place-based strategies should address housing affordability, workforce participation, and opportunities for underrepresented groups, ensuring that innovation does not create enclaves of privilege.
- *Linking Local Strengths to Global Networks.* Policies should help local ecosystems position themselves as nodes in global knowledge and value chains, leveraging international collaborations, trade links, and talent mobility to sustain competitiveness.

Closing Reflection

Innovation ecosystems are not created by chance, nor are they engineered by fiat. They evolve through deliberate choices, institutional craft, and cultural commitment. Australia's opportunity lies in turning its fragmented efforts into a coherent, adaptive system capable of generating economic prosperity, social legitimacy, and global relevance.

The task ahead is to align placemaking, economics, business logic, and governance so that they reinforce one another—delivering innovation outcomes that are resilient, inclusive, and transformative.

PART VII: SELECTED CASE STUDIES

Part VII presents detailed case studies of innovation districts, precincts, and hubs drawn from Australia and internationally. These narratives offer practical examples of how ecosystems evolve, develop, and adapt over time.

Each case study follows a common structure, addressing context, development stages, research and commercial outcomes, governance arrangements, and lessons for practice. Together, they highlight the diversity of ecosystem models, ranging from university-anchored health and biomedical precincts to urban regeneration projects and technology clusters, and reveal the conditions under which they succeed or struggle.

This part serves as both a reference and a source of inspiration, offering evidence for policymakers, planners, and institutional leaders seeking to strengthen their own innovation environments.

Introduction

The case studies included in the *Handbook* have three essential characteristics:

1. Concentration and co-location of universities, hospitals, institutes and firms to enable collaboration on R&D, research translation and commercialisation
2. Shared infrastructure and programs, including public university capital, and shared platforms, including incubators and testbeds, that reduce transaction costs for industry collaboration
3. Industry development policy frameworks that link land use, transport and investment attraction to create employment and export-facing capacity in areas such as technology, health, medical and biotech, and regional development.

Case examples have been written using publicly available information based on a reporting template developed in collaboration with people involved in the Canberra Innovation Ecosystem. Reporting was required under the following headings.

- Context
- Development stages – from inception through to maturity
- Research excellence, translation and commercial outcomes
- Challenges and future directions

- Conclusion
- Lessons for other regions

Over 70 case studies have been prepared and are included in a searchable text database. Case studies average 3,000 words (7 pages), with some longer, reflecting the amount of publicly available information.

The case studies included here have been verified by ecosystem coordinators in each of the relevant districts or precincts—the list of all case studies profiled in an Addendum to the Part.

1. The Canberra Innovation Ecosystem

Innovation ecosystems are critical for economic resilience and long-term growth. Cities worldwide are striving to create environments that support entrepreneurship, research commercialisation, and technology development.

Canberra offers a unique case study of how a city traditionally reliant on public sector employment has evolved into a vibrant innovation hub, leveraging its world-class research institutions, government support, and strategic partnerships.

This study provides important insights into how cities can transform their economies through targeted innovation strategies.

Context

Australia's capital city, Canberra, has transformed from a public sector-dominated economy into a dynamic, innovation driven hub. Home to over 466,000 residents, with a broader regional population nearing 750,000, Canberra boasts Australia's most educated workforce.

Over the past two decades, sustained efforts by the ACT Government, universities, research organisations, and the private sector have driven economic diversification and knowledge-intensive industry growth. This case study examines Canberra's innovation evolution, exploring the development of its ecosystem and identifying key challenges and future opportunities.

Studying Canberra's innovation ecosystem is important for several reasons in a broader innovation ecosystem context.

- The ecosystem has evolved " bottom-up," demonstrating how well-considered policy interventions, infrastructure investments, and collaboration between government, academia, and industry can accelerate knowledge-intensive economic development. Canberra has transformed from a city with limited private sector innovation into one that has successfully nurtured globally competitive start-ups, cutting-edge research, and new industry sectors. Its journey offers valuable lessons for other regions looking to enhance their innovation capabilities.
- Canberra's experience also highlights the importance of effective institutional alignment in advancing innovation. Universities, research organisations, and incubators have been pivotal in driving technological advancements and commercialising intellectual property. ANU, UNSW Canberra, and CSIRO have produced high impact spin-offs, attracted venture capital, and positioned the city as a leader in emerging fields such as quantum computing, cybersecurity, and sustainable technologies.
- Canberra's case illustrates the role of small but targeted funding initiatives in supporting early-stage ventures. Programs such as the Innovation Connect (ICON) grants, the Canberra Innovation Network (CBRIN), and ANU Connect Ventures have provided crucial financial and mentoring support, bridging the gap between research and commercial success. This underscores the importance of government-backed seed funding and venture capital availability in building a robust and sustainable start-up ecosystem.
- Canberra's innovation journey is also significant within the broader Australian context. As the national economy transitions towards high-tech industries, Canberra is a model of how relatively smaller, knowledge-intensive cities can drive economic diversification. Its focus on defence, space, quantum technologies, and digital health aligns with national priorities, demonstrating how regional innovation hubs can contribute to national economic strategies.

The case study is particularly relevant as it examines ongoing challenges. Despite its achievements, Canberra still faces barriers such as growth-stage capital shortages, talent retention issues, and the need for industry diversification beyond its strong defence and cybersecurity sectors. Addressing these challenges requires strategic policy adjustments, stronger links to international investment, and continued institutional collaboration.

Laying the Foundations: 2001-2008

The early 2000s marked the first significant efforts to nurture innovation in Canberra. Epicorp, a business incubator established with Commonwealth support, played a foundational role by investing in ventures such as Cohda Wireless, Windlab, and Locata.

While Epicorp has ceased operations, its legacy persists through later initiatives such as Lighthouse Innovations and Activate Capital.

National ICT Australia (NICTA), launched in 2002 as a Centre of Excellence in ICT research, significantly contributed to the innovation landscape. It cultivated expertise in artificial intelligence, computer vision, and software systems, strengthening industry-academia linkages. The organisation later merged into CSIRO's Data61, sustaining Canberra's leadership in digital research.

Funding mechanisms also emerged, notably the ACT Knowledge Fund (2003-2006) and the Canberra Business Development Fund (2005). These initiatives catalysed early-stage commercialisation, supporting start-ups like Seeing Machines and Liquid Instruments. The establishment of ANU Connect Ventures in 2005, backed by MTAA Super and the ACT Government, further bolstered seed funding, accelerating investment in research-driven start-ups.

By the late 2000s, Canberra had established core components of an innovation ecosystem: incubators, research funding, and growing investor interest.

Institutional Development and Growth: 2008-2018

During this period, universities became pro-active in commercialising their research. ANU and the University of Canberra (UC) launched InnovationACT, a student-led start-up competition to encourage and build capability in entrepreneurship. UNSW Canberra deepened its focus on defence, space, and cybersecurity, while CIT adapted its vocational offerings to the needs of a knowledge-based economy.

Lighthouse Innovations, established in 2008 as a public-private incubator, provided tailored support to tech start-ups. The ACT Government also launched the Innovation Connect (ICON) program, offering competitive grants for proof-of-concept ventures. ICON has since remained a crucial element of Canberra's early-stage innovation landscape.

A defining moment came in 2014 with the creation of the Canberra Innovation Network (CBRIN). Established through a partnership between the ACT Government, ANU, UC, UNSW Canberra, CSIRO, and industry players, CBRIN took a leadership role in innovation efforts. It cultivated collaboration, supported start-ups through the Griffin Accelerator, and facilitated knowledge sharing via networking events such as First Wednesday Connect.

During this period, Canberra emerged as a leader in defence and cybersecurity, with UNSW Canberra and ANU playing pivotal roles. High growth start-ups such as Instaclustr, Reposit Power, and Lithicon demonstrated the ecosystem's capacity to produce globally competitive companies. Instaclustr's acquisition by NetApp in 2022 for USD500 million highlighted Canberra's potential as a deep tech hub.

Beyond STEM, cultural and creative industries also flourished. Screen Canberra and digital media and gaming initiatives enriched Canberra's innovation ecosystem, demonstrating a broadening of innovation beyond traditional technology fields.

Expansion and Maturation: 2018-Present

Since 2018, Canberra's innovation ecosystem has matured, with institutions expanding their industry linkages. ANU launched the Software Innovation Institute in 2020, while UC established the Open Source Institute with foundational support from Instaclustr. The UC Health Innovation Hub emerged as a key player in digital health and aged care research.

UNSW Canberra's city precinct is becoming a co-location hub for defence and cybersecurity start-ups, reinforcing Canberra's role in national security innovation. The university's Launch on Northbourne facility has further supported deep tech entrepreneurs working on dual-use technologies.

Spin-offs from ANU, UC, CSIRO, and UNSW Canberra have proliferated, including Quantum Brilliance, Samsara, Nomad Atomics, Syenta, SkyKraft, and Vai Photonics. These ventures have collectively raised hundreds of millions of dollars, employing high-skilled professionals and reinforcing Canberra's reputation in quantum research, where ANU leads nationally.

Sustainability innovation has also flourished. Canberra achieved its 100% renewable energy target in 2020, attracting investment in hydrogen energy, circular economy solutions, and low-emission transport. Smart city initiatives integrating AI and IoT into urban planning further underscore Canberra's role as a forward-thinking city.

International engagement has grown, with the ACT Government's International Engagement Strategy (2020-2025) attracting foreign investment into technology and research sectors. Collaboration with regional innovation hubs across southern New South Wales has strengthened Canberra's broader economic footprint.

Challenges and Future Directions

Despite its success, Canberra faces significant challenges:

- Growth-stage capital: Mid-tier start-ups struggle to secure investment to scale operations. While the upcoming Activate Capital venture fund will provide early-stage funding, Canberra requires additional growth-stage capital sources.
- Talent retention: Canberra's small population limits workforce scalability, necessitating strategies to attract and retain skilled professionals.
- Industry diversification: While defence, cybersecurity, and space remain strengths, expanding into med-tech, advanced manufacturing, and creative industries is critical for economic resilience.
- Policy coordination: Stronger alignment between ACT and federal innovation strategies would optimise funding opportunities and cross-sector collaboration.

A major opportunity lies in the knowledge corridor linking UC, the AIS, the future North Canberra Hospital, CSIRO, ANU, and UNSW's new city campus. This corridor, valued at billions of dollars in innovation infrastructure, has the potential to become a world-class innovation district. However, its development requires clear governance and strategic planning.

Conclusion

Canberra's transformation into an innovation hub underscores the power of sustained investment, institutional collaboration, and policy foresight. Its experience offers key lessons for other cities seeking to build resilient innovation ecosystems:

- Leverage research institutions: Universities and research organisations serve as vital catalysts for commercialisation.
- Encourage public-private collaboration: Canberra's model demonstrates the benefits of joint government, industry, and academic engagement.
- Specialise in strategic sectors: Canberra's focus on defence, cybersecurity, quantum, and sustainability has created areas of competitive advantage.
- Secure global investment: Canberra's engagement in international partnerships has enhanced its innovation ecosystem's credibility.
- Support early-stage entrepreneurship: Targeted funding programs have been instrumental in nurturing high growth start-ups.
- Integrate smart city initiatives: Urban innovation enhances both economic and environmental sustainability.

With ongoing investment in digital technology, sustainability, and diversified industry growth, Canberra is well-positioned to maintain its status as a national innovation leader. Its experience provides a roadmap for other regions seeking to drive innovation-led economic development. The next phase of Canberra's journey will depend on strategic leadership, coordinated policy efforts, and the capacity to scale its innovation infrastructure for long-term success.

2. Tech Central, Sydney

Sydney's Tech Central has emerged as Australia's largest and most ambitious innovation district, spanning six square kilometres from Haymarket to Camperdown and South Eveleigh in the heart of the nation's most populous city. With over 5.3 million residents in Greater Sydney, the city serves as Australia's economic powerhouse, contributing approximately one-quarter of the nation's GDP.

Context

Tech Central currently hosts the highest concentration of technology businesses anywhere in Australia, including global unicorns Atlassian, Block (Afterpay), Canva, SafetyCulture and Rokt, alongside 150 research institutes and two world-leading universities. The district represents a fundamental shift in how Australia approaches innovation ecosystem development, moving from isolated tech parks to integrated urban innovation districts that combine research, industry, and lifestyle amenities.

Within Tech Central, University partners benefit from direct access to experts, labs, equipment, office space, and a vibrant community of innovators. It blends large-scale infrastructure, communication labs, and data analytics to tailor solutions for startups, SMEs, or large organisations.

Unlike traditional innovation precincts that developed organically around single institutions, Tech Central is a deliberate attempt to orchestrate innovation at a metropolitan scale. This complex ecosystem encompasses 250 hectares of established urban landscape, demanding coordination approaches that harness the collective capabilities of universities, hospitals, government authorities and industry partners while fostering authentic collaboration.

The district's development reflects Sydney's evolution from a primarily finance and trade-focused economy to one increasingly driven by knowledge-intensive industries, digital technologies, and innovation. This transformation has been accelerated by strategic government investment, university-industry partnerships, and the concentration of Australia's largest technology companies within a compact, highly connected urban area.

Historical Context and Early Foundations: 1995-2015

The origins of Sydney's current innovation landscape can be traced to the establishment of the ATP at Eveleigh in 1995. The Park was established as an innovation hub through a collaboration between The University of Sydney, UNSW, UTS and ANU. This early experiment in university-industry collaboration provided valuable lessons about the challenges of sustaining innovation ecosystems within commercial property development frameworks.

During this formative period, Sydney's universities began developing more intensive approaches to research commercialisation. The University of Sydney established its commercialisation office (the BLO) and began nurturing start-up activity, while UTS developed its reputation for industry engagement and practical technology transfer. Both institutions recognised that proximity to industry and access to urban amenities would be crucial for attracting talent and fostering innovation.

The period also saw the emergence of early startup communities and co-working spaces across the Sydney CBD. Organisations like Fishburners, founded in 2011, began creating the cultural infrastructure necessary for entrepreneurship. When Fishburners started, there were few options for start-up founders looking for support. The sense of community is in Sydney's DNA and has always been the heart and soul of what the city does.

However, the limitations of the dispersed approach became apparent when ATP was sold to property developer Mirvac in 2015. The sale led to the displacement of many smaller, innovative firms, revealing tensions between commercial property development imperatives and the needs of the innovation ecosystem. This event catalysed government thinking about how to create more sustainable innovation infrastructure.

The Sydney Start-up Hub Era: 2016-2020

The establishment of the Sydney Start-up Hub (SSH) in 2017 marked a pivotal moment in Sydney's innovation journey. Located in a 17,000 square metre space across 11 floors at 11-31 York Street, above Wynyard Station, the Hub represented the NSW Government's first major attempt to create centralised innovation infrastructure in the CBD.

The start-up hub housed four anchor tenants - Fishburners, Stone & Chalk, The Studio and Tank Stream Labs - providing 610 desks for Fishburners alone and creating a co-location environment that benefited the entire ecosystem. The Hub's strategic location provided start-ups with unprecedented access to Sydney's financial district, corporate headquarters, and transport infrastructure.

The Hub's anchor tenants each developed distinct specialisations that strengthened the overall ecosystem. Stone & Chalk became known for its curated spaces for tech start-ups, scale-ups and corporate teams, with a particular focus on fintech. Tank Stream Labs held three levels of the Sydney Start-up Hub and expanded to other locations across the Sydney CBD, including Barangaroo.

The major achievements of the Sydney Start-up Hub can be summarised as follows–

- *Community Building:* Becoming home to around 2,500 "residents" and thousands more through community and events spaces, collaborative workspaces and a Regional Landing Pad.
- *Economic Impact:* Jobs, Investment, Tourism, and Western Sydney Minister Stuart Ayres has said that the SSH had helped create over 1,000 jobs and attract more than $280 million in investment.
- *Regional Significance:* Over 40% of Australia's startups are conceived in NSW and the Hub has helped establish Sydney as a startup leader in the Asia-Pacific region.

This period also saw the rise of highly successful programs like *Antler* grow and flourish in Australia in the SSH building. Startups *Insuretech* and Upcover, founded at the Sydney Startup Hub in 2019, have grown from strength to strength, building on the ecosystem enabled at SSH and have gone on to raise a $19 million Series A in 2025.

During this period, other successful companies emerged from the ecosystem that would later become anchors for Tech Central. Notable examples include companies that received early support and connections through the Hub's networks, demonstrating the value of concentrated innovation infrastructure.

The Emergence of Tech Central: 2018-2023

While the Sydney Start-up Hub was establishing itself as a key node in Australia's start-up ecosystem, a broader vision was taking shape around Central Station.

Beginnings

The concept of Tech Central emerged from recognition that Sydney needed innovation infrastructure that could compete globally and leverage the city's unique assets, particularly its world-class universities and transport connectivity.

In 2018, prominent Australian technology business leader, David Thodey lead a panel of ecosystem leaders to evaluate the opportunity of this area as an innovation district. This panel included Scott Farquhar and representatives of Tech Sydney, Fishburners, Sydney Local Health District, both Universities and Business Sydney.

The ecosystem leadership group worked closely with Julie Wagner, a global expert on innovation districts to validate the opportunity that this area represented and catalysed key stakeholders in the district to engage in the opportunity.

The recommendations of this group formed the core commitments implemented by the NSW Government in regard to affordable space, activation and anchor tenant attraction.

Strategic Location and Infrastructure

Tech Central is Australia's largest and most dynamic tech and innovation district, spanning six diverse neighbourhoods and anchored by world-leading universities. Located in the heart of Sydney, Tech Central is the epicentre of the city's innovation and technology scene. The district's strategic advantages include the following.

Transport Connectivity: Central Station is Australia's busiest transport hub, with the new Sydney Metro providing fast connections across the metropolitan area. Tech Central is highly accessible, thanks to the new Sydney Metro, which offers fast and frequent services. The area is also well-served by buses, light rail, and nearby train stations, ensuring easy travel.

University Partnerships: The Camperdown Innovation and Health Precinct (55 hectares) is Australia's premier biomedical research and innovation ecosystem. The University of Sydney's far-reaching research infrastructure forms the foundation of the precinct, integrating with Royal Prince Alfred Hospital's clinical excellence. UTS anchors the Haymarket-Ultimo precinct with its focus on research collaboration and industry partnerships.

Industry Concentration: Tech Central features an agglomeration of businesses in digital technology (especially software and AI), deep tech, creative and gaming industries, and medical and biotech sectors. In addition, Tech Central has industry-ready talent and collaboration opportunities on its doorstep, with access to over 160,000 students and more than 150 research institutes and centres of excellence.

Atlassian as the Catalyst

The decision by Atlassian, Australia's largest software company, to establish its new global headquarters in Tech Central provided the anchor investment necessary to catalyse broader development. The UTS city campus is located at the southern border of Sydney's central business district, close to Central Station and Railway Square, within Sydney's emerging Tech Central. Atlassian's commitment represented more than just a real estate decision - it signalled confidence in Sydney's potential as a global technology hub.

University Collaboration and Research Commercialisation

Both The University of Sydney and University of Technology Sydney recognised the opportunity to leverage their proximity to create new models for research commercialisation.

Already, UTS and The University of Sydney are working on *Digital Central*, an ambitious collaboration to harness the strength of the two universities' digital sciences research and teaching. Digital Central will help anchor Tech Central and drive industry collaboration and job growth, enabling NSW to achieve its ambition of becoming a world-leading digital industries hub.

The universities developed complementary strengths that enhanced the overall ecosystem in several important ways.

- *University of Technology Sydney*

Through its UTS Startups program, UTS has access to a wide network of funders, investors and entrepreneurs and works closely with organisations in the private, public and not-for-profit sectors to commercialise UTS research and innovations. In Ultimo, as part of the UTS Startups program, UTS has established the largest incubator and coworking space operated by an Australian University, with more than $2500m^2$ and 250 desks available to its community of more than 600 active startups.

- *The University of Sydney Programs*

The Genesis program, run by the University of Sydney Business School, supports the most promising start-ups through mentoring from experts, assistance with media exposure and a final award of $25,000. At the beginning of 2020, there were around 200 start-ups actively engaged in the Genesis program. The Sydney Knowledge Hub is a University of Sydney hosted co-working and innovation ecosystem that connects researcher-led startups with industry via access to campus facilities, tailored programs, and collaborative networks to accelerate research commercialisation.

Infrastructure Investment and Government Support

The NSW Government has demonstrated sustained recognition that creating a world-class innovation district requires significant infrastructure investment across multiple budget cycles.

In 2021 the NSW Government first allocated $48.5 million to commence the work to catalyse Tech Central. Which provided funding for the establishment of the Tech Central Scaleup Hub, the Scaleup Accommodation Grant and the development of brand concepts for Tech Central.

In 2022, the government announced an $8 million Tech Central Research and Innovation Infrastructure Fund to help drive research and commercialisation activity in the Tech Central precinct, centred on Sydney's Central Station and including UTS and the University of Sydney.

The 2022 funding focused on creating shared infrastructure that could benefit multiple organisations rather than supporting individual companies. Then-Minister for Enterprise, Investment and Trade, Stuart Ayres, said the fund would help realise proposals that provide significant physical and digital infrastructure to the Tech Central precinct, expand technical expertise, collaboration and commercialisation and otherwise build capacity across its target industries and research areas.

The recent 2025-26 NSW Budget included $38.5 million for Tech Central "to flourish as a melting pot for groundbreaking innovation, research, and lifestyle." This includes $5 million for the relocation of the Sydney Start-up Hub from Wynyard to Tech Central.

The progression from the $8 million infrastructure fund in 2022 to the $38.5 million allocation in 2025 demonstrates the NSW Government's continued commitment to developing Tech Central's capabilities and infrastructure over time.

Sectoral Specialisation and Industry Development

Tech Central's development has been characterised by the emergence of distinct sectoral strengths that leverage Sydney's unique advantages and existing capabilities.

Financial Technology (FinTech): Sydney's position as Australia's financial capital has made FinTech a natural area of strength within Tech Central. The proximity to major banks, insurance companies, and financial services firms has enabled FinTech start-ups to access customers, partners, and domain expertise more readily than in other locations.

Creative and Gaming Industries: The Chippendale Creative Precinct (30 hectares) showcases the transformative potential of adaptive reuse in urban innovation. Centred around the Central Park development and UTS Faculty of Arts, the precinct demonstrates how former industrial sites can evolve into vibrant creative quarters. This creative infrastructure has supported the development of gaming, animation, and digital media companies within the broader Tech Central ecosystem.

Biomedical and Health Technology: The University of Sydney's commercialisation initiatives and industry partnerships drive breakthrough innovations in healthcare, biotechnology, and medical devices. The integration with Royal Prince Alfred Hospital and other health facilities has created unique opportunities for health technology innovation and clinical translation.

Deep Technology and Advanced Manufacturing: The prototyping facility will offer transformative manufacturing capabilities, promising to enhance Australia's semiconductor, photonics and deep-tech manufacturing ecosystem. These capabilities position Tech Central to support the advanced manufacturing and deep technology sectors that are increasingly important for national competitiveness.

Tech Central also has a "satellite" location with the UTS Tech Lab, located in Botany. This 18,000sqm state-of-the-art research facility offers bespoke industry partnerships for engineering and deep tech focused innovation. The University of Sydney operates an International Centre for Crop and Digital Agriculture at Narrabri in north-west NSW.

Coordination Challenges and Governance Evolution

The development of Tech Central has highlighted the complex governance challenges inherent in coordinating large-scale urban innovation districts. Unlike its international counterparts in North America and Europe, Tech Central operates without an integrated metropolitan coordination framework.

Instead, it functions through an intricate maze of State Government agencies, local councils, statutory authorities and NGOs.

Such institutional fragmentation poses particular challenges for coordinating innovation district development.

Multi-Jurisdictional Complexity

At the core of Tech Central's development lies an intricate governance challenge: coordinating a multifaceted innovation landscape comprising diverse property owners and major institutional stakeholders. These include Central Station, two prominent universities, a 1,200-bed tertiary hospital with associated biomedical research centres, two local government authorities, and numerous private sector interests.

The regulatory framework includes multiple approval authorities:

- The Minister for Planning is the consent authority for State Significant Development
- The Department of Planning and Environment is responsible for assessing applications
- Transport for NSW for transport-adjacent development
- City of Sydney Council for standard developments
- Heritage NSW for preservation oversight

Alliance Formation and Collaboration Models

In response to these coordination challenges, stakeholders have developed new collaborative frameworks. Alliance partners work to create the conditions that nurture research, collaboration and commercialisation outcomes in a vibrant and diverse destination. They are committed to creating a place that attracts national and international expertise and investment.

The Camperdown Ultimo Alliance, established in 2018, is one model for how universities and other institutions can coordinate their activities more effectively within innovation precincts. The Alliance brought together the University of Sydney, UTS, Sydney Local Health District, City of Sydney, Inner West Council, TAFE NSW and Cicada Innovations to provide a shared precinct voice and collaborative leadership for Tech Central.

The Alliance has demonstrated collaboration with the NSW Government on various precinct development initiatives. These have included projects designed to connect people and businesses, such as the Tech Central (Camperdown Ultimo) place-based transport strategy and the Tech Central calendar of events, which aimed to connect a global audience to the diverse array of activities happening in the precinct.

As a collaborative precinct leadership model, the Camperdown Ultimo Alliance is a key stakeholder to work with government and other stakeholders to optimise Tech Central's potential through coordinated planning and shared commitment to research, collaboration and commercialisation outcomes.

This approach reflects international best practices in innovation district governance, where institutional coordination and long-term strategic perspective are essential ingredients for success and assisted in the continued catalysation of the district.

Current Transformation and Future Directions: 2024-2025

The period from 2024 to 2025 has marked a significant transformation in how Tech Central is being developed and governed, with new strategic initiatives aimed at realising the district's full potential.

New Government Strategy and Vision

The NSW Government has announced a new plan to unlock the full potential of the Tech Central. This fresh vision will establish the district as a leading place to live and work and will play a critical role as a future driver of NSW Government key priorities, including housing, healthcare, night-time economy, visitor economy, and world-class transport infrastructure. The approach recognises that successful innovation districts must integrate multiple urban functions rather than focusing solely on technology and research.

Sydney Start-up Hub Transition

One of the most significant developments has been the decision to relocate the Sydney Start-up Hub to Tech Central. With Tech Central positioned as Australia's leading hub for technology and innovation, the Sydney Start-up Hub will move to a location in the heart of Tech Central in October 2025. This is aimed to ensure that founders, entrepreneurs, investors, corporates, and the entire innovation ecosystem can collaborate with universities and research institutes that are "right on their doorstep".

This transition signals a strategic integration of Sydney's start-up ecosystem with its research and innovation infrastructure. However, the transition has also highlighted some of the challenges in balancing different stakeholder interests and managing change in complex ecosystems.

Investment NSW Leadership Role

Investment NSW will lead the development of the strategy in collaboration with the Department of Planning, Housing and Infrastructure, and Transport for NSW. This institutional arrangement reflects recognition that successful innovation districts require coordination across multiple government agencies and policy domains.

The enhanced role for Investment NSW includes:

- Attracting investment and talent to Tech Central
- Facilitating coordination between start-ups, research institutes, and universities
- Ensuring alignment of government programs with district needs
- Brand promotion and community building

Infrastructure and Development Pipeline

The district continues to see significant infrastructure development, with Atlassian's 39-storey headquarters serving as a flagship project. The precinct also has a key technology driver as it will be anchored by Atlassian's 39-storey skyscraper, which has been "in the works" for the past five years.

Sectoral Growth and Expansion

International Connectivity and Global Positioning: Tech Central's development has been shaped by the recognition that successful innovation districts must be globally connected and competitive. Sydney's position in the Asia-Pacific region provides unique advantages for creating connections with major innovation hubs in Asia, North America, and Europe.

Asia-Pacific Gateway Role: Sydney's time zone, business culture, and established connections make it a natural hub for technology companies expanding across the Asia-Pacific region. Many global technology companies have established their regional headquarters in Sydney, bringing international expertise and connections to the Tech Central ecosystem.

International Events and Conferences: Tech Central hosts hundreds of events annually, connecting innovators and entrepreneurial businesses. These events celebrate local business success and provide robust support for the entire ecosystem, fostering a dynamic environment where ideas flourish, partnerships form, and opportunities flourish.

SXSW Sydney is a major international event that showcases Tech Central's capabilities to a global audience. The Tech & Innovation Expo provides opportunities for local companies to connect with international partners and investors.

Economic Impact and Success Metrics

While comprehensive economic impact data for Tech Central is still being developed, early indicators suggest significant positive outcomes:

Employment Growth: Sydney's colossal Tech Central is being positioned as Australia's answer to Silicon Valley. The district is expected to accommodate approximately 25,000 workers across its various developments.

Investment Attraction: The concentration of research institutions, universities, and technology companies is beginning to generate measurable innovation outputs, though comprehensive data on patents, publications, and commercialisation outcomes requires longer-term tracking.

Challenges and Critical Success Factors

Despite its promise, Tech Central faces several significant challenges that will determine its long-term success:

Coordination and Governance: The successful orchestration of these varied elements could establish Tech Central as a cornerstone of Australia's innovation capability, generating substantial economic value while positioning Sydney as a leading innovation hub in the Indo-Pacific's innovation landscape.

Achieving this coordination remains challenging given the complex stakeholder environment and multiple regulatory frameworks.

Talent Attraction and Retention: Sydney's high cost of living and housing prices create challenges for attracting and retaining talent, particularly in the early stages of their careers, when many start-up employees and researchers are most mobile.

Scale and Critical Mass: Tech Central must maintain strong connections with other innovation centres across Australia while avoiding becoming isolated from the broader start-up and research community.

Lessons for Other Innovation Districts

Tech Central's development offers several important lessons for other cities attempting to create innovation districts:

Anchor Institution Importance: The role of major universities and established companies as anchors has been crucial for providing stability and attracting additional investment.

Transport and Location Advantages: The central location and excellent transport connectivity have been fundamental to Tech Central's success, enabling it to attract talent and companies that might otherwise locate elsewhere.

Government Coordination Requirements: The experience highlights the importance of having clear governance structures and coordination mechanisms when multiple government agencies and private stakeholders are involved.

Patient Capital and Long-term Commitment: Creating world-class innovation districts requires sustained investment and commitment over many years, with returns often taking a decade or more to fully materialise.

Future Opportunities and Strategic Directions

Several opportunities could further strengthen Tech Central's position as a leading innovation district.

Quantum Computing and Advanced Technologies: Given the density of quantum research, teaching and companies in Tech Central, the precinct is globally ranked as a centre of Quantum and related activity. The precinct is also home to the Sydney Quantum Academy, a key collaborative effort to drive future workforce development for the future quantum sector. The precinct is also home to emergent Quantum Technology companies including Q-CTRL, Diraq, Quantum Brilliance, and others.

Advanced Manufacturing Integration: The Vaccine and RNA Design Centre (VRDC) will offer Australian mRNA vaccine and RNA therapeutic design capability, with the opportunity to bring RNA molecules to market. The Centre develops cloud-based tools for designing mRNA vaccines and RNA treatments through partnerships with UNSW, ANU, CSIRO and NSW Department of Primary Industries. These capabilities could position Tech Central as a leader in advanced biomanufacturing.

Sustainability and Clean Technology: Sydney's commitment to carbon neutrality and sustainable development creates opportunities for clean technology innovation within Tech Central.

International Partnership Development: Expanding formal partnerships with innovation districts in key international markets could accelerate knowledge transfer and investment flows.

Conclusion

Tech Central is one of Australia's most ambitious urban innovation experiments, demonstrating both the transformative potential and inherent complexities of modern innovation ecosystem development. Over less than a decade, it has evolved from a conceptual vision to Australia's largest concentration of technology companies, research institutions, and innovation infrastructure.

The district's success reflects several key factors: strategic location and transport connectivity, strong anchor institutions, government commitment and investment, and the development of collaborative governance mechanisms. However, significant challenges remain, particularly around coordination, talent attraction, and maintaining momentum as the district matures.

The NSW Government is committed to the success of Tech Central and seeing it flourish as a melting pot for groundbreaking innovation, research, and lifestyle. This commitment, combined with the district's strong foundations and growing international recognition, positions Tech Central to play an increasingly important role in Australia's innovation landscape.

For other cities and regions considering similar initiatives, Tech Central's experience demonstrates that creating world-class innovation districts requires sustained commitment, patient capital, strong institutional partnerships, and governance mechanisms.

Success is not guaranteed, but with the right combination of assets, strategy, and execution, innovation districts can become powerful drivers of economic growth and technological advancement.

The next phase of Tech Central's development will be crucial in determining whether it can achieve its ambition of becoming a globally competitive innovation hub. Early indicators are positive, but the true test will be whether the district can sustain its growth trajectory, attract world-class talent and companies, and generate the breakthrough innovations that define leading innovation ecosystems.

As Australia seeks to diversify its economy and strengthen its position in the global knowledge economy, Tech Central's success or failure will have implications far beyond Sydney. It is a test case for whether Australian cities can compete effectively with established global innovation centres and create the conditions necessary for sustained innovation-led growth.

3. Macquarie Park Innovation District, Sydney

Macquarie Park Innovation District (MPID) is one of Australia's most significant experiments in creating a world-class innovation ecosystem through strategic corporate-university collaboration. Located 13 kilometres north-west of Sydney's central business district, this 350-hectare precinct has emerged as one of Australia's largest non-CBD commercial districts and the fourth largest employment concentration in New South Wales, housing over 180 multinational enterprises, 200 small and medium enterprises, and anchored by Macquarie University.

Context

With over 63,000 workers and a growing residential population, Macquarie Park exemplifies the transformation from a traditional suburban office park into a dynamic innovation district. The precinct's evolution reflects broader trends in urban economic development, where proximity to research institutions, transport infrastructure, and corporate headquarters creates conditions for innovation driven growth.

The strategic location within the broader Capital Region, encompassing parts of northern Sydney and connected by world-class transport infrastructure, including the Sydney Metro Northwest, positions MPID as a critical component of Greater Sydney's innovation ecosystem. Unlike many innovation districts that emerge organically around universities, MPID is a deliberate, planned approach to creating collaborative networks between established multinational corporations and cutting-edge research capabilities.

Historical Development: From Corporate Park to Innovation District (1980s-2010s)

Early Corporate Concentration (1980s-2000s)

Macquarie Park's foundation as a business precinct began in the 1980s with the establishment of what was initially conceived as a traditional suburban office park. The area's transformation into a major corporate hub was driven by several key factors: proximity to major transport corridors, abundant land availability, and Macquarie University, established in 1964.

Innovation was at the heart of Macquarie University's establishment, with the Government seeking to develop a purpose-built precinct closely following the Stanford University model (creating the so-called 'Stanford of the South') where business and industry would cluster around research and teaching.

Throughout the 1990s and early 2000s, the precinct attracted significant corporate investment from multinational companies seeking modern facilities outside Sydney's congested CBD. Major corporations that established headquarters or significant operations during this period included Johnson & Johnson, Abbott Laboratories, Optus, Canon, Fujitsu, Sony, and Siemens. This concentration of global companies created what economists call agglomeration effects – the benefits that arise when firms locate near one another.

The pharmaceutical and biotechnology sector emerged as a particular strength during this period, with companies like AstraZeneca, Hyundai, and Douglas Hanly Moir Pathology establishing significant operations. This clustering reflected both the proximity to Macquarie University's research capabilities, cheap land with appropriate zoning for manufacturing and light-industrial work, availability of specialised facilities and a skilled workforce.

Telecommunications and technology companies also found the location attractive, with Optus's relocation from North Sydney to Macquarie Park in 2007 representing a significant milestone. The move of Australia's second largest telecommunications company signalled the precinct's growing importance as a technology hub.

Infrastructure Development and Connectivity

The period from 2000 to 2015 saw substantial investment in transport and digital infrastructure that would prove crucial to the precinct's later transformation into an innovation district. The development of modern road networks, including upgrades to the M2 Motorway and Lane Cove Tunnel, improved connectivity to Sydney's broader metropolitan area.

Macquarie University's expansion during this period also contributed to the precinct's growing capabilities. The university's investment in research facilities, particularly in areas such as biotechnology, engineering, and information technology, created research infrastructure that later became central to innovation activities.

The establishment and growth of the Macquarie Centre as a major retail hub provided essential amenities for the growing workforce, while residential development in surrounding areas began to address the traditional suburban office park challenge of maintaining vibrancy beyond business hours.

Macquarie University Hospital opened in 2010, it is the first private hospital to be located on a university campus in Australia, integrating clinical practice with research and education. The hospital integrates its clinical services with nearly 30 speciality clinics and draws in industry partners for clinical trials.

Strategic Formation: The Innovation District Concept (2015-2018)

The Vision Takes Shape

The formal conceptualisation of Macquarie Park as an innovation district began in 2015 with Macquarie University's leadership recognising the unique opportunity created by the concentration of research capabilities and corporate presence. The university initiated discussions with major corporate partners to explore how traditional corporate-university relationships could evolve into something more transformative.

The MPID was officially launched in 2016 as an industry-led collaboration, representing a new model for innovation ecosystem development in Australia. Unlike government-led initiatives, MPID emerged from the bottom-up, driven by the recognition among key stakeholders that their proximity could create value far beyond traditional transactional relationships.

The founding partnership included Macquarie University and several major corporate partners including Optus, Abbott, Konica Minolta, NAB, and AMP Capital (former operators of the Macquarie Centre). This partnership represented a deliberate strategy to leverage the existing corporate concentration while creating new mechanisms for collaboration, knowledge transfer, and innovation.

Evolution of the Collaboration Model

In 2021 MPID continued its transformative journey by merging with Connect Macquarie Park to form Connect Macquarie Park Innovation District (CMPID. Connect Macquarie Park was a non-profit transport management association established in 2013 to drive better services and act as an advocate for its members, primarily around transport and congestion issues in relation to the NSW Government's plans to convert the existing heavy rail to a Metro standard and expand it to the Sydney CBD and beyond.

Both organisations recognised the potential in the district to drive sustainable and inclusive economic growth through collaboration and innovation.

Governance and Organisation

The establishment of the merged CMPID as an independent, not-for-profit organisation represented a crucial governance innovation. CMPID was tasked with facilitating collaboration among diverse stakeholders while advocating for the district's development needs and maintaining its competitive position.

The governance model recognised that successful innovation districts require active coordination and facilitation rather than assuming that proximity alone would generate collaboration. CMPID's role encompassed strategic planning, advocacy, event coordination, facilitated collaborations and serving as a unified voice for the district's interests in policy discussions.

This organisational approach reflected international best practices in innovation district development, drawing inspiration from successful models in Boston, Barcelona, and Toronto while adapting to Australian institutional and regulatory contexts leveraging the Associations Act.

The University as Innovation Anchor

Macquarie University's role in MPID represented an evolution of the traditional university-industry relationship model. Rather than simply conducting research that might eventually be commercialised, the university positioned itself as an active facilitator of innovation networks and collaborative problem-solving.

The launch of the Macquarie University Incubator in 2017 marked a significant milestone in this evolution. The incubator adopted an "open to all" model, supporting university-affiliated start-ups and entrepreneurs from the broader community. This approach reflected the district's ambition to serve as a hub for the entire Sydney innovation ecosystem rather than a closed university-corporate collaboration.

The incubator's programs were designed around practical business development needs, including structured mentorship, investor connections, and access to specialised facilities. By 2022, the incubator had supported over 75 companies, demonstrating the demand for such services within the broader innovation ecosystem.

In 2024 the University built on this model by opening up the Macquarie University Deep Tech Incubator which is a cutting-edge comprised of laboratories, equipment, secure offices and co-working areas to support deep tech innovators.

Acceleration and Infrastructure Investment (2018-2025)

Transport Revolution and Sydney Metro

The conversion of the Epping to Chatswood heavy rail to the autonomous metro standard as part of Sydney Metro Northwest project opened in 2019. In 2024 this line was extended from Chatswood to Sydneham via the Sydney CBD and represented a transformational moment for Macquarie Park, providing direct rail connections to the Sydney CBD for the first time and in under 20 minutes. The $8.3 billion investment in automated metro technology served three stations within the district: Macquarie Park, Macquarie University, and North Ryde.

This transport infrastructure investment fundamentally altered the district's competitive position within Greater Sydney, addressing one of the primary limitations of suburban office parks: connectivity to the broader talent pool and business ecosystem. The metro connection enabled the district to attract workers who preferred to live in other parts of Sydney while working in Macquarie Park.

The transport improvements also facilitated international connectivity, with improved links to Sydney Airport enabling more frequent business travel and supporting the district's growing international business relationships.

Campus and Precinct Development

Macquarie University's ~$1 billion campus development program during this period created world-class facilities designed to support collaboration with industry partners. The development included new research facilities, Macquarie University Hospital, collaborative spaces, and student accommodation, transforming the campus into a more open, connected environment.

The university's approach to campus planning explicitly considered the needs of the broader innovation district, creating facilities and spaces that could accommodate corporate partners, start-ups, and collaborative research projects. This represented a significant departure from traditional university planning approaches that prioritised academic needs over broader economic development objectives.

Substantial private sector investment during activity during this period, created exceptional commercial, collaboration, and innovation spaces across the district. This investment reflected growing confidence in the district's long-term prospects and the effectiveness of the collaborative model.

Research Commercialisation and Start-up Formation

The period from 2018 to 2025 saw unprecedented levels of research commercialisation and start-up formation within the district. The Macquarie University Incubator supported 109 companies through its educational programs, attracting $19 million in venture capital funding, generating over $13 million in revenue and employing 296 people in 2024.

Notably, 34% of incubator start-ups had at least one female founder, demonstrating the inclusive approach to entrepreneurship development. The incubator's focus on diverse participation reflected broader recognition that innovation ecosystems perform better when they draw from diverse talent pools.

The success of companies emerging from the incubator attracted attention from national and international investors, with several firms securing significant funding rounds and expanding their operations. This success created positive feedback loops, attracting more entrepreneurs and investors to the district.

Corporate Innovation Programs

Major corporate partners began developing innovation programs during this period that moved beyond traditional R&D activities to embrace open innovation models. Several companies established innovation labs and accelerator programs that worked closely with university researchers and local start-ups.

These programs reflected the recognition that innovation increasingly occurs through networked collaboration rather than internal R&D alone. The physical proximity within the district enabled more frequent and informal collaboration, leading to joint research projects, licensing agreements, and acquisition opportunities.

The success of these programs attracted additional corporate partners to the district, with companies like Medtronic, Zeeker, Geely, and Sobi establishing operations. This expansion demonstrated the growing reputation of MPID as a preferred location for companies seeking to enhance their innovation capabilities.

The Innovation Ecosystem in Practice

Collaboration Mechanisms and Networks

MPID's success in fostering innovation stems from multiple formal and informal collaboration mechanisms that have emerged within the district. Regular networking events, including communities of practice, advisory committees, annual summits and regular knowledge sharing sessions, create opportunities for serendipitous connections between researchers, entrepreneurs, and corporate innovators.

The establishment of shared facilities and collaborative spaces has enabled project-based partnerships that might not have emerged through formal corporate relationships alone. These include specialised laboratories accessible to multiple organisations, co-working spaces for start-ups, and meeting facilities designed to accommodate diverse stakeholder groups.

The district's approach to intellectual property and collaboration agreements has evolved to support more fluid partnerships while protecting legitimate commercial interests. This has required developing new contractual models and relationship frameworks that enable rapid collaboration without the need for extensive legal negotiations.

Talent Development and Retention

One of MPID's significant advantages is its ability to attract and retain high-skilled talent through employment opportunities, educational resources, and quality of life factors. Macquarie University enables continuous learning and professional development opportunities for workers across the district.

The district's companies collectively employ thousands of knowledge workers, creating career pathways that enable professionals to move between organisations while remaining within the geographic area. This talent circulation strengthens the overall ecosystem by facilitating knowledge transfer and professional network development.

Graduate programs and internship opportunities provided by corporate partners create direct pathways for university students to enter the innovation economy, while executive education programs cater to the professional development needs of experienced professionals.

Sustainability and Smart City Initiatives

MPID has positioned itself as a leader in sustainable urban development and smart city technologies. The district's commitment to environmental sustainability includes targets for renewable energy adoption, waste reduction, and sustainable transport options.

The Zero Emission Macquarie Park Bus Depot, supported by $100 million in government investment, is a significant step towards sustainable transport within the district. Combined with the $200 million investment in the new Macquarie Park Precinct and Bus Interchange, these investments demonstrate the integration of sustainability objectives with innovation district development.

Smart city initiatives include the deployment of IoT sensors for traffic management, energy optimisation systems, and digital platforms that improve coordination among district stakeholders. These technologies serve both operational efficiency objectives and provide testbeds for innovation companies developing smart city solutions.

Economic Impact and Performance

Employment and Economic Output

By 2024, MPID housed over 63,000 workers across 180 multinational enterprises and 200 small and medium enterprises, making it the fourth largest employment concentration in New South Wales. The district's economic output contributes significantly to both the Sydney metropolitan area and the national economy.

The concentration of high-value industries, including pharmaceuticals, biotechnology, telecommunications, and professional services, generates substantial economic multiplier effects. Studies by Biointelect in 2024 highlighted the district's role as a "deep tech" hub, characterised by high levels of R&D spending, which positions it as a key economic growth engine for NSW.

The district's companies collectively invest heavily in R&D activities, with spending levels that significantly exceed national averages. This investment in innovation activities generates both immediate economic benefits and longer-term competitive advantages for Australian industry.

The example of Cochlear is an exemplar case study of industry and university collaboration and co-location. Cochlear's global headquarters and manufacturing operations are in a purpose-built facility on the Macquarie University campus. This is located opposite the Hearing Hub which brings together Australia's leading hearing and healthcare organisations to collaborate on world-leading research projects. The co-location of Cochlear and the Hearing Hub at the University unites researchers, educators, clinicians and innovators with expertise in linguistics, audiology, speech pathology, cognitive and language sciences, psychology, nanofabrication and engineering sciences

Investment Attraction and Capital Formation

MPID's success in attracting investment reflects both the quality of opportunities within the district and the effectiveness of its collaborative model. Established corporate partners provides validation for investors considering early-stage companies, while the university connection offers technical due diligence capabilities.

The formation of investment networks within the district, including angel investor groups and venture capital partnerships, has created more efficient capital allocation mechanisms. Companies emerging from the incubator and university research programs can access multiple funding sources without leaving the geographic area.

International investment attraction has become increasingly important, with several multinational companies establishing their Asia-Pacific innovation centres within the district. This international presence brings additional resources, market access, and technical expertise to the broader ecosystem.

Research Commercialisation Success

The district's approach to research commercialisation has generated significant intellectual property and technology transfer outcomes. University-industry collaborations have resulted in numerous patent applications, licensing agreements, and spin-off companies.

The success of research commercialisation efforts reflects both the quality of research conducted within the district and the effectiveness of mechanisms for translating research into commercial applications. Experienced corporate partners provide market insight and commercialisation expertise that complements university research capabilities.

Macquarie University's partnership with Uniseed, one of Australia's oldest university venture funds, provides additional commercialisation support and investment capital for promising research projects. This partnership enables companies to access funding from initial concept through to growth stages.

The University's capability growth in commercialisation is evident through success like Celosia Therapeutics, a biotech spin-out developing advanced gene therapies for neurodegenerative diseases which in 2024 completed one of the largest Series A raises for an Australian biotech start-up, raising $16.75 million to advance its program for amyotrophic lateral sclerosis (ALS) disease.

This landmark financing round was led by Uniseed, Australia's longest-running research commercialisation fund, with a major contribution from hedge fund manager and ALS patient, Mr Iomar Barrett, and support from tertiary education sector superannuation fund UniSuper.

Another spin-out, Apate.ai is partnering with the Commonwealth Bank of Australia (CommBank) to fight against cyber scams by deploying a fleet of thousands of AI-powered bot profiles. This agreement follows a successful pilot program announced delivered in late 2024 and demonstrates how spin-outs can partner with corporates to accelerate their innovations.

Challenges and Strategic Responses

Urban Planning and Development Pressures

MPID's success has created significant development pressures that require careful management to maintain the district's character and functionality. In late 2024 the NSW Government's Macquarie Park state-led rezoning as one of the Transport Oriented Development (TOD) Accelerated Precincts was completed. The TOD rezoning reflects the need to accommodate critical housing growth to meet key housing targets.

The TOD rezoning will enable:

- capacity for 9,600 new built-to-sell homes

- floor space capacity of 3.15 million sqm for a balance of between 100,00 jobs and 10,000 built-to-rent (BTR) homes based on market conditions
- affordable housing contributions of between 3% to 10% for all new residential development
- up to 14 hectares of open space, with new or improved parks and sport fields
- more vibrant streets with new amenities, commercial, office and retail spaces
- better walking and cycling connections to transport and public open spaces

The challenge involves balancing residential development needs with the space requirements of innovation activities. Residential development can enhance district vibrancy and reduce commuting, but it must be planned to complement rather than compete with innovation activities for scarce land resources.

The master planning process led by AJC Architects and Tract Consultants has developed a framework for managing this growth through integrated urban design that creates public spaces, walkable streets, and mixed-use development while maintaining the district's innovation focus.

CMPID continues to advocate for a viable mixed-use outcome that preserves MPID's fundamental role as an innovation and jobs engine along with housing and amenity outcomes.

Talent Competition and Retention

Despite its advantages, MPID faces increasing competition for high-skilled talent from other innovation hubs within Sydney and internationally. The district's relatively suburban character, while offering certain lifestyle advantages, may be less attractive to some talent segments compared to inner city alternatives.

The district's stakeholders have responded by providing enhanced amenities, improving transport connections, and developing more diverse housing options. The goal is to create an "18-hour district" that provides attractive lifestyle options beyond traditional business hours.

Professional development opportunities, including access to university programs and corporate training initiatives, help differentiate the district's value proposition for knowledge workers. The ability to work on cutting-edge projects while accessing continuous learning opportunities appeals to many

Future Opportunities

MPID's growth and success will be factored across four key areas.

Specialist Innovation Infrastructure

The new RNA Research and Pilot Manufacturing Facility will conduct small-scale GMP facility production of messenger RNA (mRNA), plasmid DNA (pDNA) and lipid nanoparticles (LNP) encapsulation at a scale that will enable phase 1-3 clinical trials of new therapeutics. At Macquarie University and operated by Aurora Bioscience it is expected to open in 2026.

The Australian Genome Foundry, at the University, is part of the ARC Centre of Excellence in Synthetic Biology (CoESB), an open-access cutting-edge facility designed to accelerate the design, construction, and testing of synthetic biology solutions using high-throughput robotics and automation.

The district will need to leverage this infrastructure to build more talent and industry density to build the next generation of companies, such as Cochlear.

Commercial

Following the TOD rezoning process there will be a need to ensure MPID can provide the right commercial, innovation and advanced manufacturing growth space for the start-ups, spin-outs and SMEs in a mixed-use environment with residential developments.

Close collaboration between industry and all levels of government across investment attraction, planning levers and infrastructure delivery will be critical to ensure innovation outcomes are achieved.

Governance

As of July 2025, CMPID's independent governance model does not include any Government representation and sponsorship which limits its ability to be strategically coordinate with government. Representation and support from local and state government would result in accelerated outcomes and would align with best practice internationally with more than one-third of international district/science parks being funded predominantly by government.

Transport

Building on the success of Metro CMPID is advocating for an accelerated expansion to connect up the district to the new Western Sydney Airport which will further increase the district's talent catchment and provide direct links to a 24/7 international airport.

Other public transport improvements such as new rapid bus links and active transport infrastructure, delivered as part of the urban renewal, will further enhance the accessibility and attractiveness of the district to industry and talent.

4. Bradfield City and Aerotropolis, Western Sydney

The Bradfield City and Western Sydney Aerotropolis is Australia's most ambitious greenfield urban development project, centred on the new Western Sydney International (Nancy-Bird Walton) Airport. This coordinated development spans 11,200 hectares of rezoned land, with the 114-hectare Bradfield City Centre serving as the metropolitan hub of a new innovation driven economic region.

Context

The Aerotropolis development is a fundamental departure from traditional airport-adjacent planning, designed from inception as an integrated innovation ecosystem anchored by advanced manufacturing, aerospace and defence, agribusiness, and emerging technologies. The project's scale and coordination distinguish it from conventional urban expansion projects across Australia.

Bradfield City, named after John Bradfield, who designed the Sydney Harbour Bridge, is positioned to become Sydney's third city alongside Sydney CBD and Parramatta. The development will deliver 20,000 direct jobs and 10,000 homes while serving as the innovation hub for the broader Aerotropolis region, which targets 200,000 new jobs over the next 20 years. by 2036.

The Western Sydney City Deal, signed in March 2018, established the governance framework for this transformation through a 20-year agreement between the Australian Government, New South Wales Government, and eight local councils. This deal commits over $20 billion in coordinated infrastructure investment, representing one of the largest planned urban developments in Australian history. Since then, additional announcements have pushed this commitment to around $27 billion

The project's strategic positioning provides access to half the world's population within a 14-hour flight radius, designed to leverage the new 24-hour international airport as an economic catalyst for export-oriented industries and global supply chain integration.

Foundation and Early Planning: 2018-2020

The Western Sydney City Deal established the institutional framework for coordinated development, moving beyond previous fragmented approaches to create a unified planning authority. The Deal's 38 specific commitments outlined infrastructure delivery timelines, employment targets, and governance structures essential for managing the project's complexity.

The Western Parkland City Authority was formed under the WPCA Act in 2018 as part of the City Deal commitments. On 1 July 2024 the WPCA became the Bradfield Development Authority with a refocused, tighter remit - to deliver Bradfield City Centre, attract investment to Bradfield and the Aerotropolis, and drive industry growth through the AMRF.

The creation of the Bradfield Development Authority reflected international best practice in large scale urban development, where dedicated agencies with clear mandates and adequate resources have proven essential for coordinating complex, multidecade projects. The Authority was given specific responsibilities for master planning, infrastructure coordination, land development, investment attraction, and placemaking activities within the 114-hectare city centre.

Early planning decisions established the Aerotropolis as comprising ten precincts, each designed around specific industry clusters and infrastructure requirements. The Aerotropolis Core precinct, containing Bradfield City Centre, was positioned as the metropolitan hub, with surrounding precincts dedicated to agribusiness, advanced manufacturing, and airport-related logistics.

The NSW Government's rezoning of 6,500 hectares in 2020 represented one of the largest single land release exercises in NSW history. This rezoning process established planning controls that prioritised employment-generating land uses over traditional residential development, reflecting the project's focus on-job creation rather than housing supply.

During this period, the first major private sector commitments emerged, led by Celestino's $5 billion Sydney Science Park announcement. This 280-hectare development validated private sector confidence in the precinct's potential and demonstrated market appetite for large scale innovation-focused developments.

Infrastructure commitments during this phase included the $5.3 billion Western Sydney International Airport, the $11 billion Sydney Metro Western Sydney Airport rail line, and the $2 billion M12 motorway. These investments provided the essential connectivity infrastructure required to attract business investment and establish the precinct's viability.

Institutional Innovation in City Building

The Bradfield Development Authority is a distinctive institutional model for large scale urban development in Australia, with the BDA granted specific powers and responsibilities for creating Bradfield City Centre from inception through to operational maturity.

The BDA is a government agency, with development capabilities enabled through the Act. BDA's mandate encompasses master planning, infrastructure development, land assembly and release, investment attraction, placemaking, ongoing management, and the establishment of a new innovation ecosystem through the creation and operation of the Advanced Manufacturing Readiness Facility (AMRF).

This comprehensive scope distinguishes it from traditional development authorities that typically focus on single aspects of urban development such as land release or infrastructure provision.

The Authority's approach emphasises integrated development around the central metro station, creating a compact, walkable urban centre designed to support 24-hour activity. This planning philosophy reflects international best practice in transit-oriented development while adapting these principles to Australian conditions and the specific requirements of an innovation-focused city.

The BDA's master planning process prioritised community engagement and stakeholder consultation, ensuring that the city's development reflected both technical requirements and community aspirations. The Authority conducted extensive consultation during master plan development, incorporating feedback from potential residents, businesses, and regional stakeholders.

Infrastructure Development: 2020-2023

The period from 2020 to 2023 witnessed the Bradfield Development Authority's transition from planning to active development management. The Authority coordinated infrastructure delivery across multiple agencies in Bradfield City, while maintaining oversight of the integrated development vision.

Infrastructure delivery has been approached with strong Designing with Country principles and commitments of the Authority. The Authority has established the following targets:

- Net zero for Scope 1 and 2 emissions from BDA corporate activities by 2025
- Achieve net zero emissions by 2025 for Scope 1 and 2 emissions
- Provide on-site renewable energy generation and storage
- Allocate space for last mile logistics, e-vehicle charging and shared resource options
- Install two EV charging stations in all buildings, or one EV charging station per 30 car parking spaces
- Targeting the future inclusion of Scope 3 emissions through the development of a comprehensive net zero roadmap by 2026

The BDA's infrastructure coordination role proved critical during this phase, as the city centre's development required synchronisation between transport infrastructure (managed by Transport for NSW), utilities provision (coordinated with Sydney Water and energy providers), and digital infrastructure (delivered through private partnerships). The Authority served as the central coordination point for these multiple infrastructure streams.

The Bradfield Development Authority pioneered several innovative approaches to development management. The Authority's procurement strategies emphasised design excellence and sustainability outcomes, establishing requirements that exceeded standard development industry practices. The selection of cross laminated timber construction and rammed earth walls for the First Building reflected the Authority's commitment to environmental leadership.

The Authority's land development strategy focused on development-ready sites with coordinated infrastructure provision, reducing risk and complexity for private sector partners. This approach contrasted with traditional greenfield development where infrastructure provision often lags behind development demand.

Master planning activities undertaken by the BDA during this period established the spatial framework for over 2 million square metres of mixed-use development. The Authority's planning approach emphasised flexibility and adaptability, creating development frameworks that could accommodate changing market conditions and industry requirements over the multidecade development timeline.

The BDA's engagement with research institutions during this period established the partnerships that would become central to Bradfield's innovation ecosystem. The Authority's role in facilitating partnerships with its AMRF demonstrated the capability to broker complex relationships between government, industry, and academia.

From Planning to Delivery: 2023-Present

The period since 2023 has demonstrated the Bradfield Development Authority's transition from planning and infrastructure coordination to active city delivery and management. The opening of the First Building in March 2025 marked a critical milestone in the Authority's evolution from development agency to city operator.

The BDA's management of the First Building project exemplified its integrated development approach. The Authority coordinated building design, tenant attraction, and operational management while ensuring alignment with broader city centre objectives. The building's incorporation of the AMRF, commercial tenancies, and public spaces demonstrated the Authority's placemaking capabilities.

The Authority's strategy for the First Building established operational models that will guide future development phases. The BDA's approach emphasised synergies between tenant, facilitating collaboration opportunities, and maintaining the building's role as a catalyst for broader innovation ecosystem development.

The finalisation of the master plan in September 2024 represented a significant achievement in translating the city centre vision into development-ready frameworks. The Master Plan has been approved by the Minister for Planning and Public Spaces. It reflects effective stakeholder engagement and regulatory navigation capabilities.

The Authority's management of Stage 2 enabling works through 2025 established operational capabilities in infrastructure delivery and project management. These works, covering roads, utilities, and digital connectivity, were delivered through the BDA's coordination of multiple contractors and service providers while maintaining quality and schedule requirements.

The BDA's approach to digital infrastructure, delivered in partnership with Superloop, demonstrated innovative thinking about smart city capabilities. The Authority's requirements for high-capacity fibre connectivity and extensive mobile coverage reflected its understanding of technology industry requirements and future city operational needs.

The Authority's management of the Second Building development approval process showcased its regulatory navigation capabilities and stakeholder management skills. The building's design as an innovation incubator reflects the BDA's strategic thinking about ecosystem development and its role in facilitating industry-research collaboration.

Central Park's design and approval process, managed by the Authority, demonstrated its placemaking capabilities and community engagement approach. The 2-hectare civic space will serve as the city's community focal point, reflecting the BDA's understanding of the social infrastructure required for successful city development.

The Authority's management of Superlot 1, the first major private sector land release, established frameworks for ongoing private development partnerships. The BDA's approach to land release emphasised development quality and alignment with city centre objectives rather than simple revenue maximisation.

Governance and Institutional Model

The BDA's governance structure includes a board with expertise spanning urban development, finance, and industry sectors, providing strategic oversight while enabling operational flexibility. This board structure balances commercial development requirements with public sector accountability, creating capability for complex decision-making across multiple stakeholder groups.

The Authority's relationship with the parent Department of Housing, Planning & Infrastructure creates coordination mechanisms for regional scale planning while maintaining focused delivery capability for the city centre. This institutional arrangement enables both strategic alignment and operational effectiveness, addressing challenges that have affected previous large scale development projects.

The BDA's powers include land acquisition and assembly, development control, infrastructure provision, and place management, providing comprehensive authority over city development. These powers enable integrated development approaches that would be difficult to achieve through traditional planning mechanisms or fragmented agency responsibilities.

The Authority's financial model combines government funding for public infrastructure with revenue generation through land sales, development partnerships, and commercial tenancies. This approach enables cross-subsidisation between public good outcomes and commercial returns, supporting development quality and community infrastructure provision.

The BDA's procurement and contracting approaches emphasise design excellence, sustainability outcomes, local economic development, and First Nations design engagement. This the way in which development authorities can drive broader policy objectives through their purchasing power and development requirements.

The Authority's stakeholder engagement model includes regular community consultation, industry advisory groups, and government liaison mechanisms, ensuring that development decisions reflect diverse interests while maintaining project momentum and delivery capability.

Innovation Ecosystem Development and Authority Role

The Aerotropolis has developed as an integrated innovation ecosystem spanning multiple industry sectors and research institutions. The AMRF serves as the central hub, providing shared access to advanced manufacturing technologies and expertise that individual companies could not economically justify independently.

Research partnerships have been established across aerospace and defence, advanced manufacturing, agribusiness, and emerging technologies. The collaboration with the AMRC at the University of Sheffield provides international benchmarking and access to global manufacturing networks, while domestic university partnerships focus on skills development and research commercialisation.

The ecosystem approach recognises that innovation occurs through networks and collaboration rather than isolated facilities. The physical design of Bradfield City Centre prioritises proximity and connectivity, with research facilities, commercial offices, and residential areas integrated within walking distance.

Government investment in shared infrastructure and facilities has de-risked private sector participation and enabled smaller companies to access capabilities typically available only to large corporations. The

AMRF's shared-use model provides access to expensive equipment and expertise on a fee-for-service basis.

Industry cluster development has focused on sectors where Australia has competitive advantages and export potential. The proximity to the 24-hour airport enables participation in global supply chains and access to Asian markets, particularly important for advanced manufacturing and agribusiness sectors.

Investment Attraction and Economic Model

The Aerotropolis development model emphasises job creation and export-oriented industries rather than traditional residential-led development. Government infrastructure investment has provided confidence for private sector participation by establishing essential connectivity and shared facilities before requiring commercial commitments.

The Bradfield Development Authority's Investor Concierge Service provides comprehensive support for businesses considering establishment in the region, including site selection, regulatory guidance, and workforce development assistance. This coordinated approach contrasts with traditional fragmented investment attraction efforts.

Private sector investment has responded positively to government infrastructure commitments and the integrated planning approach. Beyond major planned developments like Sydney Science Park, multiple smaller companies have committed to establishing operations to access the research facilities and supply chain opportunities. At present, there is around $14.8 billion of private development in the planning pipeline for the Aerotropolis

The focus on advanced manufacturing and technology industries reflects strategic positioning to capture higher-value economic activity rather than competing on labour costs. Integrating research facilities and skills development programs supports this positioning by providing access to capabilities and expertise.

Government funding mechanisms have supported early-stage business development and research commercialisation. The AMRF's shared-use model provides access to advanced manufacturing capabilities without requiring individual companies to make large capital investments.

Current Challenges and Development Trajectory

Despite substantial progress, the Aerotropolis faces ongoing challenges in scaling from initial facilities to a fully functioning innovation ecosystem. The transition from government-led infrastructure development to private sector-driven growth requires sustained business confidence and market validation.

Skills development is both an opportunity and a constraint, with the region hosting significant manufacturing expertise but requiring workforce development for advanced manufacturing and emerging technologies. The university partnerships and paid internship programs address this challenge but require ongoing coordination and funding.

The scale of planned job creation requires successful attraction of major employers across multiple industry sectors. While early commitments provide validation, achieving the target of 200,000 new jobs will require sustained economic growth and continued infrastructure investment.

Transport connectivity beyond the immediate airport area will influence the precinct's integration with Greater Sydney. The success of the metro line and potential future transport investments will determine accessibility for workers and residents across the broader region.

Environmental sustainability and climate adaptation present ongoing challenges, particularly managing urban heat and water resources in Western Sydney's climate. Integrating green infrastructure and sustainable design practices will be critical for long-term livability and competitiveness.

Conclusion

The Bradfield Development Authority model demonstrates how dedicated institutional capacity can enable large scale urban development of unprecedented complexity in Australia. The Authority's comprehensive mandate and hybrid governance structure have proven essential for coordinating the multiple dimensions of creating a new city from greenfield sites.

The BDA's approach to master planning, infrastructure coordination, and stakeholder engagement offers a template for other jurisdictions seeking to undertake similar large-scale developments. The Authority's emphasis on integrated development, design excellence, and innovation ecosystem creation distinguishes its work from conventional development authority approaches.

The success of the Bradfield model will depend on the Authority's ability to maintain development quality and strategic vision while scaling from initial facilities to a fully functioning city. Early results demonstrate the viability of the institutional approach, but sustained execution over the multidecade development timeline will determine long-term success.

Lessons for Oher Regions

The Authority's experience provides lessons about the importance of dedicated institutional capacity, comprehensive planning powers, and adequate financial resources for large scale urban development. The BDA model addresses many of the coordination failures and capacity constraints that have affected previous attempts at planned city development in Australia.

The Bradfield Development Authority's integration of economic development, placemaking, and infrastructure delivery within a single institutional framework is a significant innovation in Australian urban governance. This model's potential application to other major development projects across Australia could enhance the nation's capability for managing complex urban development challenges.

The international partnerships facilitated by the Authority, particularly with the University of Sheffield, demonstrate how development agencies can access global expertise and networks to accelerate local capability development. This approach to international engagement offers lessons for other development authorities seeking to build innovation ecosystems.

The Authority's emphasis on advanced manufacturing and technology industries, supported by purpose-built facilities and research partnerships, positions Bradfield to contribute to Australia's economic diversification and competitiveness in sectors where global competition is intensifying.

5. Newcastle Innovation Ecosystem

Newcastle, the second largest city in New South Wales, has emerged as Australia's leading regional innovation hub, with a dynamic ecosystem centred around the University of Newcastle's Integrated Innovation Network (I2N) and supported by a collaborative network of research institutions, incubators, and industry partners.

As home to over 320,000 residents in the greater Newcastle area and serving as the capital of the Hunter region with a population approaching 650,000, Newcastle has successfully leveraged its industrial heritage and research strengths to build a thriving knowledge-based economy.

Context

The city's transformation from a traditional coal and steel centre to an innovation driven economy is one of Australia's most successful regional diversification stories. Newcastle's strategic location between Sydney and Brisbane, combined with world-class universities, established research institutions, and a growing start-up ecosystem, has positioned it as a magnet for entrepreneurs, researchers, and investors seeking alternatives to expensive capital city markets.

The Hunter region's innovation ecosystem encompasses a broad geographic area, including Newcastle, the Central Coast, and surrounding local government areas, with strong connections to defence installations at Williamtown and emerging renewable energy zones across the broader region. This geographic diversity provides unique advantages for start-ups and scale-ups across multiple industry sectors, from defence technology to clean energy and advanced manufacturing.

Between 2016 and 2021, census data demonstrate the region's economic evolution. The proportion of workers in professional, scientific, and technical services increased significantly, while traditional manufacturing employment stabilised around new high-tech and advanced manufacturing activities. Educational attainment has risen substantially, with nearly 40% of the adult population now holding tertiary qualifications, reflecting the city's growing knowledge intensity.

Newcastle's innovation journey has been characterised by strong institutional collaboration between the University of Newcastle, Hunter Medical Research Institute (HMRI), CSIRO, local government, and private sector partners. This collaborative approach has created an integrated ecosystem that supports entrepreneurs at every stage of their journey, from initial concept through to successful exit.

Early Foundations: 2000-2015

Newcastle's innovation ecosystem began taking shape in the early 2000s, building on the city's existing strengths in research and engineering. The Hunter Medical Research Institute (HMRI), established in 1998 as a partnership between the University of Newcastle, Hunter New England Local Health District, and the community, became Australia's largest regional medical research institute. With 1,600 researchers and support staff, HMRI specialises in translational research moving discoveries from laboratory to clinical application, providing the crucial foundation for Newcastle's emerging biotechnology sector.

The University of Newcastle played a crucial role during this foundational period, with its strong engineering, science, and medical research capabilities providing the intellectual foundation for technology transfer and commercialisation activities. Newcastle Innovation, the university's technology transfer arm, was established to facilitate the movement of research outputs into commercial applications.

A pivotal early success was the development of oncolytic virus technology by Associate Professor Darren Shafren at HMRI. Beginning in the early 2000s, Shafren's research into using common cold viruses to target cancer cells laid the groundwork for what would become **Viralytics**, one of Australia's most successful biotechnology companies. The work demonstrated Newcastle's capacity to translate fundamental research into globally competitive commercial applications.

The establishment of the *Newcastle Institute for Energy and Resources (NIER)* in 2010 was a significant institutional development. Built with $30 million from the Australian Government's Education Investment Fund and $2.2 million from NSW Government funding, NIER was constructed on the site of BHP's former Newcastle Technology Centre, symbolically linking Newcastle's industrial heritage with its innovation future. The institute's focus on multidisciplinary research in energy, resources, food, and water sectors provided crucial infrastructure for industry-academia collaboration and pilot-scale technology development.

During this period, the region also began developing capabilities in defence and underwater technologies. *BlueZone Group*, founded in 2000 as ATSA Defence Services by Darren and Gunilla Burrowes, emerged from the Royal Australian Navy's Minehunter Coastal Project. The company represented Newcastle's first significant defence technology start-up, leveraging the region's naval heritage and proximity to defence installations to create innovative underwater technology solutions.

The establishment of *Eighteen04* in 2015 was a significant milestone in Newcastle's evolution as a clean technology hub. As Australia's first dedicated climate tech incubator, Eighteen04 provided workspace, mentorship, and community for sustainability-focused start-ups. The organisation's location in Newcastle's creative precinct symbolised the city's broader transformation from heavy industry to knowledge-intensive sectors.

Eighteen04's early success in attracting and nurturing clean tech start-ups demonstrated the viability of Newcastle as a location for sustainability innovation. The incubator's focus on connecting local entrepreneurs with global networks helped establish Newcastle's reputation beyond traditional economic sectors.

This period also saw the emergence of university-based entrepreneurship programs, with student competitions and early-stage funding mechanisms beginning to create a culture of innovation and commercial thinking among researchers and students. These initiatives laid important groundwork for the more structured ecosystem development that would follow.

Institutional Development and Ecosystem Formation: 2015-2020

The period from 2015 to 2020 was a decisive phase in Newcastle's innovation ecosystem development, characterised by the establishment of formal support structures and the achievement of significant commercial milestones that validated the region's potential as an innovation hub.

The formation of *Hunter iF* (Hunter Innovation Foundation) in 2018 marked a significant step towards creating a unified regional innovation network. Established as a not-for-profit organisation with initial funding from the New South Wales Government's Local Innovation Network grant, Hunter iF connected entrepreneurs, researchers, and businesses across the greater Newcastle and Hunter region.

Hunter iF's development of regular networking events, including the monthly *Innovators Lunch* series, created consistent opportunities for collaboration and knowledge sharing among ecosystem participants. The organisation's focus on connecting regional innovators with opportunities and resources across New South Wales and beyond helped address one of the key challenges facing regional innovation hubs: access to networks and expertise.

The *Hunter Innovation Project* launched during this period with backing from the New South Wales Government and core partners including Newcastle City Council, the University of Newcastle, and Newcastle NOW. The project aimed to deploy smart city infrastructure throughout Newcastle's city centre, creating the digital connectivity and innovation-enabling infrastructure necessary to support technology-intensive businesses.

This initiative was significant for its immediate infrastructure benefits and for demonstrating government commitment to Newcastle's transformation into a knowledge-based economy. The project's focus on creating a digitally connected innovation precinct helped attract businesses and entrepreneurs in the technology and digital-creative sectors.

A watershed moment was in 2018 when *Viralytics* was acquired by Merck for $502 million. It was one of the largest biotechnology transactions in Australian history. The acquisition validated Newcastle's research capabilities on a global scale and demonstrated that world-class companies could emerge from regional Australian innovation ecosystems.

The success of Viralytics had multiple positive effects on the broader ecosystem. Beyond the direct financial returns to investors and researchers, the transaction provided proof-of-concept for the viability of Newcastle-based biotech ventures, attracted attention from national and international investors, and inspired a new generation of researchers to consider commercialisation pathways for their work.

During this period, the University of Newcastle also began developing more sophisticated commercialisation capabilities. The establishment of partnerships with national venture capital firms and the development of structured pathways for research commercialisation helped create the foundations for the more formal accelerator programs that would emerge in subsequent years.

The region's defence and aerospace capabilities were strengthened through continued collaboration with CSIRO and expanded relationships with defence contractors and research organisations. The proximity to Williamtown Aerospace Centre and ongoing defence industry activities provided a stable foundation for technology companies serving government and commercial markets.

MGA Thermal emerged as another significant success story. Spun out from research at the University of Newcastle by Professor Erich Kisi and Dr Alexander Post, the company developed revolutionary thermal energy storage technology using "miscibility gap alloys" that store energy as heat while remaining solid. MGA Thermal's technology addressed a critical challenge in renewable energy storage, offering a cost-effective alternative to battery storage for large scale applications.

The company's early success in raising capital and attracting commercial partnerships demonstrated Newcastle's growing capabilities in deep technology commercialisation beyond biotech and defence sectors. MGA Thermal's development also illustrated the importance of sustained university research investment in creating breakthrough commercial opportunities.

Acceleration and Maturation: 2020-Present

Since 2020, Newcastle's innovation ecosystem has undergone rapid acceleration and maturation, marked by the establishment of sophisticated support infrastructure and the achievement of scale in start-up creation and capital formation.

The formal launch of the *University of Newcastle's Integrated Innovation Network (I2N)* represented a significant escalation in institutional commitment to innovation and entrepreneurship. Anchored by the purpose-built *Q Building* in Honeysuckle, I2N created Australia's most comprehensive regional innovation facility, combining co-working spaces, accelerator programs, maker facilities, and event venues in a single integrated complex.

The *I2N Hub Honeysuckle* facility comprises three levels serving different stages of venture development. The Innovation Common on the ground floor provides free co-working space and innovation concierge services for prestart-ups and early-stage entrepreneurs. Level 2 houses seminar and program delivery spaces, while Level 3 offers dedicated office space for start-ups and scale-ups requiring more sophisticated facilities.

I2N's accelerator programs have become central to Newcastle's start-up creation activities. The **I2N Accelerator** operates three specialised streams: Cleantech, Health tech, and All-Industries, each tailored to specific sectoral needs and opportunities. The programs provide structured 13-week experiences combining mentorship, expert workshops, industry connections, and access to funding networks.

The *I2N Cleantech Accelerator*, developed in partnership with the Trailblazer for Recycling and Clean Energy (TRaCE), has emerged as a nationally significant program for sustainability-focused start-ups. The accelerator's focus on clean energy, water management, recycling, supply chain optimisation, and sustainable manufacturing aligns with Newcastle's broader positioning as a clean technology hub.

Recent cohorts from I2N accelerators have demonstrated the program's effectiveness in creating viable businesses. Companies such as *Wild Yeast Zoo*, which has foraged over 500 wild yeasts for applications in biofuels and pharmaceuticals, and *Cloud Grow*, which provides intelligent irrigation solutions for urban green infrastructure, exemplify the innovative solutions emerging from the ecosystem.

Hone Agriculture is also a significant success story from this period. Founded by PhD students Antony Martin, William Palmer, and Jamie Flynn, the company developed handheld testing devices that revolutionise agricultural analysis by providing instant chemical property assessments of soil, crops, and grain samples. Hone's growth from academic research to a commercial company with team members across multiple cities demonstrates the ecosystem's capacity to support scaling ventures.

Newcastle's position as a renewable energy hub has been reinforced by the region's achievement of 100% renewable electricity supply and the development of renewable energy zones across the broader Hunter region. This infrastructure advantage has attracted clean technology companies and provided real-world testing environments for energy innovation.

The establishment of formal venture capital partnerships has addressed one of the traditional challenges facing regional innovation ecosystems. The University of Newcastle's partnership with *Uniseed*, Australia's longest-running university venture fund, provides structured pathways for research commercialisation and start-up funding.

Through Uniseed, Newcastle-based ventures gain access to a network spanning eight major Australian universities and CSIRO, collectively responsible for over 60% of Australia's research expenditure. This partnership significantly expands the funding and collaboration opportunities available to local start-ups while maintaining Newcastle's regional advantages.

The region has also developed strength in agritech and advanced manufacturing. Companies such as *Diffuse Energy*, which develops small-scale wind turbines for distributed energy generation, demonstrate Newcastle's capacity to create solutions addressing both regional and global challenges.

The Port of Newcastle is pursuing an ambitious $2.4 billion diversification strategy to reduce coal dependency from 72% to 50% of revenue by 2030. Key projects include a deepwater container terminal, clean energy precinct for hydrogen and ammonia production, and renewable electricity transmission infrastructure, supporting regional economic resilience.

The Melt Accelerator is Australia's first integrated industrial prototyping lab and hardware accelerator, based in Warners Bay near Newcastle. It provides programs, investment, and facilities to help clean technology and advanced manufacturing start-ups scale from idea to global markets using their proven 4P process: prove, prototype, pilot, and produce.

The *Future Industries Facility* was recently awarded $20.7 million through the Federal Government's regional Precincts and Partnerships Program. This University of Newcastle facility will provide two industrial-scale innovation spaces and accommodate 16 academic-industry partnerships for rapid technology development across energy, resources, and manufacturing sectors, contributing an estimated 740 direct and indirect jobs.

Energy Industry Jobs Plan (NZEA): The Energy Industry Jobs Plan from the Net Zero Economy Authority provides structured support to workers when coal or gas-fired power stations close. It includes career planning, education and training, financial advice, paid time off for interviews, and connections with new employers, ensuring coordinated transition assistance during the energy sector transformation.

Current Challenges and Future Opportunities

Despite its substantial progress, Newcastle's innovation ecosystem faces several challenges that require continued attention and strategic intervention. Access to growth-stage capital remains a constraint, with successful start-ups often requiring investment levels that exceed regional capacity. While early-stage funding has improved significantly through programs like I2N and partnerships with national venture capital, the transition from start-up to scale-up continues to present challenges.

Talent acquisition and retention represent ongoing concerns, particularly for deep technology companies requiring specialised skills. Newcastle's relatively small professional population and competition from Sydney and Melbourne-based opportunities can limit the talent pool for scaling companies. However, the region's lifestyle advantages and lower costs provide competitive benefits that many companies leverage.

The need for continued diversification beyond traditional strength areas remains important. While defence, clean energy, and biotechnology have proven successful, developing capabilities in areas such as advanced manufacturing, space technology, and digital services will provide additional growth opportunities and economic resilience.

International connectivity is both a challenge and an opportunity. Newcastle's international airport provides direct access to regional markets, but expanding global connections will support the ecosystem's continued development. The region's proximity to Sydney provides access to international markets while maintaining cost advantages.

The development of Newcastle's *innovation corridor* presents significant future opportunities. The geographic concentration of the University of Newcastle, HMRI, CSIRO facilities, I2N hubs, and supporting infrastructure creates the potential for a world-class innovation district. Strategic planning and continued infrastructure investment could enhance this natural clustering and attract additional research and commercial activities.

The region's renewable energy infrastructure provides unique opportunities for companies developing clean technology solutions. The combination of renewable energy generation, storage, and transmission infrastructure creates real-world testing environments that are increasingly attractive to global clean technology companies.

Newcastle's defence and aerospace capabilities, anchored by Williamtown Aerospace Centre and supported by university research strengths, provide opportunities for continued growth in these strategically important sectors. The development of space industry capabilities, building on existing aerospace expertise, is a promising growth area.

Conclusion

Newcastle's transformation into Australia's leading regional innovation hub demonstrates the potential for strategic, collaborative ecosystem development to drive economic diversification and growth. The city's success in leveraging institutional strengths while building new capabilities provides valuable lessons for other regions seeking to develop knowledge-based economies.

The integrated approach adopted in Newcastle, combining university research excellence, government support, private sector engagement, and community collaboration, has created a sustainable foundation for continued innovation and entrepreneurship. The ecosystem's ability to support companies from initial concept through to successful exit, as demonstrated by Viralytics and other success stories, validates the model's effectiveness.

Newcastle's focus on developing distinctive sectoral strengths in defence, clean technology, biotechnology, and advanced manufacturing has enabled the region to compete effectively with larger urban centres while maintaining its unique advantages. This strategic specialisation, combined with general innovation support infrastructure, provides a template for other regions seeking to build competitive innovation ecosystems.

The ecosystem's emphasis on collaboration and knowledge sharing, exemplified by organisations such as Hunter iF and the I2N network, has created a supportive environment that encourages entrepreneurship and facilitates connections for start-up success. These soft infrastructure elements are often as important as formal programs and facilities in determining ecosystem vitality.

Lessons for Other Regions

Newcastle's experience demonstrates that regional innovation hubs can achieve significant scale and impact when supported by appropriate institutional frameworks, funding mechanisms, and collaborative partnerships. The city's continued evolution towards becoming a globally recognised innovation destination reflects the cumulative effect of sustained investment and strategic development over two decades.

For other regions seeking to develop innovation capabilities, Newcastle's experience highlights the importance of building on existing strengths while remaining open to new opportunities. The combination of patient capital, institutional support, collaborative culture, and geographic advantages has enabled Newcastle to punch above its weight in the national innovation landscape.

The region's continued commitment to infrastructure development, talent development, and international connectivity will be crucial for maintaining momentum and achieving the next phase of growth. Newcastle's innovation ecosystem provides a compelling example of how midsized cities can create world-class innovation environments that support both local prosperity and global impact.

As Newcastle looks towards the future, the foundation established over the past two decades provides a robust platform for continued growth and diversification. The city's experience proves that with strategy, investment, and collaboration, regional centres can build innovation ecosystems that rival those of major metropolitan areas while offering unique advantages for entrepreneurs, researchers, and investors.

6. UNE Smart Farms and Incubator

The University of New England (UNE) SMART Farms is one of Australia's most significant examples of how regional universities can lead agricultural innovation and technology development. Located in Armidale, New South Wales, this world first integrated agricultural research facility has transformed from a traditional university farm operation into a globally recognised centre for precision agriculture, sustainable farming practices, and agricultural technology development.

Context

Spanning 3,655 hectares across eight properties, the UNE SMART Farms are both commercial agricultural operations and cutting-edge research facilities, supporting disciplines including natural and agricultural ecosystems, poultry, livestock and canine sciences, precision agriculture and zoology. The SMART Farms provide the context for problem-based experiential learning for undergraduates and research and community activities with most of the Farms within a five-minute drive of the main UNE campus.

Tullimbah Feedlot at Kingstown is a major site for methane work. Clarks Farm, which is opposite the UNE campus, was backed through drought funding to be a demonstrator of climate resilience. There has also been significant pasture development work with DLF Seeds on other plots of land close to campus and in glasshouses on campus. Also, on campus at the livestock facility, UNE has the largest number of research methane chambers in the world.

The development of the SMART Farms ecosystem demonstrates how regional institutions can leverage their existing strengths in agricultural education and research to create innovation platforms that benefit both academic research and commercial agricultural practice. Over the past two decades, the facility has evolved from a traditional teaching farm into a sophisticated "landscape laboratory" that showcases the integration of technology, science and farming practice.

The SMART Farms approach combines commercial viability with research excellence, ensuring that innovations developed on-site are tested under real-world conditions. This model has attracted significant industry collaboration, government support, and international recognition, positioning the New England region as a leader in agricultural technology development.

Early Foundations: Building Research Capabilities (1990s-2007)

The origins of the UNE SMART Farms can be traced to the university's long-standing commitment to agricultural education and research. UNE had maintained working farms for teaching and research purposes for decades, but the transformation into a technology-focused innovation platform began in the late 1990s when precision agriculture emerged as a distinct discipline.

Professor David Lamb, a physicist whose research interests include applied optics and precision agriculture, has worked in precision agriculture for more than 25 years and has led more than 40 industry-funded R&D projects. After completing his Bachelor of Science, David did Honours and then his PhD at UNE. He left for CSU, a plasma physics graduate, "big on arcs and sparks and lightning", just when the US military was making GPS technology available to the public.

At CSU, David got hooked on precision agriculture, and many opportunities soon opened up. "It was like the wild west, and I was blessed because I was at ground zero," David said. "When I returned to UNE eight years later, the notion of smart farming and the NBN was just firing up".

The early 2000s saw UNE systematically building its research capabilities in precision agriculture. David established the University of New England's Precision Agriculture Research Group, and the internationally renowned SMART Farm (an acronym which stood for smart, measurable, accessible, rural technologies) project. UNE offers Australia's only Graduate Certificate and Master of Science degree dedicated to precision agriculture, establishing the institution as the national leader in formal precision agriculture education.

This period was characterised by the development of fundamental research capabilities and the attraction of industry funding for applied research projects. The university's focus on precision agriculture aligned with broader national trends towards technology adoption in farming, driven by the need for increased productivity and environmental sustainability.

Establishing the SMART Farms Concept (2007-2015)

David led the launch of Australia's first Smart Farm at the University of New England, marking a significant milestone in agricultural innovation.

The SMART Farm was conceived as a world first "landscape laboratory" that would integrate commercial farming operations with advanced research and technology demonstration.

Using precision ag was dominated by crop farmers until then – yield monitors on harvesters, soil survey technologies and satellite remote sensing of crops. UNE kicked off precision livestock management, including sensors for estimating pasture biomass, GPS livestock tracking and virtual fencing.

The farm's design incorporated multiple scales of operation, from individual paddocks to entire farm systems, allowing researchers to test technologies and management practices across different contexts. This approach recognised that agricultural innovation needed to be tested under real commercial conditions to ensure practical applicability.

During this formative period, the SMART Farm began establishing partnerships with industry and government organisations. An array of multidisciplinary research is being undertaken across the Farms with a range of partners and collaborators, including projects on pasture monitoring, livestock tracking, and environmental sensing.

The farm's commercial operations remained central to its research mission. "It is a real living and breathing farm with all the same problems that farmers have," says David. "We try not to get in the way of the farming, because it is got to be a real farm. But the beauty of it is we get wonderful feedback and wonderful performance metrics that are real-world based".

Key research areas during this period included remote sensing technologies, precision livestock management, and the development of sensor networks for environmental monitoring. The farm became a testing ground for emerging technologies including unmanned aerial vehicles (UAVs), satellite-based monitoring systems, and automated livestock monitoring equipment.

Innovation Centre and Technology Integration (2015-2020)

The SMART Farms are home to the SMART Farm Innovation Centre (SFIC), launched in 2015. The SFIC is a research, education and outreach facility built to showcase the integration of technology, science and farming.

The $2 million centre opened in March 2015 and operates as a technological and educational hub for UNE. Up to 200 students a year pass through the centre, and not just those studying agriculture: students from science, mathematics, business and education degrees are visiting the centre to learn about the technology being used.

The SFIC, located between Kirby and Newholme, is the physical hub of a heavily connected landscape with a network of on-farm connected devices. On-farm connectivity comprises the latest and emerging methods, including store-and-forward and mesh telemetry, Long Range Wide Area Network (LoRA WAN), mobile phone network and satellite-direct.

The facility is linked via AARNet, NBN (fibre, terrestrial wireless and satellite) and a farm-wide telemetry architecture facilitating a landscape laboratory. SMART Farms is a national demonstrator site showcasing the latest on-farm technologies.

The Innovation Centre transformed the SMART Farms from a research facility into a comprehensive technology demonstration and education platform. Networked tree sensors track carbon storage levels above and below ground. Motion-triggered cameras provide a camera trapping grid used to monitor wildlife. Virtual fencing & ear-tag tracking involve R&D on controlling and monitoring livestock movements, behaviour and health via virtual, real-time systems.

Research projects during this period expanded significantly in scope and complexity. Areas of research currently being undertaken on the UNE SMART Farms include the National Resource Flock Project, on-animal sensors for monitoring animal health and welfare, UAV automated surveillance of in-field hotspots for improved management, and low-cost remote water tank monitoring using satellites to directly access sensor data.

Despite the research and education focus, the SMART Farms still operate as a commercial farm: to do otherwise would jeopardise the practicality of the research. The lambs and cattle produced are sold to abattoirs and local saleyards, while the wool is sold through the Sydney wool markets.

The period also saw increased industry collaboration and government recognition. The UNE SMART Farm, which integrated cropping and livestock advances, was a world first under David's leadership of the Precision Agricultural Research Group and earned him the Vice-Chancellor's Award for Research Excellence in 2007.

Ecosystem Development and Regional Impact (2018-Present)

The success of the SMART Farms led to broader ecosystem development initiatives in the New England North West region. The UNE SMART Region Incubator (UNE SRI) working across the region and has physical locations in Moree, Tamworth, Armidale and Narrabri. The UNE SRI has focused on benefits to the regional community, values diversity of ideas, and is open and inclusive.

UNE SMART Region Incubator founding Director, Professor Lou Conway developed entrepreneurship programs and attracted over $10m in funding during the first eight years of delivery. A focus on the development of a community of founders in the region to bring to market solutions for agriculture, health, the environment and creative industries ensured a broad representation of businesses and an inclusive foundation. With 51% of UNE SRI founders women, the incubator's success in supporting diverse entrepreneurship can be demonstrated.

The SMART Region Incubator received significant government support. The fit-out of NOVA in the old Armidale Library building on Faulkner Street was funded by a $900,000 commitment by UNE and supported by a peppercorn lease arrangement for the building with Armidale Regional Council. With support across all levels of Government, the UNE SRI was able to secure a strong base for growth. Since inception the UNE SRI captured founder data which showed that over $30m in business investment was realised and over 300 jobs created.

In terms of support for agricultural technology-focused start-ups, early funding support of $320,000 from the Federal Government for the *AgTech Gateway program* established the roadmap for success of AgTech start-ups in the region. The well-known annual hackathon Agmentation, held on-site at the UNE SMART Farm Innovation Centre was the breeding ground for launching a number of new start-ups.

The AgTech Gateway program offers tailored, one-on-one support to AgTech start-ups and scale-ups in the New England North West region of New South Wales. The SMART Farms provide an ideal environment for innovation in areas including: spatially-enabled livestock management; telecommunication systems and Internet of Things (IoT); earth observation systems monitoring feed base and land use sustainability; datasets of soil, land types, pasture systems, elevation, weather and livestock health; remote & automated monitoring systems;

and data integration to assess the effectiveness of management decisions and optimise planning of new infrastructure.

Research partnerships have expanded to include major industry collaborators such as Myriota Global (MG) a nanosatellite and IoT connectivity solutions developer. MG has developed an ultra-low-cost remote IoT connectivity platform, demonstrating the farm's role in testing emerging commercial technologies.

Professor David Lamb's leadership in the field gained international recognition. In 2022, the Pierre C. Robert Precision Agriculture Award from the International Society of Precision Agriculture made David the first Australian to receive the honour. David was appointed Chief Scientist of the Food Agility CRC, extending UNE's influence in national agricultural innovation policy.

UNE SMART Farms was a founding member of the nascent Global Smart Farms Network which was launched at the Banff Innovation Summit, Canada 2024. The global network is now slowly expanding, and UNE SMART Farms continues to mentor other like Australian entities including the Charles Sturt University Global Digital Farm and the Longerenong College Data Farm.

UNE SMART Farms is going through a period of further upgrade and investment thanks to the tireless efforts of Justin Bailey (SMART Farms Director) and Lou Hogan (Smart Farms industry engagement manager).

Current Research Focus and Technology Development

Today's SMART Farms operates as a comprehensive agricultural technology testing and development platform. A collaboration across seven universities and 15 industry and AgTech companies, the SMART Farm Learning Hub provides learning modules for secondary and tertiary students that use real farm data, including from the UNE SMART Farms.

Tullimba Feedlot uses GrowSafe feeders to monitor individual feed intake and allow calculation of breeding values for feed efficiency in beef cattle. GreenFeed units allow measurement of methane emissions from individual animals, demonstrating the farm's focus on sustainability and climate change mitigation.

The facility's research infrastructure continues to expand. There are several facilities suitable for large scale poultry studies on UNE SMART

Farms. At Kirby, there are three poultry sheds. Two of these are older-style sheds with no temperature control, and these are suitable for layer floor pen work, supporting diverse research across multiple agricultural sectors.

Water resource management is a significant research focus. Dr Debbie Bower is testing whether farm dam management that benefits production through cleaner water has dual benefits for the frogs, birds, reptiles and bugs that also live there as part of a $425,469 Australian Research Council grant.

Challenges and Future Directions

Despite its successes, the UNE SMART Farms ecosystem faces several ongoing challenges. The regional location, while providing authentic agricultural conditions, limits access to some technology companies and investors who remain concentrated in major cities. Maintaining the balance between commercial viability and research objectives requires constant attention to ensure the farm remains representative of industry conditions.

The rapid pace of technological change in agriculture presents both opportunities and challenges. New technologies such as artificial intelligence, machine learning, and autonomous systems require significant investment in infrastructure and expertise. The farm must continuously evolve its capabilities to remain at the forefront of agricultural innovation.

Funding sustainability remains a constant challenge. While the facility has successfully attracted industry and government support, the long-term nature of agricultural research requires sustained investment commitments that can be difficult to secure in rapidly changing policy environments.

Integrating climate change adaptation and mitigation strategies is a growing focus area. The region is responsible for about 20% of New South Wales's agricultural produce and plays a significant role in exports. These include cotton, wool, cattle and crops such as barley, wheat and sorghum, making climate resilience research increasingly critical.

Workforce development presents both challenges and opportunities. While the region produces highly skilled agricultural graduates, retaining talent in regional areas requires continued investment in career opportunities and research infrastructure. The SMART Farms role in demonstrating career pathways in agricultural technology becomes increasingly important.

Broader Regional Impact and Innovation Ecosystem

The SMART Farms success has contributed to broader regional development initiatives. Industry growth is supported by the region's higher education, agricultural technology expertise and ICT investments, including a SMART Region Incubator, breeding and genetics research organisations, and Cooperative Research Centres.

The SMART Region Incubator through its distributed model of support with Regional Connectors on-site at Armidale and Tamworth continues to support agribusiness innovation.

The facility's educational impact extends beyond traditional university boundaries. "One of the key challenges we have got in agriculture and food production systems is a lack of interest from the very brightest students in this sector," says Mark. "And what we find when we open up some of these technologies to some of the very brightest students is that they are really engaged".

Regional economic development benefits from the SMART Farms research activities. Continued agricultural innovation and technology, led by R&D facilities in Narrabri, Glen Innes, Tamworth and Armidale, offer the potential to diversify and grow broadacre agriculture and grazing that will make the region a leader in 21st-century agricultural production.

Conclusion

The UNE SMART Farms is a successful model for how regional universities can lead agricultural innovation while maintaining strong connections to industry and community needs. The facility's evolution from a traditional teaching farm to a world-recognised agricultural technology centre demonstrates the potential for regional institutions to compete globally in specialised research areas.

Key factors in the SMART Farms success include strong leadership in precision agriculture research, sustained commitment to maintaining commercial farming operations alongside research activities, strategic investment in technology infrastructure, and the development of comprehensive industry partnerships. The facility's emphasis on real-world testing and validation has enhanced the credibility and applicability of its research outcomes.

The SMART Farms integration with broader regional development initiatives, particularly through the SMART Region Incubator, illustrates how research institutions can serve as catalysts for innovation ecosystems that extend beyond university boundaries. This approach has created employment opportunities, supported start-up development, and attracted investment to the region.

Lessons for Other Regions

Other regions seeking to develop agricultural innovation capabilities can draw several lessons from the UNE SMART Farms experience. The importance of combining research excellence with commercial viability ensures that innovations developed are practically applicable. Strong industry partnerships provide both funding and real-world validation opportunities. Investment in technology infrastructure creates demonstration capabilities that attract further collaboration and investment.

The SMART Farms success in precision livestock management and sustainable agriculture practices demonstrates the potential for Australian agricultural research to lead global innovation in these critical areas. As agriculture faces increasing pressure to improve productivity while reducing environmental impact, the research and technology development conducted at facilities like the UNE SMART Farms becomes increasingly valuable.

The facility's emphasis on interdisciplinary collaboration, bringing together expertise from physics, agriculture, engineering, and business, provides a model for addressing complex agricultural challenges that require diverse perspectives and skills. This approach has enhanced the quality and applicability of research outcomes while providing students with exposure to real-world problem-solving approaches.

The UNE SMART Farms continued evolution will depend on maintaining its balance between fundamental research and practical application, adapting to emerging technologies and agricultural challenges, and sustaining the partnerships that have been central to its success. The facility's role in addressing climate change, food security, and sustainable agriculture positions it to remain relevant and influential in global agricultural innovation.

The UNE SMART Farms experience demonstrates that regional Australia can compete successfully in high-technology industries when supported by investment, leadership, and strategic vision. This success provides a template for other regional centres seeking to develop innovation capabilities that build on existing industry strengths while embracing emerging technologies and practices.

7. Parkville Biomedical Precinct, Melbourne

The Parkville Biomedical Precinct is one of Australia's most significant knowledge-intensive urban developments, encompassing the southern hemisphere's largest biomedical precinct alongside cutting-edge engineering and creative innovation facilities. Located immediately north of Melbourne's central business district, this 2.5-square-kilometre district spans from East Melbourne through Parkville and Carlton to Royal Park, creating an unparalleled concentration of research institutions, hospitals, universities, and technology companies.

Context

The Parkville Precinct comprises over 40 hospitals, medical research institutes, biotechnology organisations and universities, employing 49,000 people and educating over 7,000 biomedical, health and medical students each year, while attracting more annual competitive biomedical research funding than any other area in Australia.

When combined with the adjacent innovation infrastructure, including Melbourne Connect, RMIT University's city campus, and the Parkville area is a globally significant innovation ecosystem.

The district's evolution from a collection of separate institutions to an integrated innovation ecosystem reflects Melbourne's strategic approach to knowledge-based economic development. Unlike traditional business districts that developed organically around commercial activity, the Parkville Precinct has been deliberately cultivated through coordinated investment in research infrastructure, innovation facilities, and ecosystem building initiatives spanning multiple decades.

This transformation has positioned Melbourne as a leading global centre for biomedical research and innovation, while establishing new models for university-industry collaboration and research commercialisation. The district's success demonstrates how strategic precinct development can leverage existing institutional strengths to create dynamic innovation hubs that drive both economic growth and societal impact.

Historical Foundations: Building Australia's Biomedical Centre (1850s-2000)

The foundations of the Parkville Biomedical Precinct were laid in the 19th century with the establishment of Melbourne's premier educational and healthcare institutions. The University of Melbourne was founded in 1853 on a large area north of Grattan Street in Parkville, just north of Melbourne's central business district. This strategic location, close to the city centre yet with sufficient space for expansion, would prove crucial to the district's later development.

The medical and healthcare cluster began forming in the early 20th century as hospitals and research institutions established themselves in proximity to the university. The Royal Melbourne Hospital, Royal Women's Hospital, and other major healthcare facilities created a natural concentration of medical expertise and infrastructure.

The Parkville Biomedical Precinct has touched the lives of millions of people around the world with the class of clinical care its hospitals provide, the outstanding education its universities deliver and the world-renowned research of its institutions.

Throughout the mid-20th century, this concentration deepened as research institutes and specialty medical facilities co-located to take advantage of shared expertise and infrastructure. The establishment of research centres like the Walter and Eliza Hall Institute of Medical Research, the Murdoch Children's Research Institute, and the Melbourne Research Institute created a critical mass of biomedical research capability.

The Parkville Biomedical Precinct consistently attracts around 23% of NHMRC funding annually, more than any other state in Australia, and has a publication citation rate twice the international average. This research excellence, combined with the co-location of clinical care and education, established the foundation for what would become a globally recognised innovation ecosystem.

By 2000, the area had developed into Australia's premier biomedical cluster but remained primarily focused on research and clinical care rather than commercialisation and innovation. The transformation into a comprehensive innovation district would require deliberate intervention and strategic investment over the following two decades.

The Innovation Turn: Strategic Development (2000-2015)

The early 2000s marked a strategic shift towards innovation-focused development in the Parkville area. This period saw coordinated efforts by government, universities, and industry to transform the existing research cluster into a comprehensive innovation ecosystem capable of translating research into commercial outcomes.

The strategic transformation was enabled by the Victorian Government's Science, Technology and Innovation (STI) initiative, launched in 1999 by Premier Steve Bracks and Treasurer John Brumby. Backed by over $620 million, this decade-long program supported biomedical research alongside other technology sectors across Victoria.

Central to the biomedical strategy was the creation of the Bio21 Cluster in 1999, which evolved into Bio21 Australia Limited in 2001 under the leadership of Chairman David Penington. This initiative directly resulted in the establishment of the Bio21 Institute in 2002, with John Brumby's approval of the precinct's five-year business plan marking a pivotal moment in the development of the biomedical cluster.

The Bio21 Institute's David Penington Building, officially launched in 2005, became the physical manifestation of this strategic vision for university-industry collaboration.

The University of Melbourne played a central role in driving this transformation, recognising that its research excellence needed to be coupled with stronger industry engagement and commercialisation capabilities. The university's expansion into Carlton, particularly through acquisitions of strategic sites, demonstrated a commitment to creating integrated innovation facilities that could bridge academic research and commercial application.

During this period, the university established key partnerships with industry and government to create shared innovation infrastructure. The development of research collaboration agreements with major pharmaceutical companies and biotechnology firms brought industry expertise and investment directly into the precinct, creating new pathways for research commercialisation.

The Victorian Government's Melbourne Biomedical Strategic Plan provided a framework for coordinated development across the precinct, aligning infrastructure investment, research priorities, and commercialisation objectives. This strategic approach recognised that individual institutions, no matter how excellent, could not create a comprehensive innovation ecosystem without deliberate coordination and shared vision.

RMIT University's presence in the adjacent city area added complementary capabilities in engineering, technology, and design, creating opportunities for interdisciplinary collaboration that would become increasingly important as the innovation district matured. RMIT's Activator centre for entrepreneurship was designed to support aspiring founders, start-ups and organisations driven to create meaningful change through entrepreneurship.

By 2015, the groundwork had been laid for the comprehensive innovation district that would emerge over the following decade, with strategic investments in infrastructure, partnerships, and ecosystem development beginning to create visible results in terms of start-up formation, industry collaboration, and research commercialisation.

Infrastructure Development and Ecosystem Building (2015-2020)

The period from 2015 to 2020 saw unprecedented infrastructure development that transformed the Parkville area into a truly integrated innovation district. The centrepiece of this transformation was Melbourne Connect, a world-class innovation precinct developed by the University of Melbourne on the site of the former Royal Women's Hospital.

The University acquired the 8,000-square-metre site in 2012 and announced plans for Melbourne Connect (then known as the Carlton Connect Initiative) in 2014. The project embodies a forward-looking University of Melbourne initiative, developed in partnership with Lendlease, to create an innovation precinct centred on collaboration through the co-location of the University's engineering faculty and industry partners.

Melbourne Connect represented a new model for university-industry collaboration, designed around the concept of co-location and shared facilities. Behind the precinct's façade is a 'superfloor' that provides meeting, theatre, and workshop spaces, and connects three separate buildings: a five-storey cross laminated timber building, a 10-storey building housing the Melbourne University School of Engineering and commercial spaces, and a 13-storey student accommodation building.

The facility incorporated several innovative elements that would become models for other innovation districts globally. The Science Gallery Melbourne, Australia's first Science Gallery, created a new form of public engagement with research and innovation. The Creator Space provided access to advanced manufacturing and prototyping facilities for both students and industry partners.

Melbourne Connect provides access to world-class amenities, including high-speed internet, meeting rooms, conference facilities, and on-site cafés, creating an ideal hub for networking and growth.

Parallel to Melbourne Connect's development, the broader *Melbourne Innovation Districts* (MID) initiative was launched as a partnership between the City of Melbourne, the University of Melbourne, and RMIT University. The initiative is set to drive investment in the knowledge economy and help shape the city's future, developing a world-class urban district that supports and develops next-generation Melbourne through emerging technologies and innovation.

This period also saw significant investment in research infrastructure across the broader biomedical precinct. The Victorian Government's commitment to major hospital redevelopments, combined with new research institutes and expanded university facilities, created a comprehensive ecosystem of innovation infrastructure.

The precinct's maturation has been supported by unprecedented investment levels. Since 2014, the Victorian Government has invested more than $1.3 billion in medical research alone, while total capital investment in research and healthcare facilities across the Melbourne Biomedical Precinct exceeded $3 billion in the decade leading up to 2022. This investment wave accelerated significantly under the Andrews Government, culminating in the announcement of a $5-6 billion expansion of hospitals and research facilities in the district.

The infrastructure developments were designed as individual buildings *and* as components of an integrated ecosystem. Deliberate attention was paid to connectivity—both physical and programmatic—between different facilities and institutions.

The approach reflects international best practices in innovation district development and recognition that successful innovation requires networks and collaborations rather than isolated centres of excellence.

Research Commercialisation and Industry Engagement (2020-Present)

Since 2020, the Parkville Biomedical Precinct has demonstrated its maturity as an innovation ecosystem through accelerated research commercialisation, major industry partnerships, and the emergence of a vibrant start-up community. This period has been characterised by the translation of decades of research investment into tangible commercial outcomes and global industry engagement.

The biomedical sector has led this commercialisation wave, building on the precinct's established research strengths. The Parkville Biomedical Precinct's five clinical research strengths—innovative diseases, neurosciences including mental health, cancer, child health, and healthy ageing—represent areas where it will generate the strongest health and economic impact over the next decade. These research strengths have attracted major international pharmaceutical and biotechnology companies to establish operations within the district.

Notable industry partnerships have included CSL's decision to locate its new global headquarters within the precinct, which will bring biotech incubator space for up to 40 start-ups. This development exemplifies how established industry leaders are utilising the district for their own operations and as platforms for ecosystem development and start-up support.

The COVID-19 pandemic demonstrated the global significance of the district's research capabilities through rapid scientific breakthroughs that shaped the international response. Scientists at the Peter Doherty Institute for Infection and Immunity became the first researchers outside China to successfully isolate and culture SARS-CoV-2, sharing this critical achievement with the World Health Organization and laboratories worldwide.

This breakthrough enabled the development of diagnostic tests, treatments, and vaccines globally, demonstrating how the precinct's concentrated research infrastructure could deliver outcomes of immediate international impact during health emergencies.

Melbourne Connect has emerged as a central hub for university-industry collaboration and the development of start-ups. The facility hosts a growing number of technology companies, research institutes, and innovation programs that utilise the co-location model to accelerate the translation of research.

The entrepreneurship ecosystem has been strengthened through programs across multiple institutions. RMIT's Activator has launched over 170 start-ups into market through its programs, with 49% founded or co-founded by females, and has awarded over $3.1 million in pre-seed funding to start-ups through Activator Capital Fund and Jacques Nasser Fund for Innovation.

The district's success in research commercialisation reflects not just the quality of its research base but the deliberate development of innovation infrastructure and support systems. The co-location of research institutes, industry partners, start-up incubators, and support services has created an ecosystem where ideas can move efficiently from laboratory to market.

International recognition has followed these developments, with the district increasingly viewed as a global hub for biomedical innovation and a model for knowledge-intensive urban development. Major international pharmaceutical companies, research institutions, and government agencies now see the Parkville Biomedical Precinct as an essential partner for accessing Australian research capabilities and Asia-Pacific markets.

Institutional Architecture and Governance

The success of the Parkville Biomedical Precinct reflects an institutional architecture that strikes a balance between institutional autonomy and coordinated development. Unlike innovation districts managed by a single entity, this district operates through a network of partnerships and collaborative agreements that leverage the strengths of multiple institutions while maintaining their distinct identities and missions.

The University of Melbourne serves as a central anchor institution, but its approach has been to create platforms for collaboration rather than to dominate district development. Melbourne Connect exemplifies this approach, functioning as a shared innovation infrastructure that serves the university's research and education missions while providing a platform for industry engagement and start-up development.

The close proximity of the Precinct Partners has been key to building a culture of collaboration over many years. This culture extends beyond formal partnerships to include informal networks, shared events, and cross-institutional initiatives that strengthen ecosystem connectivity. This is reflected in the *Melbourne Innovation District* initiative, which embodies a broader governance framework that coordinates development across multiple institutions and levels of government.

This MID governance model recognises that successful innovation districts require both institutional leadership and ecosystem-level coordination. Individual institutions provide the foundational capabilities and resources, while collaborative frameworks enable ecosystem-level initiatives that no single institution could achieve on its own.

The Victorian Government has played a crucial strategic role, providing funding and policy frameworks that support innovation district development. Parkville is a designated Priority Precinct and National Employment and Innovation Cluster, recognised globally as a leading biomedical and research precinct, walking distance from Melbourne's central business district.

The private sector has been engaged as a customer for university research and as a partner in ecosystem development. Major companies have made strategic investments in district infrastructure and programming, recognising that their own innovation capabilities are enhanced by participation in a broader ecosystem.

Economic Impact and Industry Development

The economic impact of the Parkville Biomedical Precinct extends far beyond its geographical boundaries, representing a significant driver of Victoria's knowledge economy and Australia's position in global innovation networks. The district's economic contribution reflects both direct employment and economic activity within the precinct and the broader spillover effects of innovation and industry development.

The Parkville Biomedical Precinct employs 49,000 people and educates over 7,000 biomedical, health and medical students each year, attracting more annual competitive biomedical research funding than any other area in Australia. These direct employment figures represent high-value knowledge work that generates significant economic multiplier effects throughout the broader economy.

The district has become a magnet for international investment in biotechnology and medical technology, with major global companies establishing regional headquarters and research operations within the precinct. This international engagement has strengthened Australia's position in global biotechnology supply chains and created pathways for Australian research to reach international markets.

Research commercialisation has generated significant economic returns, with university spin-offs and start-ups from the district raising substantial venture capital investment and achieving significant valuations. The success of companies that have emerged from district research institutes has created a demonstration effect that attracts further investment and talent to the ecosystem.

The district's industry development impact extends beyond biotechnology to encompass engineering, advanced manufacturing, and digital technologies. Melbourne Connect's engineering focus has created new capabilities in areas such as advanced manufacturing, artificial intelligence, and sustainable technologies that complement the biomedical core.

International students also provide a significant economic contribution, with the district's universities attracting thousands of students annually who contribute both to the local economy and to global knowledge networks. Parkville forms part of Melbourne's 'International Education Gateway', with the City North area accommodating over 40,000 international students in 2019.

The economic impact model demonstrates how strategic investment in innovation infrastructure can generate returns that far exceed initial investment. The district's success has validated Melbourne's approach to knowledge-intensive economic development and provided a model that other cities seek to replicate.

Challenges and Future Directions

Despite its significant achievements, the Parkville Biomedical Precinct faces several challenges that will require strategic attention to maintain its global competitiveness and continue its growth trajectory. These challenges reflect both the success of the district and the evolving nature of global innovation competition.

Space constraints represent perhaps the most immediate challenge, as the district's success has created demand for expansion that may exceed available land within the existing boundaries. The high cost of real estate in inner Melbourne presents particular challenges for start-ups and growing companies that may be priced out of the district despite benefiting from its ecosystem.

Talent retention and attraction remain an ongoing challenge, particularly in competition with international innovation hubs that may offer different career opportunities or lifestyle advantages. While the district has been successful in attracting international talent, retaining graduates and experienced professionals requires continued attention to career pathways and quality of life factors.

The integration between different components of the district could be strengthened further. While significant progress has been made in creating connections between institutions and sectors, opportunities remain to deepen collaboration and create more seamless pathways for knowledge transfer and commercial development.

Infrastructure demands continue to grow as the district evolves. The construction of Parkville Metro Station as part of the Metro Tunnel Project will provide fast and easy access across the city by 2025. However, continued investment in both physical and digital infrastructure will be necessary to support the district's growth and maintain its global competitiveness.

The district has significant opportunities for expansion and development. It is expected that by 2030, Parkville Biomedical Precinct will be one of the world's premier biomedical and research hubs, and by 2036, the precinct will have generated many thousands of new jobs. Realising this vision will require continued strategic investment and coordination across multiple institutions and levels of government.

The district's future development will need to strike a balance between growth and sustainability, ensuring that expansion enhances rather than undermines the collaborative culture and ecosystem dynamics that have driven its success. This will require thoughtful planning that preserves the district's essential characteristics while enabling continued evolution and adaptation.

Global Significance and Replication Models

The Parkville Biomedical Precinct has already achieved global recognition as a leading model for knowledge-intensive urban development and biomedical innovation. Its success provides valuable insights for other cities and regions seeking to develop innovation districts and strengthen their knowledge economies.

The Precinct has established partnerships with global health institutions, including the World Health Organization, the Bill & Melinda Gates Foundation, the American Foundation for AIDS Research, the National Institutes of Health and the Wellcome Trust. These international partnerships demonstrate how local innovation districts can become nodes in global knowledge networks.

The Precinct's approach to university-industry collaboration has attracted international attention. According to innovation districts expert Julie Wagner from the Global Institute for Innovation Districts (GIID), who studied Melbourne's development in 2017, the city appeared to be creating an 'innovation spine' running through its centre, comprising at least six innovation precincts:

1. Fishermans Bend Employment Precinct - advanced manufacturing, 230-hectare area adjacent to CBD with anchors like Boeing and Defence Science and Technology
2. Melbourne Innovation District (City North) - engineering, design, and a mix of industry
3. Parkville biomedical precinct
4. City North/RMIT Social Innovation Precinct - RMIT's vision is for an innovation precinct in Melbourne's city north, bordered by Lygon, Victoria, Swanston and Queensberry streets
5. Docklands - Part of the broader central Melbourne innovation corridor
6. Carlton Connect/Melbourne Connect - University of Melbourne's engineering and innovation facilities

International delegations regularly visit the district to study its development model and governance approaches. The combination of institutional excellence, strategic government support, and private sector engagement provides a template that other regions seek to adapt to their own contexts.

The district's success in biomedical innovation has global significance given the increasing importance of health technologies and biotechnology in the global economy. As cities worldwide seek to develop capabilities in these sectors, the Parkville-Carlton model provides proven approaches to building research-intensive innovation ecosystems.

Replication requires careful attention to local context and existing institutional capabilities. The Precinct's success builds on decades of investment in research infrastructure and the presence of world-class anchor institutions that cannot be replicated quickly in other locations.

The governance model that balances institutional autonomy with ecosystem coordination provides insights for other multi-institutional innovation districts. The experience demonstrates how universities, research institutes, hospitals, and private companies can collaborate effectively while maintaining their distinct missions and competitive positions.

Conclusion

The Parkville Biomedical Precinct is one of Australia's most significant achievements in knowledge-intensive urban development. Its evolution from a collection of separate research institutions to an integrated innovation ecosystem demonstrates how strategic investment, institutional collaboration, and ecosystem thinking can create globally competitive innovation hubs.

The Precinct's success can be attributed to several key factors: the presence of world-class anchor institutions, strategic government investment in innovation infrastructure, a commitment to collaboration across institutional boundaries, and the development of shared facilities and programs that facilitate knowledge transfer and commercial development.

Melbourne has biomedical capabilities unparalleled in Australia and among the world's best, with an exceptional network of skilled workers, quality education providers, leading research institutes and an advanced health system. The district has successfully translated these capabilities into economic outcomes and global impact.

The district's model provides valuable insights for innovation policy and urban development more broadly. Its success demonstrates how cities can leverage existing institutional strengths to create new forms of economic development that generate both local prosperity and global impact.

As the precinct continues to evolve, it faces the challenge of maintaining its collaborative culture and innovation dynamics while accommodating growth and expansion. Success in meeting these challenges will determine whether the district can maintain its global leadership position and continue to serve as a model for knowledge-intensive development.

The Parkville Biomedical Precinct's experience demonstrates that with appropriate strategy, investment, and governance, cities can create innovation ecosystems that compete globally while generating significant local economic and social benefits. Its continuing evolution will provide important lessons for the future of urban innovation and knowledge-based economic development.

8. Innovation Districts (City North), Melbourne

Melbourne, Australia's cultural capital and fastest growing city, has emerged as a global leader in innovation driven economic development through the establishment of Melbourne Innovation Districts (MID). The MID partnership between the City of Melbourne, University of Melbourne, and RMIT University is a collaborative approach to urban innovation that leverages the concentration of knowledge institutions and investment in the area immediately north of Melbourne's CBD.

Context

City North, selected as the first innovation district, is home to 21 per cent (60,260) of all knowledge sector jobs in Melbourne and encompasses many of Victoria's formative institutions, including the central campuses of RMIT and the University of Melbourne, Queen Victoria Market, State Library of Victoria, Royal Exhibition Buildings, and the Melbourne Biomedical Precinct.

Melbourne's innovation ecosystem has experienced remarkable growth, with Victoria now home to over 3,500 start-ups with a total enterprise value of $132 billion. The city has produced several globally recognised companies including Airwallex, Atlassian (with significant Melbourne operations), and Culture Amp, demonstrating the maturation of an ecosystem that has evolved from traditional manufacturing to knowledge-intensive industries.

The geographic concentration of universities, research institutes, hospitals, and corporate headquarters within a compact urban area has created unique opportunities for collaboration and knowledge transfer. Melbourne's population of over 5 million provides a substantial talent pool, with the city consistently ranking among the world's most liveable cities, attracting international talent and investment.

Melbourne is growing from a medium-sized metropolis to a world city of 8 million people, creating both opportunities and challenges for innovation-led development. The establishment of the Melbourne Innovation Districts is a strategic response to this growth, positioning the city to compete globally while addressing local social and environmental challenges.

Building Foundations: 1990s to 2010

Melbourne's innovation ecosystem has deep historical foundations, particularly in the biomedical and advanced manufacturing sectors. The Melbourne Biomedical Precinct has a rich history spanning over 160 years, with the University of Melbourne established in 1853 and the Royal Melbourne Hospital founded in 1848 creating early anchors for knowledge-intensive development.

The emergence of software companies such as Seek and REA Group during the dotcom boom of the mid-1990s demonstrated Melbourne's potential as a technology hub. Companies such as Microsoft, Google and AWS began opening offices in Sydney between the 1980s and early 2000s, drawn to the shared language and proximity to Asian markets, making foundational investments in cloud infrastructure. This infrastructure development created an environment for Melbourne-based SaaS platforms to emerge.

Atlassian, founded by Mike Cannon-Brookes and Scott Farquhar in Sydney in 2002 but with significant Melbourne operations, demonstrated the potential for Australian software companies to achieve global scale. Atlassian went public on the Nasdaq in 2015 at a $4.4 billion valuation, raising $1.6 billion at IPO. The company's success inspired a generation of entrepreneurs and provided both capital and experienced talent for the broader ecosystem.

During this period, RMIT University established itself as a leader in applied research and industry collaboration. RMIT began as a night school offering classes in art, science and technology in response to the Industrial Revolution in Australia, operating as a private college for more than a hundred years before merging with the Phillip Institute of Technology to become a public university in 1992.

The University of Melbourne strengthened its research capabilities and industry connections, particularly in the biomedical sciences. The Melbourne Biomedical Precinct consistently attracts around 23% of Australia's annual competitive research funding, more than any other cluster in Australia.

Early venture capital activity began to emerge, with funds such as Starfish Ventures (established in 2001) providing early-stage capital to technology companies. The ecosystem began to develop institutional foundations for sustained innovation, including incubators, accelerators, and angel investor networks.

Institutional Development and Early Success: 2010-2016

The period from 2010 to 2016 witnessed significant institutional development and the emergence of Melbourne's first major start-up successes. This phase was characterised by increased government support, university commercialisation efforts, and the establishment of key infrastructure.

The Melbourne Accelerator Program (MAP) was launched by the University of Melbourne in 2012, becoming Australia's leading university-based start-up accelerator. MAP offers a range of programs including the Velocity Program (a part-time pre-accelerator program for early-stage founders), the Franklins (a female founders network), and the Master Class series. MAP has supported over 100 start-ups and played a crucial role in establishing Melbourne's reputation as a start-up hub.

During this period, Melbourne began to see the emergence of significant technology companies that later became global success stories. The ecosystem was developing the foundations that later supported companies like Airwallex and Culture Amp, while companies like Canva (based in Sydney) demonstrated the potential for Australian start-ups to achieve global scale and provided inspiration for the broader national ecosystem.

RMIT University expanded its focus on innovation and entrepreneurship, building on its historical strengths in applied research and industry collaboration. The university began developing programs that bridged traditional vocational education with emerging technology sectors and intensified its commitment to impact-focused translational research.

The Victorian Government began recognising the potential of the innovation economy, though comprehensive policy frameworks were still in development. Early initiatives focused on supporting research commercialisation and developing the venture capital ecosystem, along with active support for industry-research infrastructure in such sectors as healthcare and network computing.

During this period, Airwallex was founded in 2015 by university friends Jack Zhang, Lucy Liu, Max Li, and Xijing Dai at a Melbourne coffee shop. The start-up's financial pain of making payments overseas for supplies sparked the idea for Airwallex, offering SMEs access to interbank mid-market exchange rates when making cross-border transactions.

The biomedical sector continued to strengthen, with significant investments in research infrastructure and the development of spin-out companies. CSL, Australia's largest pharmaceutical company, expanded its operations in the Melbourne Biomedical Precinct, creating demonstration effects for other companies.

The period concluded with the emergence of several successful Melbourne-based companies and the recognition that Melbourne had the potential to develop a globally competitive innovation ecosystem. However, challenges remained around access to capital, talent retention, and coordination between different ecosystem actors.

Strategic Development and LaunchVic: 2016-2020

The establishment of LaunchVic in 2016 marked a turning point for Melbourne's innovation ecosystem. LaunchVic was created by the Victorian Government as an independent start-up agency to oversee the allocation of its $60 million innovation fund. This represented the first comprehensive government strategy to support start-up development in Victoria.

The Melbourne Innovation Districts partnership was formally established in 2016 when RMIT University began discussions with the City of Melbourne and University of Melbourne about the potential to jointly transform the City North area as a shared Innovation District.

RMIT's experience in Barcelona in the heart of the 22@ Innovation District provided stimulus and inspiration for the potential impact that networked collaboration in a strategically integrated innovation district could have. The partners recognised that the Victorian governments' commitment to building large scale transport infrastructure, along with Melbourne's lively demographic growth, would create both demand and opportunity innovation.

LaunchVic's early funding rounds supported the development of ecosystem infrastructure. The Round Two grants contributed $4.9 million in funding to seven local projects, including Sydney-based accelerator program Startmate running a two-year program in Victoria, Girl Geek Academy bringing the SheHacks program to the state, and several regional innovation hubs.

Airwallex achieved significant milestones during this period, demonstrating the potential for Melbourne-based start-ups to scale globally. In 2017, Airwallex completed a $US13 million Series A funding round led by Tencent Holdings, Sequoia Capital China and MasterCard, less than 12 months after raising $4.5 million pre-launch. By 2019, Airwallex had raised $100 million at a valuation above $1 billion, joining other successful Australian tech companies as a unicorn.

The Melbourne Innovation Districts initiative began taking shape with the development of planning frameworks and the selection of City North as the first innovation district. City North was selected because of its long and successful history of innovation, and the concentration of knowledge institutions and investment in the area.

RMIT began developing its vision for a distinctive precinct in City North on a city block historically used for workforce training, with the potential to connect strongly with wider district. The Victorian Government funded RMIT through the Victorian Higher Education State Investment Fund (VHESIF) to develop a detailed proposal to create a social innovation precinct in Melbourne's city north.

The University of Melbourne planned for Melbourne Connect, recognising the need for a purpose-built innovation precinct, which would be built in the converted former women's hospital and include a multi-level development bringing together start-ups, maker spaces, governments, academics and bigger businesses.

Melbourne Connect and the RMIT's Social Innovation Precinct site are both on Cardigan St, helping to connect a 'spine' across the wider district which, with the addition of new public transport options, widened opportunities for talent, capital and new ventures to cluster and grow in City North.

During this period, several other notable companies emerged or expanded significantly. Culture Amp, founded by Didier Elzinga, became a leader in employee feedback and analytics software. SafetyCulture, developed by Luke Anear, established itself as a global leader in workplace safety technology.

The biomedical sector continued to strengthen with major investments in research infrastructure and the attraction of international companies. With over $2.8 billion investment in research and healthcare facilities over the past decade, the Parkville Precinct produces 20 per cent of Australia's patent citations.

Acceleration and Global Recognition: 2020-Present

The period from 2020 to the present has been characterised by rapid growth, international recognition, and the physical realisation of major infrastructure projects. The COVID-19 pandemic, while presenting challenges, also accelerated digital transformation and highlighted the importance of innovation ecosystems.

Melbourne Connect opened in 2021 as Australia's newest purpose-built innovation precinct. Powered by the University of Melbourne in partnership with a consortium led by Lendlease, Melbourne Connect features a 75,800sqm precinct with interconnecting buildings arranged around a central landscaped open space known as Womin Djerring (Come Together). The precinct features circa 35,000 square metres of A-Grade office space, the Telstra Creator Space, Science Gallery Melbourne, retail and a 3,000 square metre superfloor shared meeting and events space.

Following City North being selected as the first innovation district in 2016, the Melbourne Innovation Districts City North Opportunities Plan was developed and endorsed at Future Melbourne Committee on 12 November 2019. Since Council's adoption of the Plan in 2020, Council and partners have made substantial investment in public realm upgrades within the precinct, and MID has delivered over 40 projects that align with the 'Urban Realm Objectives'.

LaunchVic expanded its programs significantly, supporting the growth of the start-up ecosystem. Since 2017, LaunchVic has supported 1,150 budding entrepreneurs and start-ups through its funded programs, with the ecosystem growing to over 3,500 start-ups valued at more than $129 billion. LaunchVic has funded 15 accelerator programs supporting over 600 start-ups and established nine new venture capital funds that collectively target to raise over $300 million in capital for Victorian start-ups.

Airwallex secured $US300 million in funding, becoming valued at $US6.2 billion ($9.6 billion AUD) and growing to $US720 million in annualised revenue with 150,000 business customers and 1700 staff over 26 offices. The company has become a flagship example of Melbourne's fintech capabilities. CSL moved into new purpose-built headquarters on the Haymarket Roundabout at City North.

RMIT progressed its City North Social Innovation Precinct plans. The precinct will train tens of thousands of highly skilled workers for the

Victorian economy over the next two decades, contributing to major workforce challenges such as electrification, mental health and social care. Recent activation initiatives focus on student and community health and wellbeing, using the City North built and urban environments as a living-lab.

The Melbourne Biomedical Precinct continued to expand significantly. Parkville has close to 73,000 jobs and could be home to over 80,000 jobs by the mid-2030s. The Victorian Government announced a $5-6 billion expansion of hospitals and research facilities at Arden and Parkville, including a new dedicated Clinical Trials Centre and Digital Health Command.

The start-up ecosystem achieved significant scale and diversity. Melbourne is home to start-up sectors including Life Sciences and Biotech, Fintech, SaaS, AI, Blockchain and advanced manufacturing, with early-stage start-up value more than doubling in a year to $23.6B (AUD). According to LaunchVic's latest research, there are over 3,400 start-ups in the ecosystem, with its value jumping by 56% to $36.9 billion AUD in just one year.

Success Stories and Economic Impact

Melbourne's innovation ecosystem has produced several globally significant companies that demonstrate the maturation of the local ecosystem. These success stories have created demonstration effects, provided experienced talent, and generated capital for reinvestment in new ventures.

Airwallex is one of the most significant success stories of the Melbourne ecosystem. Founded in Melbourne in 2015 by four founders, Airwallex provides cross-border payment services and has become a unicorn within three years of its launch. The company has raised hundreds of millions in venture capital and employs over 800 staff based across 12 offices globally. The founders have launched a $200 million VC fund to support other tech companies on their global growth journeys.

Atlassian, while founded in Sydney, has maintained significant Melbourne operations and has become a model for Australian technology companies. Atlassian achieved a market cap of US$48B at time of writing, peaking at over $100B in 2021. In September, Atlassian launched a $70 million fund for investment into early-stage start-ups building apps for its cloud products.

Culture Amp has emerged as a global leader in employee experience technology, demonstrating Melbourne's strength in software-as-a-service solutions. The company has raised significant venture capital and employs hundreds of people globally while its Melbourne headquarters.

SafetyCulture, founded by Luke Anear, has become a global leader in workplace safety technology, serving millions of users worldwide. The company's success has highlighted Melbourne's capabilities in developing solutions for traditional industries through technology innovation.

The fintech sector has been strong, building on Melbourne's historical strengths in financial services. The top-funded segment in 2022 at $1.3 billion raised was fintech, one of Australia's strongest exports, with companies leveraging Australia's well-established financial infrastructure.

The economic impact extends beyond individual company success stories. Deloitte Access Economics revealed that the Victorian start-up sector supports almost 19,000 jobs and could contribute an additional 15,000 jobs each year over the next 20 years. Successful start-ups are job multipliers, with every tech job creating an additional five jobs in non-technical areas.

The biomedical sector has also generated significant economic impact. The Melbourne Biomedical Precinct employs 49,000 people and educates over 7,000 biomedical, health and medical students each year, attracting more annual competitive biomedical research funding than any other area in Australia.

Government Policy and Support Mechanisms

The Victorian Government's approach to innovation support has evolved from ad hoc initiatives to comprehensive ecosystem development strategies. LaunchVic is the centrepiece of this approach, providing targeted support across the start-up lifecycle. This has included:

- Grants for program providers and investor groups to establish in Victoria, with funding focusing on pre-accelerator programs, accelerator programs, angel networks, and VC funds. The CivVic Labs program provides $12,500 equity-free funding through a six-week bootcamp experience, with two teams securing further $25,000 funding to bring their prototype to life.
- Two targeted investment funds: the $10 million Alice Anderson Fund (a sidecar fund that co-invests in women-led start-ups) and the

Hugh Victor McKay Fund (for Victorian AgTech start-ups). These funds address specific market gaps while leveraging private sector investment.
- Support for the growth of specialised accelerator programs like the Melbourne-based Med-tech Actuator, which now successfully runs start-up cohort programs in Singapore and Japan as well as Victoria, and by promoting the establishment of new early-stage VC funds in Melbourne.

The government has supported infrastructure development through significant investments in major precincts. The Melbourne Metro Tunnel project, due for completion in 2025, will enhance connectivity between innovation districts. The Arden Metro Station will connect Arden to Parkville (2 minutes) and the CBD (5 minutes), enabling increased collaboration between innovation precincts.

Regulatory support has been important, particularly in fintech where Australia's supportive regulatory environment and tech-savvy population helped start-ups revolutionise solutions like buy-now, pay-later services, peer-to-peer lending platforms and digital banking alternatives. Interaction or alignment between jurisdictions, including the Commonwealth Government's growing focus on sovereign capability, advanced manufacturing and integrated urban precincts, is a work in progress.

The partnership model adopted for Melbourne Innovation Districts is an innovative approach to governance, bringing together local government, universities, and industry in a collaborative framework. This approach promotes and advance technology innovation as a driver of new jobs and enterprise, with equal weight given to social needs and the public realm alongside enterprise activation and new technology.

Challenges and Future Directions

Despite significant progress, Melbourne's innovation ecosystem faces several challenges that require ongoing attention and strategic responses.

Access to growth-stage capital remains a constraint, particularly for companies seeking to scale internationally. While early-stage funding has increased significantly, the availability of Series A and later-stage funding from domestic sources remains limited. This has led many successful companies to seek offshore investment or relocate operations to access global capital markets.

Talent retention and attraction is an ongoing challenge. While Melbourne's livability and quality of life attract international talent, competition from global technology centres and the relatively small-scale of some local companies can make it difficult to retain senior technical and commercial talent.

The concentration of innovation activity in inner Melbourne creates both opportunities and risks. While clustering effects generate significant benefits, there is a risk of excluding communities and businesses in outer suburban areas from innovation-led economic development.

International connectivity and market access remain important considerations. RMIT's international locations in Barcelona and Hanoi offer platforms to partner with community and industry across the City North ecosystem and into South-East Asia on workforce training needs and digital transformation.

Infrastructure development continues to be critical. The Arden Precinct presents a landmark urban renewal opportunity that will complement Parkville and become a globally recognised innovation precinct, attracting occupants in life sciences, research, digital technology, and educational technology sectors.

Climate change and sustainability present both challenges and opportunities. Melbourne's innovation ecosystem is well-positioned to develop solutions for the clean energy transition, with strengths in energy technology, smart cities, and sustainable manufacturing.

The future development of Melbourne Innovation Districts will require continued collaboration between government, universities, and industry. The imperative for innovation in innovation districts oscillates between primarily economic objectives and aligning innovation investments and outcomes with social and environmental needs. International collaboration and peer exchange with the growing number of urban innovation districts established around the world may help influence this process.

Conclusion

Melbourne Innovation Districts City North is a successful model of collaborative ecosystem development that leverages existing institutional strengths while addressing contemporary urban challenges. The partnership between the City of Melbourne, University of Melbourne, and RMIT University has created a framework for innovation that extends beyond traditional technology transfer to encompass social innovation and urban development.

The ecosystem's success can be attributed to several key factors: strong anchor institutions, supportive government policy, access to capital through LaunchVic and private investors, a culture of collaboration between universities and industry, and the demonstration effects of early success stories like Airwallex and Culture Amp.

Melbourne's approach offers several lessons for other cities seeking to develop innovation ecosystems. The importance of anchor institutions, particularly universities, in providing both research capabilities and talent cannot be overstated. The collaborative governance model adopted for Melbourne Innovation Districts provides a framework for coordinating actions across different sectors while maintaining institutional autonomy.

The strategic focus on specific sectors where Melbourne has competitive advantages, including fintech, biomedical technologies, and software-as-a-service, has enabled the ecosystem to develop depth and specialisation. This approach has created knowledge spillovers and supply chain effects that strengthen the overall ecosystem.

Government support through LaunchVic demonstrates the value of dedicated start-up agencies that can provide targeted support across the innovation lifecycle. The combination of direct funding, ecosystem development programs, and infrastructure investment has created conditions for sustained growth.

Melbourne Innovation Districts City North has the potential to become one of the world's leading innovation ecosystems. The combination of world-class universities, growing start-up community, supportive government policy, and strategic location in the Asia-Pacific region provides a strong foundation for continued growth.

The challenge will be maintaining the collaborative culture and inclusive approach that has characterised the ecosystem's development while scaling to accommodate rapid growth. Integrating social innovation and environmental sustainability into the innovation agenda is an opportunity to create a model for innovation-led development that addresses the challenges of the 21st century.

Lessons for Other Regions

Other cities can learn from Melbourne's experience the importance of patience and long-term thinking in ecosystem development. The success of Melbourne Innovation Districts builds on decades of institutional development and is the culmination of sustained effort rather than rapid transformation. The collaborative approach adopted in Melbourne provides a template for other cities seeking to leverage their institutional assets for innovation-led economic development.

Melbourne's innovation ecosystem demonstrates that mid-sized cities can compete globally by leveraging their unique assets and fostering collaboration between key institutions. The continued development of Melbourne Innovation Districts City North will provide valuable insights into how innovation ecosystems can evolve to address the challenges and opportunities of rapid urban growth in the 21st century.

9. Gold Coast Health and Knowledge Precinct

The Gold Coast Health and Knowledge Precinct (GCHKP) spans 200 hectares and brings together more than $5 billion in health, research, education and transport infrastructure to drive world-class discovery and development. With over 16,000 workers, including 1000 researchers, within a thriving community of clinicians, entrepreneurs, students and global partners, the precinct is where research, industry and care come together to solve global challenges, improve lives and create jobs of the future.

Context

The GCHKP contributes $3.4 billion value-add to the Queensland economy and is home to almost 20,000 students, 2,500 residents and 50 SMEs and tech companies. The precinct is a unique model of urban transformation, conceived as vehicle for economic diversification and emerging from the legacy of a major international sporting event in the 2018 Commonwealth Games, to become Asia-Pacific's newest innovation hub.

Located in Southport on the Gold Coast, the precinct is transforming what was historically known as a tourism and lifestyle destination, into a knowledge-intensive economy centred on health, medical research, and advanced technologies. Formerly based on tourism and construction, the Gold Coast is now Australia's fastest growing economy.

The Precinct is a partnership between the City of Gold Coast, Griffith University, Gold Coast Health, and the Queensland Government. This collaborative governance model has enabled rapid development and positioned the Gold Coast as a serious contender in Australia's innovation landscape, demonstrating how strategic partnerships between government, academia, and health services can catalyse economic transformation.

Foundation and Strategic Development: 2008-2018

With the original vision for an innovation precinct dating back to 2001, the genesis of the GCHKP can be traced to strategic planning decisions made in the lead-up to the 2018 Commonwealth Games. The precinct is identified in the South-East Queensland Regional Plan (2031) as a major residential and mixed-use development node. This designation reflected forward-thinking by policymakers who recognised the potential to leverage major event infrastructure for long-term economic benefit.

The precinct's foundation was built on significant public investment in health and education infrastructure. Griffith University's presence was crucial during this foundational period. In 1990, the Gold Coast College of Advanced Education (est. 1987) became an official campus of Griffith University, with the Gold Coast campus growing over two decades to become the university's largest campus with over 19,000 students.

Griffith's Institute for Glycomics, now the Institute for Biomedicine and Glycomics, was established at the Gold Coast campus in 2001, focusing on glycans and glycan-binding proteins to develop novel drugs and vaccines. Founded by Emeritus Professor Mark von Itzstein, who also discovered the anti-influenza drug relenza, its research targets medical conditions such as cancer, infectious diseases and neurological conditions, and has now expanded into the university's largest research institute.

In December 2008, then Premier Anna Bligh turned the first sod to commence construction of the Gold Coast University Hospital, at a cost of approximately $1.8 billion. It became the cornerstone investment that would anchor the broader precinct development.

The hospital was built on a greenfield site adjacent to Griffith University's Gold Coast campus, demonstrating the strategic co-location approach that would become central to the precinct's model. This proximity between health services, research capability, and education infrastructure created the foundation for what would become a world-class innovation ecosystem.

Opened in July 2013, by The Honourable Dame Quentin Bryce AD CVO, the $150 million Griffith Health Centre (Ian O'Connor Building) is now a key facility for health services, housing community clinics alongside world-class teaching and learning spaces.

The Gold Coast University Hospital opened three months later, on September 28, 2013, in what was the largest planned hospital relocation in Australia.

Commonwealth Games Legacy and Transformation: 2018-2020

The 2018 Commonwealth Games provided a unique catalyst for the precinct's development, demonstrating how major sporting events can generate lasting economic and infrastructure benefits. In 2016, the then Deputy Premier Jackie Trad alongside the Gold Coast Mayor Tom Tate unveiled the masterplan for this new precinct to grow the jobs of the future. The Masterplan had 200 hectares to work with and a vision of employing 20,000 people when complete.

The new health and knowledge commercial hub, inside what was established as the Commonwealth Games Athletes Village would be an important part of the precinct in years to come, as it was strategically designed for post-Games transformation. Known as the Parklands Project and valued at $550 million, it hosted 6,600 athletes and officials.

Thanks to the legacy project, the Precinct is set to support thousands of jobs for Queenslanders and generate $2.9 billion for the City's economy once fully developed. This is one of the most successful examples of Games legacy planning in Australia, with infrastructure designed from the outset for long-term economic benefit rather than single-use sporting purposes.

The transformation was swift and comprehensive. Following the Games, the site underwent a remarkable transformation, continuing the evolution of the GCHKP. The previous Athlete Village apartments were transformed into the Smith Collective, containing 1252 one-, two- and three-bedroom dwellings, the largest and first 'Build to Rent' development in Australia.

A Griffith Institute for Tourism and Griffith Business School report modelling the economic impacts of the 2018 Commonwealth Games, produced for the Office of the Commonwealth Games, predicts a $2 billion increase in Queensland Gross State Product. It predicts the Games will continue to improve the local economy to 2028.

South East Queensland (SEQ) City Deal, 2022 (20 Years)

A regionwide compact across 11 councils, aligning with Olympic legacy opportunities. It includes discrete, published amounts, notably:

- $150 million SEQ Innovation Economy Fund to "ignite" significant innovation precincts.
- $30 million for SEQ rail corridor digital connectivity.
- $40 million Smart Region Digital Plan/local digital priorities.
- Major place-based items such as Brisbane Metro Woolloongabba Station ($450 million), Kangaroo Point Green Bridge ($190.2 million), a $285.8 million Livability Fund, $105 million for resource recovery infrastructure, and multiple planning/business-case lines.

In relation to the above SEQ Innovation Economy Fund, in 2024 Griffith University, with a $10m contribution from City of Gold Coast, successfully bid for $25m in funding from the SEQ Innovation Economy Fund to develop the Health and Advanced Technologies Research and Innovation Centre (HATRIC), set to be complete by 2027. This will be a transformational building for the Precinct focused on drug and vaccine development, clinical trials, advanced design and additive manufacturing, and co-locate scale-up companies.

Research Excellence and Commercial Development: 2020-Present

The GCHKP is home to the Institute for Biomedicine and Glycomics, which boasts 400 researchers and staff and has a host of vaccine, drug and diagnostic technologies in commercial licensing, co-development or early pipeline stages, including vaccines for Malaria and Strep A in clinical trials. The Australian Centre for Precision Health Technologies (PRECISE) brings together Griffith University's substantial medical technology capabilities with a range of clinical collaborators.

Medical device development, along with advanced prototyping across a range of industry sectors is supported by Griffiths Advanced Design and Prototyping Technologies Institute (ADaPT), a key facility for design, testing, 3D printing and prototyping. Leading research holds the promise of a breakthrough treatment for spinal cord injury, through a novel cell transplantation therapy, now in human clinical trials, and the BioSpine advanced rehabilitation system incorporating robotics and electrical stimulation paired with a brain–computer interface and virtual reality technology, controlled by a personalised digital twin.

Other key research expertise includes AI, machine learning and computer vision processing, along with sensor technologies, micro and nano electronics and new materials science. The recent formation of Griffith's Quantum and Advanced Research Technologies Institute

(QATRI) builds on two decades of Quantum science research and brings capability together, including the Queensland Micro and Nanotechnology Centre, and the Queensland Node of the Australian National Fabrication Facility. This concentration of research institutes is one of the most significant clusters of health and technology research capability in the Asia-Pacific region.

The precinct has attracted significant international attention and investment through partnerships with global companies including Sanofi and Philips. As part of Griffith University's involvement as a partner in the Sanofi Translational Science Hub, the global pharma leader opened a new office within the GCHKP in 2023. This decision by a global pharmaceutical leader validated the precinct's research credentials and commercial potential.

Also in 2023, healthcare technology leader Philips partnered to establish a specialised facility for Image-guided Surgical Training and Robotics with two precinct clinicians who are global leaders in treatments and world first clinical trials for stroke and neurovascular conditions, and pioneers in surgical robotics in Dr Hal Rice and Dr Laetitia de Villers.

Early-stage R&D is supported by an ecosystem that includes the Cohort Innovation Space, home to the precinct's start-up community, and various programs to drive commercialisation and entrepreneurship, including Cohort's signature LuminaX Healthtech Accelerator.

Recent major developments, as outlined earlier include the Health and Technology Research Innovation Centre (HATRIC) will further support technology translation and scale-up companies. This is forecast to begin construction in late 2025.

The precinct's research has produced significant breakthroughs across multiple disciplines. From world first breakthroughs in spinal cord repair, artificial heart technology and stroke intervention to next-generation ventures in biotech, sportstech, clean energy and AI, the GCHKP is building a connected community where you can collaborate, grow and lead real change.

Lumina Commercial Cluster: The Heart of Innovation

Lumina, the Queensland Government's cluster of 9.5 hectares of development-ready land in the GCHKP, provides the only dedicated location for businesses and enterprises to be part of this cutting-edge Precinct.

As landowner, master developer and planning authority, Economic Development Queensland (EDQ), as part of the Queensland Government, is facilitating the development of Lumina over the next 10. This long-term approach ensures coordinated development and strategic curation of tenants that align with the precinct's innovation objectives.

Lumina is home to the Cohort Innovation Space, opened in 2019, which provides co-working desks, office, labs and innovation programs. The first private commercial development, the $80m Proxima building, opened in mid-2024.

Major developments within Lumina include the RDX building. Northwest Healthcare Properties ("Northwest"), through its NZX-listed fund, Vital Healthcare Property Trust, a new state-of-the-art, $154 million 6-Star Green Star, Life Sciences Centre of Excellence

Set to open in early 2026, RDX is Lumina's flagship facility designed for cutting-edge health, medical research and digital technology enterprises.

Leading Australian data centre owner NextDC is developing a data centre within Lumina, which will supercharge innovation and AI adoption by precinct partners and companies.

Companies establishing operations within Lumina demonstrate the diversity of innovation taking place. The precinct is home to the international office of BiVACOR, a medical device company developing a world first rotary total artificial heart which has recently received breakthrough device designation by the United States FDA following successful trials with six heart-failure patients in the US and Australia.

BiVACOR has plans to manufacture the device for the Asia-Pacific and other global markets from an expanded facility in the GCHKP, while continuing R&D alongside Griffith University mechanobiology researchers. In an adjoining lab at Cohort, Southern RNA has established as an Australian-first manufacturer of RNA for personalised cancer therapeutics.

Governance and Leadership Structure

The Precinct Office is a collaboration between the Queensland Government, City of Gold Coast, Griffith University and Gold Cost Health. A Precinct Partnership Group, at CEO/Vice-Chancellor level, provides overall direction, while a Strategic Advisory Group, at senior management level, provides partner guidance and decision-making.

A third level of governance in the Precinct Team Working Group brings together around 30 operational level staff across the Precinct from partners to coordinate on the ground activities. The focus is to bring coordination across the Precinct, to raise the profile, attract investment, and develop strategic projects and governance.

The precinct's leadership demonstrates the importance of dedicated management in complex multi-stakeholder environments. The governance model emphasises collaboration and coordination across multiple entities.

As Director of the GCHKP, Craig Rowsell brings extensive experience in economic development and strategic planning. After completing a Bachelor of Applied Science at the University of Canberra, Craig began his career at the Australian Sports Commission. He was heavily involved in the Australian Government's policy and programs in high performance and community sport.

Mr Rowsell relocated to the Gold Coast in 2011 and has been involved in major economic development projects in the sport and film sectors, before being drawn to the transformational project that is the GCHKP in 2019, focusing on health, education and technology. He is highly supported by the Precinct Office team consisting of five staff, including his long-term colleague Ms Kathy Kruger who leads Strategic Projects and Engagement.

Economic Impact and Future Projections

The economic impact of the GCHKP extends well beyond its physical boundaries. The precinct is intended to support 12,400 new high-value jobs to ultimately support over 26,000 full-time equivalent positions. These projections represent a transformation of the Gold Coast's economic base from tourism and construction towards knowledge-intensive industries. The GCHKP continues to meet their annual KPIs in economic growth and job creation.

The vision appears to be working, with the Gold Coast last month recognised as Australia's fastest growing economy, outperforming state and national projections. This performance demonstrates the broader economic impact of strategic investment in innovation infrastructure.

To date, the LuminaX (LX Health) program has charged up 52 start-ups in 5 years, raising over $33 million in funding, supported by more than 1600 hours of mentorship, 60 workshops, 50+ mentors, with 15+ venture capital firms engaged across the life of the program.

Strategic Alignment and Policy Integration

The GCHKP operates within broader state and national policy frameworks that enhance its effectiveness and reach. Aligned with the Queensland Innovation Precincts and Places Strategy 2022-32, which aims to create jobs, foster industry growth, and attract significant investment, GCHKP is playing a pivotal role in advancing Queensland's innovation agenda.

> *The Precinct is focused on collaboration with other precincts. This strategic alignment will ensure that the precinct benefits from statewide coordination while contributing to broader innovation objectives.*

The precinct is positioned to leverage the Brisbane 2032 Olympic and Paralympic Games. The Brisbane 2032 Olympic and Paralympic Games is expected to bring a $4.6 billion boost to the Queensland economy in tourism and trade alone. But this event also presents a golden opportunity for Lumina and the organisations who call Lumina home, through a broader plan for capitalising on this opportunity to become a lasting legacy.

Challenges and Opportunities

Despite significant achievements, the GCHKP faces challenges common to emerging innovation precincts. The transition from research to commercialisation requires ongoing support and investment, particularly in bridging the gap between academic research and market-ready products and services.

This is not a fast process; it highly depends on economic cycles to generate and maximise opportunities. The precinct's development requires sustained commitment across multiple economic and political cycles, demanding robust governance structures and long-term strategic planning.

The precinct's success creates opportunities for broader regional development. Do not underestimate the importance of a good multimodal public transport network in growing the geographic area from which students and employees will be prepared to travel. This will grow access to a greater skills pool and continue to elevate the precinct as a place of knowledge and employment desirability.

Craig has bold ambitions for the GCHKP and would love to see the Gold Coast recognised as a global innovation leader, based on its output and impact. This ambition reflects the precinct's potential to transcend its regional origins and achieve international recognition. He is focused on enabling people to "live in Australia's best lifestyle city, while progressing your dream career".

The GCHKP has emerged as a national leader in health and innovation precinct development. The precinct's selection to host the 2024 National Health & Innovation Precincts Summit marked a significant milestone in its recognition as a leading Australian innovation hub.

A new group was established off the back of this annual conference, in the Australian Innovation District Professionals Network, to begin information sharing to build an ecosystem of trust among innovation precincts. This collaborative approach demonstrates the precinct's commitment to sharing knowledge and best practices with other innovation hubs across Australia.

Conclusion and Lessons for Other Regions

The GCHKP is one of Australia's most successful examples of strategic urban transformation driven by innovation infrastructure. The precinct's evolution from Commonwealth Games legacy site to Asia-Pacific innovation hub demonstrates the potential for coordinated public and private investment to generate sustained economic transformation.

Several key lessons emerge from the GCHKP experience.

The importance of strategic planning that anticipates post-event legacy from major sporting events, designing infrastructure for long-term economic benefit rather than single-use purposes.

The critical role of anchor institutions, particularly universities and major hospitals, in providing the foundation for innovation ecosystems.

The value of dedicated governance structures that can coordinate multiple stakeholders while maintaining strategic focus on long-term objectives. The appointment of professional precinct management has been crucial in navigating complex partnerships and maintaining momentum across political and economic cycles.

An emphasis on co-location and physical proximity between research, clinical services, and commercial development has proven essential in generating the collaboration and knowledge transfer that drives innovation. The precinct's design facilitates both planned and serendipitous interactions that accelerate research translation and commercialisation.

The GCHKP *experience highlights the importance of aligning local precinct development with broader state and national innovation strategies.* This alignment has enhanced access to funding, policy support, and collaborative opportunities that amplify the precinct's impact.

For other regions considering similar development, the GCHKP model offers a blueprint for leveraging existing strengths in health and education while building new capabilities in advanced technologies and commercialisation.

The precinct's success demonstrates that with investment, governance, and strategic vision, regional cities can compete globally in knowledge-intensive industries.

The transformation of the Gold Coast from a tourism-dependent economy to a diversified innovation hub anchored by the GCHKP provides compelling evidence that strategic investment in innovation infrastructure can generate profound and lasting economic transformation. As the precinct continues to evolve towards its vision of supporting 26,000 jobs and contributing $2.9 billion to the local economy, it is a testament to the power of collaborative, long-term strategic planning in driving innovation-led growth.

10. Sunshine Coast Innovation District

The Sunshine Coast region in Queensland is one of Australia's fastest growing innovation ecosystems, anchored by a unique combination of lifestyle appeal, educational excellence, and entrepreneurial spirit. With a population approaching 934,500 in 2023 and projected to reach 1.3 million by 2041, the region has evolved from a tourism and retail-dependent economy into a knowledge-intensive hub that attracts entrepreneurs, researchers, and technology companies seeking both commercial success and quality of life.

Context

Located approximately 100 kilometres north of Brisbane, the Sunshine Coast benefits from proximity to Queensland's capital while maintaining its distinct character as a coastal innovation destination. The region's transformation reflects a deliberate strategy to leverage its natural advantages while building the infrastructure, institutions, and networks necessary to support a modern knowledge economy.

The Sunshine Coast's innovation ecosystem is distinctive for its collaborative, community-driven approach. Unlike traditional top-down innovation districts, the region's development has been characterised by grassroots entrepreneurship, strong university-industry partnerships, and a culture of cooperation that spans sectors and organisational boundaries. This ecosystem has attracted national recognition as a model for regional innovation development, particularly in how smaller cities can compete globally while maintaining their unique identity and livability.

The economic base has diversified significantly over the past two decades. While tourism remains important, the region now hosts substantial activity in digital technologies, advanced manufacturing, health innovation, agribusiness, clean technologies, and creative industries. High-speed digital infrastructure, including direct submarine cable connections, has positioned the Sunshine Coast as a viable alternative to major metropolitan centres for technology-intensive businesses.

Foundational Years: 1996-2010

The contemporary Sunshine Coast innovation ecosystem traces its origins to the establishment of the University of the Sunshine Coast in

1996, which marked a pivotal moment in the region's economic development trajectory. As Australia's newest university, USC was designed from inception to serve as a catalyst for regional transformation, with ambitious plans to produce 60,000 graduates over two decades in a region that previously lacked significant higher education infrastructure.

The university's establishment addressed a critical gap in the regional economy, which at the time relied heavily on tourism, retail, and construction, providing limited opportunities for high-quality careers in knowledge-based industries. This recognition that sustainable economic growth required technological progress and innovation would become a central theme in the region's development strategy.

A crucial early initiative was the establishment of the Innovation Centre Sunshine Coast in 2000, which began operations in January 2002. At the USC campus at Sippy Downs, the Innovation Centre was conceived as a not-for-profit company designed to serve as a major hub for mature entrepreneurial start-ups and high growth companies. The centre's 1,700 square metre purpose-built facility provided essential infrastructure that was often lacking in regional settings, including high-speed connectivity, meeting spaces, prototyping capabilities, and access to mentorship networks.

The Innovation Centre's comprehensive support model included incubator and accelerator programs, a 30-plus mentor panel, professional partnerships, and access to funding opportunities, creating what would become a template for regional innovation support in Australia. This early investment in innovation infrastructure reflected an understanding that successful regional development required more than just physical space - it demanded the creation of entire ecosystems that could nurture entrepreneurship and facilitate connections between researchers, entrepreneurs, and industry.

During this foundational period, the region began developing the digital infrastructure that later proved crucial to its success. Recognition of the importance of high-speed connectivity led to early investments in telecommunications infrastructure, setting the stage for the region's later emergence as a digital hub.

The period also saw the beginning of collaborative relationships between the university, local government, and emerging business networks. These early partnerships established patterns of cooperation that would become hallmarks of the Sunshine Coast approach to innovation development.

Building Momentum: 2010-2018

The second phase of the Sunshine Coast innovation ecosystem's development was characterised by the emergence of community-driven organisations and the strengthening of collaborative networks. A pivotal development was the formation of Silicon Coast around 2014, which represented a shift towards grassroots organising and community building within the innovation ecosystem.

Silicon Coast emerged as a not-for-profit community organisation focused on "igniting collaboration" and serving as "the leading body for entrepreneurship and innovation in Queensland Australia". Unlike many innovation initiatives that operate primarily through government or institutional structures, Silicon Coast developed as a genuinely community-driven organisation that brought together entrepreneurs, researchers, students, and industry professionals in informal yet effective networks.

The organisation grew to encompass almost 2,000 members, creating a substantial network that connected various elements of the innovation ecosystem including co-working spaces, incubators, accelerators, networking groups, and educational institutions. This community building approach proved effective in a regional setting where personal relationships and trust networks play crucial roles in business development.

During this period, the Innovation Centre Sunshine Coast continued to mature as a regional hub. The centre reported assisting over 200 resident members since 2002 and creating over $60 million in economic value for the Sunshine Coast, demonstrating the tangible impact of sustained investment in innovation infrastructure.

The University of the Sunshine Coast also expanded its innovation activities during this period, developing stronger connections with industry and establishing research capabilities that would support commercial applications. The university's growth provided both talent and research capacity that proved essential to the region's innovation development.

This period saw the emergence of various sector-specific networks and initiatives. The ecosystem began to include specialised groups such as the Sunshine Coast Blockchain network, Digital Entrepreneurs Network of Noosa, Sunshine Coast.NET User Group, and various hackathons and pitch competitions. These sector-specific communities allowed for deeper collaboration within particular technology areas while maintaining connections to the broader ecosystem.

The region also began developing capabilities in key sectors that later became strategic priorities. Health innovation, advanced manufacturing, agribusiness, and digital technologies all began to show signs of clustering during this period, supported by university research capabilities and growing industry presence.

Recent Expansion and Strategic Development: 2018-Present

Since 2018, the Sunshine Coast innovation ecosystem has entered a phase of rapid expansion and strategic development, characterised by major infrastructure investments, formal recognition of innovation priorities, and integration into state and national innovation strategies.

A transformative development has been the construction of the new Maroochydore City Centre, representing a $2.5 billion investment in creating Australia's largest greenfield central business district. This development provides modern commercial infrastructure specifically designed to support knowledge-intensive industries and positions the Sunshine Coast as a serious alternative to traditional metropolitan centres for technology companies and professional services.

The region has benefited from high-speed digital infrastructure, including a submarine internet cable that lands directly in Maroochydore CBD. The project includes a 550km undersea fibre optic cable which connects the Sunshine Coast to the 7000km Japan-Guam-Australia South submarine cable.

Construction of the cable landing station commenced in 2018, and the network became operational in March 2020. It provides the fastest international connection point from Queensland and east Australia to Asia and is the only international cable landing on the Australian east coast outside of Sydney.

In November 2024, Sunshine Coast Council entered into a partnership with Google, NEXTDC and others on Australia Connect, a new cable initiative to further the reach, reliability and resilience of digital connectivity in Australia and the Indo-Pacific. The Tabua transpacific subsea cable that will connect the United States and Australia to Fiji is expected to land at NEXTDC's SC1 data centre in Maroochydore in 2025.

This digital infrastructure supports the region's development as a significant hub for innovation in digital technology and related sectors.

Sunshine Coast university has expanded its innovation activities through the launch of UniSC Innovation, which encompasses multiple initiatives designed to bridge research and commercialisation. Key developments include the establishment of Collider as "the Sunshine Coast's hub for entrepreneurial skills, start-up success and community venture creation" and Launch Pad as a hub for innovative businesses committed to future-focused development.

The university has also announced plans for five new specialist laboratories, including a final assembly hub for electric vertical take-off and landing vehicles, which will anchor a flagship Innovation Precinct at the USC campus. These developments represent a significant scaling up of the region's innovation infrastructure and its capacity to support advanced technology companies.

The region has gained recognition for specific sector strengths, particularly in areas aligned with Queensland's strategic priorities. The Sunshine Coast has been selected as the location for the Critical Infrastructure Information Sharing and Analysis Centre (CI-ISAC), a national organisation focused on cybersecurity, which has established operations in Maroochydore.

The Food and Agribusiness Network (FAN), described as "Australia's fastest growing food industry group," has become a significant cluster connecting and growing the region's food and agriculture businesses through collaboration, innovation acceleration, and trade promotion. This development builds on the region's natural advantages in agriculture while incorporating technology and innovation to create higher-value opportunities.

A major upcoming development is TURBINE, a $134 million project to build "Australia's first collaborative food and beverage manufacturing, education, and innovation centre" which will create a national blueprint for collaborative manufacturing and assist businesses to scale from product ideation through to export.

The ecosystem has also developed strong connections to state and national innovation strategies. The region has benefited from Queensland's Innovation Precincts and Places Strategy, which includes funding streams for strategic initiatives, infrastructure development, and technology projects that support future industry-focused collaborations.

Silicon Coast has evolved during this period to become more strategically focused, developing program offerings and events that align with broader economic development objectives. Recent initiatives have included the FWD series with events focused on climate technology, artificial intelligence, and other emerging sectors, demonstrating the organisation's capacity to respond to evolving innovation priorities.

Innovation in Key Sectors

The Sunshine Coast ecosystem has developed particular strengths in several strategic sectors that reflect both natural advantages and deliberate development strategies.

Digital Technologies and Immersive Media. The region has positioned itself as a potential world leader in digital immersive entertainment, supported by high-speed digital infrastructure and growing creative industries capabilities. This includes gaming, virtual reality, and other digital media applications that benefit from the region's lifestyle appeal in attracting creative talent.

Health and Biotechnology. The University of the Sunshine Coast's Centre for Bioinnovation is significant capability in biotechnology applications, including rapid point-of-care diagnostics, vaccine therapeutics, sustainable aquaculture tools, and bioinsecticides.

Advanced Manufacturing. Recent initiatives include the Manufacturing Excellence Forum and advanced composites being developed by Zone RV and Cure Marine, which utilise the largest 3D printer in the southern hemisphere at Coolum Business Park. Sunshine Coast Maker Space's academy specialises in advanced manufacturing techniques, including computerised numerical control machining, aimed at equipping businesses with skills and knowledge to expand manufacturing capabilities.

Geospatial and Space Technologies. The University of the Sunshine Coast's Space to Sea Accelerator program focuses on unlocking commercial potential in the geospatial industry, developing sustainable solutions for environmental challenges and disaster management through partnerships with Smart Sat CRC and industry leaders.

Clean Technology and Sustainability. The region has developed capabilities in clean technology and sustainable development, supported by both research institutions and a community culture that values environmental sustainability.

Governance and Collaborative Structures

The Sunshine Coast innovation ecosystem is distinctive for its collaborative governance approach, which relies more heavily on community networks and informal coordination than formal institutional structures. This approach has both advantages and limitations compared to more centrally managed innovation districts.

The ecosystem includes multiple industry groups and networks including the Sunshine Coast Creative Alliance, Food and Agribusiness Network, Clean Tech Network, Sunshine Coast Tech Industry Alliance (SCTechIA), and various manufacturing and creative sector organisations. These groups operate with considerable autonomy while maintaining connections through Silicon Coast and other coordinating mechanisms.

Sunshine Coast Council plays an important convening and strategic role, particularly through its economic development activities and infrastructure investments. In 2023, the council brought industry together to discuss ecosystem development to identify gaps and growth opportunities, leading to plans for a centralised innovation hub at Maroochydore and upgraded facilities at the University of the Sunshine Coast.

The collaborative approach has enabled rapid adaptation and responsiveness to opportunities but also creates challenges in terms of strategic coordination and resource allocation across multiple priorities and initiatives.

Challenges and Future Directions

Despite significant progress, the Sunshine Coast innovation ecosystem faces several challenges that must be addressed to sustain growth and competitiveness.

Scale and Critical Mass. As a regional centre, the Sunshine Coast faces ongoing challenges in achieving the scale and critical mass that characterise larger metropolitan innovation hubs. While the region's collaborative culture partially compensates for smaller scale, continued growth in population and economic activity will be necessary to support increasingly complex innovation activities.

Access to Capital. Like many regional centres, the Sunshine Coast has limited access to venture capital and other forms of growth financing compared to major metropolitan areas. While government funding and programs provide important support, the development of regional investment capabilities remains a priority.

Talent Attraction and Retention. The region's lifestyle appeal helps attract talent, but competition with major metropolitan centres for specialised skills remains challenging. Continued development of career pathways and opportunities for professional growth will be essential to talent retention.

Infrastructure Development. While significant progress has been made in digital infrastructure, continued investment in transport connections, research facilities, and commercial infrastructure will be necessary to support growth.

Coordination and Strategic Focus. The collaborative, community-driven approach that has been a strength of the ecosystem also creates challenges in terms of strategic coordination and priority setting. Current plans for coordinated development, including the innovation hub at Maroochydore and upgraded university facilities, represent attempts to address these coordination challenges.

Global Connectivity. Developing stronger connections to international markets and innovation networks remains important for regional companies seeking to scale globally.

Future Opportunities and Strategic Directions

The Sunshine Coast innovation ecosystem is well-positioned to capitalise on several emerging opportunities that align with its existing strengths and strategic investments.

Proximity to SE Asian/ASEAN economies. The Sunshine Coast Region can leverage its proximity to Southeast Asia's middle class of 200 million people as the bloc becomes the world's fourth largest economy by 2040. Priority collaboration areas identified in Australia's Southeast Asia Economic Strategy include green energy transition, healthcare, and digital economy creating pathways for two-way knowledge transfer and joint ventures that advance mutual prosperity.

Olympic Games Legacy. The region's role in the Brisbane 2032 Olympic and Paralympic Games presents opportunities to showcase innovation capabilities and attract international attention to the ecosystem. Sunshine Coast Council has identified Olympics-related preparation as a driver for accelerated infrastructure and ecosystem development.

Innovation Hub Development. The planned centralised innovation hub at Maroochydore, combined with upgraded university facilities and the TURBINE food and beverage centre, is a significant opportunity to create greater coordination and critical mass within the ecosystem.

Sector Leadership. The region has opportunities to develop leadership positions in specific sectors where it has developed particular strengths, including immersive digital technologies, sustainable manufacturing, biotechnology, and geospatial applications.

Regional Network Development. Strengthening connections with other regional innovation centres and with the broader South East Queensland innovation ecosystem could provide access to larger markets and greater specialisation opportunities.

International Partnerships. Developing strategic international partnerships, particularly with regions that share similar characteristics or complementary capabilities, could provide access to markets and knowledge that support ecosystem growth.

Lessons for Regional Innovation Development

The Sunshine Coast experience offers several insights for other regions seeking to develop innovation ecosystems.

Community-Driven Development. The success of Silicon Coast and similar community organisations demonstrates the importance of grassroots engagement and community building in regional innovation development. Formal institutional structures are important, but sustainable ecosystems require genuine community engagement and ownership.

Collaboration Over Competition. The Sunshine Coast's collaborative culture, where organisations and individuals actively support each other rather than competing for limited resources, has been crucial to ecosystem development. This approach may be important in regional settings where scale limitations make collaboration essential.

Leveraging Natural Advantages. The region's success in attracting talent and companies despite its smaller scale demonstrates the importance of leveraging natural advantages, including lifestyle appeal, cost advantages, and unique capabilities.

Infrastructure Investment. Strategic investments in digital infrastructure, research facilities, and innovation hubs have been crucial enablers of ecosystem development. These investments require long-term thinking and often involve coordination across multiple levels of government and private sector partners.

University-Industry Integration. The close integration between the University of the Sunshine Coast and the broader innovation ecosystem has been essential to the region's development. This integration goes beyond traditional technology transfer to encompass shared facilities, collaborative programs, and joint initiatives.

Incremental but Persistent Development. The Sunshine Coast's development has been characterised by incremental but persistent progress over two decades. This long-term approach may be more realistic for regional centres than attempting to create innovation ecosystems rapidly.

Sector Focus with Diversification. While the region has developed strengths in particular sectors, it has maintained diversification across multiple areas rather than concentrating on a single industry. This approach provides resilience and opportunities for cross-sector collaboration.

Conclusion

The Sunshine Coast innovation ecosystem is a distinctive model for regional innovation development that emphasises community collaboration, strategic infrastructure investment, and leveraging of natural advantages. Over two decades, the region has evolved from a tourism-dependent economy to a diversified innovation hub that competes successfully for talent and investment while maintaining its character and livability.

The ecosystem's development demonstrates that smaller regions can create competitive innovation environments through sustained effort, strategic investment, and effective collaboration. The community-driven approach exemplified by Silicon Coast and other organisations has created networks and relationships that enable rapid response to opportunities and effective resource sharing.

Key success factors have included early investment in university and innovation infrastructure, development of high-speed digital connectivity, creation of community networks that span sectors and organisations, strategic focus on sectors that align with regional advantages, and sustained support from local government and other institutional partners.

Current developments, including the Maroochydore City Centre, expanded university innovation facilities, and sector-specific initiatives like TURBINE, position the region for continued growth and increasing sophistication of innovation activities. Integrating these developments with existing community networks and collaborative structures should enable the ecosystem to achieve greater scale while maintaining its collaborative character.

The Sunshine Coast experience offers valuable lessons for other regional centres seeking to develop innovation capabilities. The emphasis on community building, collaborative culture, and leveraging natural advantages provides a template that may be relevant for regions that cannot rely solely on scale or traditional metropolitan advantages.

As the ecosystem continues to mature, challenges related to scale, capital access, and coordination will require ongoing attention. However, the foundation of collaborative relationships, strategic infrastructure, and community engagement provides a strong platform for addressing these challenges and capitalising on emerging opportunities.

The Brisbane 2032 Olympics and Paralympics represent a significant opportunity to showcase the region's innovation capabilities and accelerate ecosystem development. With appropriate strategic focus and continued investment, the Sunshine Coast is well-positioned to serve as a model for how regional centres can compete in the global innovation economy while maintaining their unique identity and advantages.

11. Henderson Defence Precinct, Perth

Henderson, located 23 kilometres south of Perth's central business district on Western Australia's Cockburn Sound, has emerged as one of Australia's most significant defence and maritime precincts. The Australian Marine Complex (AMC) at Henderson is home to approximately 150 businesses across five designated zones, serving as a world-class centre of excellence for marine industry manufacturing, fabrication, assembly, technology, repair and maintenance activities.

Context

Perth, with over 2.2 million residents, anchors Western Australia's economy and provides the skilled workforce foundation for Henderson's transformation. The state's strong engineering and technical capabilities, developed through decades in the resources sector, have proven readily transferable to defence applications.

Henderson's defence heritage traces back to 1915, when the Commonwealth Government resumed the land for defence purposes, with a large naval base originally planned between 1913-1920 before being cancelled and eventually replaced by HMAS Stirling in 1978. This historical connection to naval defence has provided the foundation for Henderson's current role as Australia's western defence hub.

The precinct has undergone remarkable transformation since the early 2000s, evolving from a general maritime facility to become central to Australia's national shipbuilding enterprise and the AUKUS partnership. The Commonwealth Government is now establishing a consolidated Defence Precinct at Henderson that will serve as a world-class shipbuilding and sustainment facility vital for Australia's national security.

Foundations and Early Development: 1988-2008

Austal commenced operations in Henderson in 1988 with a vision to build high-quality commercial vessels for the international market. This marked the beginning of Henderson's modern shipbuilding era, with the company establishing itself as a pioneer in aluminium vessel construction and advanced maritime technologies.

Austal's early success in commercial markets provided crucial foundations for defence applications. Since 1988, the company has produced over 260 commercial and defence vessels for over 100 customers in 50 countries, growing to comprise ship building facilities in Australia, the USA and Philippines. This international expansion demonstrated the global competitiveness of Henderson's capabilities.

The Australian Marine Complex was formally established in 2003 when the Common User Facility, owned by the Western Australian Government, commenced operations. This provided the essential infrastructure foundation that enabled multiple companies to co-locate and share facilities, creating the collaborative environment that would become Henderson's signature strength.

The complex was strategically designed with five designated zones: shipbuilding, technology, support industry, fabrication, and recreational boating precincts. This thoughtful planning created synergies between different maritime activities while establishing clear pathways for expansion.

During this period, Henderson also became the preferred location for Royal Australian Navy sustainment activities. The shipbuilding precinct serves as a maintenance facility for the Royal Australian Navy's Collins-class submarines and Anzac-class frigates, being located across Cockburn Sound from HMAS Stirling on Garden Island. This proximity to the fleet base created natural efficiencies and reinforced Henderson's strategic importance.

Infrastructure Development and Industry Consolidation: 2008-2020

A critical infrastructure milestone occurred in 2010 with the commissioning of the A$60 million floating dock Yargan, capable of lifting up to 12,000 tonnes from the water and transferring up to 3,500 tonnes from water to land. This capability positioned Henderson to handle increasingly complex naval sustainment requirements.

The Anzac-class frigates of the Royal Australian Navy undergo their Midlife Capability Assurance Program (AMCAP) at the Australian Marine Complex, with BAE Systems Australia leading this critical upgrade program through the Warship Asset Management Agreement Alliance. This established Henderson as the western hub for major naval sustainment operations.

Following the establishment of Austal Philippines and the re-alignment of commercial ship building operations to Balamban, Cebu in 2012, Austal Australia's focus became defence vessel design and construction for the Australian and global export markets. This strategic pivot towards defence work aligned with growing government emphasis on sovereign capabilities.

The period also saw Henderson emerge as a significant export hub for regional naval capabilities. Austal built and delivered Guardian-class patrol boats for Pacific Island nations and Cape-class patrol boats for international customers, including the Trinidad and Tobago Coast Guard.

Education and training infrastructure expanded during this period to support the growing workforce. South Metro TAFE established partnerships with the defence industry spanning over two decades, delivering qualifications and courses throughout Western Australia, culminating in the $2.4 million Naval Base Campus opening in June 2019.

The Western Australian Government recognised the strategic importance of developing Henderson through comprehensive planning. The Strategic Infrastructure and Land Use Plan for the Australian Marine Complex highlighted short, medium and long-term development required to meet Defence and commercial sector needs.

Strategic Partnership and Continuous Shipbuilding: 2020-Present

A watershed moment occurred in March 2022 when the Morrison Government announced up to $4.3 billion investment to deliver Western Australia's first large-vessel dry berth, creating a world-class precinct at Henderson and supporting thousands of local jobs. This represented the largest single infrastructure commitment to Henderson and signalled government intent to establish sovereign shipbuilding capabilities.

The Albanese Government built upon this foundation with the announcement in November 2023 of a strategic partnership between Defence and Austal Limited at Henderson Shipyard, addressing the Defence Strategic Review's observation that consolidation was needed to sustain the number of shipbuilders at Henderson.

The Strategic Shipbuilding Agreement was formally signed in August 2025, establishing Austal Defence Australia as the Commonwealth's strategic shipbuilder for Tier 2 surface combatants at Henderson. This landmark agreement provides the industrial certainty needed for long-term workforce and infrastructure investment.

The most significant announcement came in October 2024 when the Australian Government committed to establishing a consolidated Commonwealth-owned Defence Precinct at Henderson to underpin tens of billions of dollars of investment in defence capabilities over the next two decades and support approximately 10,000 well-paid, high-skilled local jobs.

Critically, the Government determined that Henderson will be the home of depot-level maintenance and contingency docking for Australia's future conventionally-armed, nuclear-powered submarines. This designation positions Henderson as central to the AUKUS partnership and Australia's future submarine capabilities.

The precinct will support construction of multiple vessel classes, including 18 Landing Craft Medium and 8 Landing Craft Heavy over the next decade, along with new general-purpose frigates for the Navy. This provides the continuous pipeline of work essential for maintaining skilled workforces and industrial capabilities.

In September 2025 the Commonwealth Government announced that it would provide a further $12 to deliver continuous naval shipbuilding and AUKUS in Western Australia. This initial investment will underpin the delivery of:

- The construction of surface vessels for the Australian Defence Force, starting with Army's Landing Craft and, pending successful consolidation, the domestic build element of Australia's future general-purpose frigates.
- Facilities to support the sustainment of Australia's surface combatant vessels.
- Contingency docking capabilities for Australia's future conventionally-armed, nuclear- powered submarine fleet from the early 2030s.
- Depot-level maintenance, including graving docks.

The Research and Education Ecosystem

Western Australia has developed strong institutional support for defence innovation. The Defence Science Centre creates connections across universities, industries and Defence to develop Western Australia's science and research capabilities, serving as the state's contribution to the Australian Defence Science and Universities Network.

The University of Western Australia established the Defence and Security Institute to unify and focus UWA's expertise and build partnerships in defence and security research, policy, engagement, and education. The university also operates an International Space Centre combining 21 speciality areas of research.

Edith Cowan University partners with the Defence Science Centre and Naval Shipbuilding College, offering seven engineering degrees that serve as pathways to careers in Defence industry. ECU has particular strengths in cyber security research and is a lead partner in the Cyber Security Cooperative Research Centre.

The Defence Science Centre has facilitated 50 industry-academic connections and coordinates visitors such as prime contractors and global research agencies. This collaborative approach ensures Western Australia's research capabilities directly support defence industry needs.

Workforce development has become increasingly sophisticated, with Western Australia's 'lower fees, local skills' initiative halving the price of key trade qualifications required by the defence sector. The state is developing a Maritime Defence Industry Workforce Development Plan to systematically address skills requirements.

Current Capabilities and Industry Base

Henderson now hosts major defence contractors including Austal, BAE Systems Australia, ASC, and Luerssen Australia, creating a concentrated ecosystem of complementary capabilities. The new infrastructure will be constructed at the southern end of the Henderson shipyard, where these companies have adjacent facilities.

Birdon, an Australian-owned maritime engineering company, has expanded its footprint at Henderson with 42 employees working from a new technical hub, drawing on global capabilities to build local capacity for Australian projects. This demonstrates how international companies are investing in Henderson's long-term potential.

Western Australia features advanced manufacturing capabilities supported by tertiary institutions that are world-class and at the forefront of next-generation defence technologies and innovation, working across defence's five operational domains: information and cyber; maritime; air; space; and land.

The state's broader industrial base provides crucial support for defence activities. Defence industry employees work alongside people with complementary skills from industries such as oil and gas, mining, manufacturing and construction, providing a resilient workforce with transferable skills.

Challenges and Future Directions

Despite remarkable progress, Henderson faces significant challenges in realising its full potential. Questions arise over where sustained funding will come from, as the Royal Australian Navy's future fleet of nuclear-powered submarines and infrastructure will consume a predicted $53-63 billion over the next decade.

There are concerns about where the projected 10,000 skilled workers will come from, with workforce development requiring urgent attention. The scale of skills requirements spans from PhD-level research through engineering to trades training, requiring coordinated effort across education and industry.

The 2023 Defence Strategic Review highlighted that challenges around workforce, layout and development of the Henderson Precinct needed to be addressed, along with relationships between infrastructure, shipbuilders, contractors and workforce providers.

Infrastructure complexity presents ongoing challenges. Much of the land at Henderson is owned by the Western Australia Government and managed through leases with multiple entities, requiring careful coordination as the precinct evolves. The transition to consolidated Commonwealth ownership will require sensitive management of existing operations.

The scale of the dry dock infrastructure required is enormous, with comparable US Navy facilities costing billions of dollars and taking years to construct. Australia has not built a military dry dock since World War II, presenting both technical and project management challenges.

Future opportunities centre on Henderson's potential to become a true innovation district. The AMC features a dedicated 24-hectare Technology Precinct to support R&D activities critical to the defence, resources and marine sectors. Expanding this capability could position Henderson as a hub for next-generation maritime technologies.

The precinct's strategic location provides natural advantages for supporting Indo-Pacific operations. Henderson will coordinate AUKUS submarine support activities with the Submarine Rotational Force-West at HMAS Stirling from 2027, with the Australian Government investing $8 billion to support this capability.

Governance and Coordination

The Australian and Western Australian governments have signed a Cooperation Agreement providing a foundation for collaboration to deliver the Defence Precinct, including developing a dedicated joint forum to deliver skilled workforce requirements and establishing necessary infrastructure.

This whole-of-government approach addresses one of the key lessons from other defence precincts: the need for sustained coordination across multiple levels of government and diverse stakeholders. The agreement demonstrates mutual commitment to collaboration over the coming decade and beyond.

Defence and the WA Government have commenced consultation with leaseholders, local industry and the commercial sector about the future of the shipyard, emphasising the importance of ensuring current activities can continue as the project progresses.

Conclusion

Henderson's transformation from a general maritime facility to Australia's western defence powerhouse demonstrates the potential for strategic government investment and industry partnership to create world-class capabilities. The precinct's success reflects careful building upon existing strengths - Western Australia's engineering capabilities, maritime heritage, and proximity to key operational areas.

The strategic partnership model between government and industry, exemplified by the Austal agreement, provides a template for sustaining long-term capability development. By securing continuous pipelines of work, Henderson has overcome the traditional boom-bust cycle that has plagued Australian shipbuilding.

The emphasis on education and training infrastructure ensures Henderson's workforce will evolve with technology and requirements. The integration of universities, TAFE institutes, and industry training demonstrates the ecosystem approach necessary for sustained success.

Henderson's designation as the AUKUS submarine maintenance hub positions the precinct at the centre of Australia's most significant defence capability program. This role requires the highest standards of technical competence, security, and safety - areas where Henderson's track record provides confidence.

Lessons for Other Regions

For other regions seeking to develop defence capabilities, Henderson offers several key lessons.

- *Building on existing industrial strengths and infrastructure* provides more sustainable *foundations than creating capabilities from scratch.*
- *International partnerships and export success validate capabilities* and provide revenue streams that support domestic requirements.
- *Sustained government commitment across political cycles is essential* for long-term success.

Henderson's experience also highlights the importance of geographic advantages - proximity to operational areas, access to skilled workforces, and suitable natural harbours. These advantages only translate to success through strategic planning, sustained investment, and collaborative governance arrangements.

The precinct's future success will depend on successfully managing the transition to consolidated Commonwealth ownership while maintaining existing capabilities and relationships. The scale of infrastructure development required presents significant challenges, but Henderson's track record of delivering complex projects provides grounds for optimism.

Most significantly, Henderson demonstrates that middle-power nations can develop world-class defence capabilities through strategic focus, international partnerships, and sustained commitment. As Australia faces an increasingly complex strategic environment, Henderson's role in providing sovereign capabilities and supporting alliance relationships will only grow in importance.

The Henderson experience proves that with the right combination of natural advantages, industrial foundation, government support, and strategic vision, defence precincts can drive both economic development and national security outcomes. For policymakers and industry leaders seeking to build resilient defence capabilities, Henderson provides a compelling case study in successful ecosystem development.

12. Tonsley Innovation Precinct, Adelaide

The Tonsley Innovation District, situated 10 kilometres south of Adelaide's central business district, is recognised as Australia's most honoured innovation district and serves as a global standard for transforming postindustrial areas. The 61-hectare area, which was previously used for car manufacturing, now accommodates about 2,000 workers across many organisations in addition to a growing residential community and provides educational services to 8,500 students each year.

Context

Adelaide, the capital of South Australia, has a metropolitan population of over 1.3 million people and has historically combined public sector jobs with a robust manufacturing industry. The end of car manufacturing in the late 2000s posed a challenge but also created a chance for economic diversification.

Tonsley is now seen as a global model for transforming traditional manufacturing sites, showing how effective government support, collaboration among institutions, and targeted industry growth can turn former industrial areas into thriving centres of innovation.

The precinct focuses on four rapidly growing sectors in which South Australia has demonstrated expertise: health and medical devices, assistive technologies, clean technology and renewable energy, mining and energy services, and automation, software, and simulation. As a founding member of the Global Institute on Innovation Districts, Tonsley can utilise frameworks that enhance its value proposition and create opportunities to advance in sustainability, economic resilience, and attracting industries.

Industrial Heritage and Transformation: 1955-2008

The history of Tonsley starts in 1955 when Chrysler acquired 73 hectares of land to develop its motor parts and accessories division and to build an engine plant. The facility, which opened in 1964, played a key role in South Australia's manufacturing economy for almost 50 years.

In 1979, Mitsubishi Motors Corporation bought a stake in Chrysler Australia, and in the next year, it acquired the rest of the shares, renaming the company Mitsubishi Motors Australia Limited. During its peak years, the plant produced vehicles including the Mitsubishi Magna, Verada, and later the ill-fated 380 model.

Shifting global trends in the automotive industry and decreasing sales presented significant challenges. On Tuesday, February 5, 2008, during a meeting at Mitsubishi's Tonsley Park car manufacturing facility, 1,000 employees were informed that they would be losing their jobs. Mitsubishi announced that it would close its manufacturing operations in Australia. The final Mitsubishi 380 sedan was produced on March 27, 2008.

This closure signalled the start of a nationwide decrease in car manufacturing and indicated that traditional Australian manufacturing could no longer compete globally. The economic and social effects reached well beyond the factory boundaries, influencing suppliers, local businesses, and the overall community's confidence in the industrial future of South Australia.

Vision and Early Development: 2010-2018

In 2010, the South Australian Government bought the old car manufacturing plant and began working on a new plan for the site, aiming to transform it into a sustainable economic centre for future industries. This decision reflected a strong commitment to changing the economy instead of permitting a decline in industry.

The government's plan focused on establishing an innovation district that would:

- Leverage existing infrastructure and skilled workforce
- Foster collaboration between research, education, and industry
- Focus on high-value manufacturing and knowledge-intensive sectors
- Create a mixed-use precinct integrating employment, education, and residential development

The site started to evolve into a globally recognised innovation district. Renewal SA has kept the 5-hectare outer shell of the Main Assembly Building (MAB) from its days as an assembly line and transformed it into an inviting and accessible area filled with vibrant urban forests, community spaces, services, and convenient connections to Adelaide's rail network and greenways.

A crucial turning point was when anchor institutions made their commitment. In 2015, Flinders University established a presence in Tonsley that accommodates the College of Science and Engineering, College of Business, Government and Law, as well as the Medical Device Research Institute and the Centre for Nanoscale Science and Technology. TAFE SA also played a prominent role in building a connected educational system.

The SAGE Group emerged as one of the first success stories, with Managing Director Andrew Downs transforming his small backyard electrical contracting business into the largest independent automation and control systems integrator in Australia. SAGE partnered with Flinders University, TAFE SA, and several other technology companies, such as Siemens, Signostics, Micro-X, Innovyz, Simulation Australia, and ZEN Energy Systems, at the Tonsley site.

Other notable early tenants included international firms such as Siemens Energy, ZEISS, and Tesla, as well as innovative Australian companies like Micro-X, which specialises in ultra-lightweight, mobile X-ray imaging equipment that utilises carbon nanotube field emission technology.

Expansion and Maturation: 2018 to Present

Since 2018, Tonsley has undergone significant growth and gained more recognition on a global scale. Since its opening in 2012, the district has made significant progress, evolving from its original goal of revitalising a struggling manufacturing sector in Adelaide's southern suburbs to now addressing some of the world's most complex challenges through innovative solutions.

Currently, approximately 2,000 people are employed at Tonsley through over 100 organisations, which is more than twice the number of employees when Mitsubishi shut down in 2008. Each year 8,500 students learn alongside this skilled and specialised workforce at the prominent academic institutions, Flinders University and TAFE SA as well as private training organisations (RTOs) such as Skillslab.

Major Success Stories and Exits

The district has achieved multiple important successes in commerce. In May 2025, the US consulting and engineering company Tetra Tech purchased the Adelaide-based SAGE Group in a significant deal estimated to be valued at $150 million. The SAGE Group began as a small backyard business in the early 1990s and has since expanded into a global company, now employing nearly 800 individuals in offices throughout Australia and the Asia-Pacific region.

In early 2025, SAGE Group was recognised as the 11th largest system integrator globally in the System Integrator Giants Report, having reported over $200 million in gross revenue the previous year. This acquisition is one of the most important technology exits in South Australia, confirming the effectiveness of the innovation district model.

Research and Development Excellence

Flinders University at Tonsley has emerged as a leader in applied research and collaboration with industry. The $120 million innovation hub serves as a central location for the University's teaching and research in areas such as computer science, artificial intelligence, defence and national security, autonomous vehicles, information technology, engineering, and mathematics.

The Medical Device Research Institute is dedicated to conducting high-quality research and improving health outcomes by applying technology in the medical and allied health fields. The Medical Device Partnering Program offers a distinctive approach for collaboration among researchers, clinicians, end users, and industry partners.

Infrastructure and Sustainability Leadership

Tonsley is home to the Flinders Factory of the Future, a specially designed facility for large scale testing of new technologies and processes before they are used in advanced manufacturing, such as shipbuilding. The Flinders Factory of the Future shares its room with the Tonsley Technical College, positioning South Australia as a leader in creating dynamic research, education, and training facilities connected to industry.

The district has also adopted mixed-use development that includes residential elements. Tonsley Village offers a novel lifestyle that integrates both starter homes and long-term residences, along with apartments and terrace houses, all located just 20 minutes from the city. The opening of the 88-room Mantra Tonsley Adelaide hotel, along with its bar and bistro, in 2024 improved the amenities in the area.

Global Recognition and Networking

Tonsley is an inaugural member of the Global Network of Innovation Districts, operated by The Global Institute on Innovation Districts

(GIID). As of 2024, the network has grown to include 45 innovation districts globally, with Australia having a number of participating districts across several states.

Challenges and Future Directions

Even with significant achievements, Tonsley continues to encounter common challenges that innovation districts face worldwide. Access to funding for growth-stage companies is still a challenge, especially for those trying to expand beyond the start-up phase. Although successful exits, such as those from the SAGE Group, show the potential for generating profits, the ecosystem would improve with an increased presence of local venture capital and private equity.

Retaining and attracting talented individuals is a continuing priority. The combination of education and industry fosters the growth of local talent pools, but to compete with the job markets in Sydney and Melbourne, we must consistently prioritise career development opportunities and improve quality of life benefits.

Statewide Innovation Integration

The new statewide Innovation District Framework will create additional opportunities for Tonsley to enhance its impact, utilise advantages, and draw more investment, as the entire state benefits from its multibillion-dollar AUKUS nuclear submarine deal. The new model will enhance the connections among Tonsley, Lot Fourteen, Adelaide BioMed City, Osborne Naval Shipyard, Edinburgh Defence Precinct, Waite Research Institute, and the upcoming Australian Space Park.

This coordinated strategy offers a significant chance to apply Tonsley's successful model to a larger innovation network, especially in the defence, space, and advanced manufacturing industries, where South Australia holds competitive strengths.

Future Development Opportunities

Recent development opportunities feature the 2.578-hectare Tonsley Administration Building site located at the district's entrance. This presents developers with a remarkable chance to undertake a significant redevelopment project at the entrance of this world-class area.

The ongoing growth of university programs, especially in new technologies like artificial intelligence, and sustainable manufacturing, is crucial for keeping Tonsley competitive in fast-changing global markets.

Conclusion

The Tonsley Innovation District is one of the most successful cases of postindustrial transformation in Australia. The State Government aims to convert the old Mitsubishi car factory into a leading high-tech manufacturing centre. This initiative has surpassed expectations by drawing in many innovators, businesses, and students, earning recognition as Australia's most awarded innovation district.

Lessons for Other Regions

The Tonsley model provides important insights for other areas implementing similar changes.

Strategic Government Leadership: Choosing to buy and develop the site instead of letting the market decide its future was essential. Government investment supplied the essential funding and forward-thinking approach required for the development of ecosystems in partnership with the private sector.

The partnership between Flinders University and TAFE SA: This established instant credibility and generated lasting demand for facilities and services. The partnership between education and industry has created significant advantages.

Concentrate on Competitive Advantages: Instead of trying to compete in every sector, Tonsley focused on areas where South Australia already excelled, including advanced manufacturing, medical devices, automation, and energy systems.

Reimagining the MAB and developing "collision spaces" have encouraged the casual interactions that are vital for fostering innovation ecosystems.

Global Connectivity: Being part of international networks and partnering with multinational companies has helped avoid isolation and provided access to worldwide markets and knowledge.

Mixed-Use Integration: By combining housing, shops, and recreational areas with places for work and learning, we can create vibrant communities instead of lifeless industrial zones.

The success of companies such as SAGE Group, which was acquired by a US firm for $150 million, shows that regional innovation districts can create businesses that compete on a global scale. As one business owner remarked, "Tonsley is a wonderland for someone like myself because it has all the ingredients in the petri dish that I need to be successful, and I've not found that anywhere else".

For cities undergoing changes after industrial decline, Tonsley demonstrates that with clear planning, ongoing investment, and teamwork in governance, old industrial areas can transform into hubs of innovation and economic development. The change from automotive assembly line to innovation district shows how areas can utilise their current resources while adjusting to changes in the global economy.

As Tonsley begins its second decade, its incorporation into South Australia's larger innovation framework and ongoing emphasis on new technologies enable it to continue being a leader in Australia's innovation sector. The district demonstrates how careful urban renewal can generate enduring economic and social benefits from industrial history.

13. 22@Barcelona

Barcelona, Spain's second largest city with over 1.6 million residents, has emerged as one of Europe's most dynamic innovation hubs through the transformative 22@Barcelona project. The district, spanning 198.26 hectares (489.9 acres) across 115 blocks in the formerly industrial Poblenou neighbourhood, is one of Europe's most ambitious urban regeneration schemes of the 21st century.

Context

The 22@ district has successfully transformed from "the Catalan Manchester" of the 19th century into a thriving knowledge-based economy, housing over 93,000 workers and generating approximately €10.3 billion in annual business turnover (Pique et al, 2019). This remarkable transformation began in 2000 when the Barcelona City Council approved the revolutionary 22@ urban planning designation, replacing the traditional industrial 22a zoning classification.

The strategic vision was clear: convert Poblenou's declining industrial landscape into Barcelona's technological and innovation district while preserving the area's industrial heritage and creating a mixed-use environment that balances economic activity with residential and recreational spaces. Since 2001, more than 8,223 new companies have established offices in the district, including major international corporations such as Yahoo! I+D, Mediapro, Microsoft, Sanofi-Aventis, Capgemini, Schneider Electric, SAP, NTT Data, and Indra.

The project's success is a paradigm shift in urban development, demonstrating how strategic planning, public-private collaboration, and adaptive governance can revitalise postindustrial areas for the knowledge economy while maintaining social cohesion and cultural identity.

The Genesis of Transformation: 1990s-2000

The seeds of Barcelona's innovation renaissance were planted in the late 1990s, grounded in two complementary visions for the city's future. The first concept, "Barcelona, City of Knowledge," was promoted by Professor Trullén and endorsed by former mayor Joan Clos, emphasising that new technologies had made information cheaper and more widely available, making knowledge the differentiating element that generates additional value and increases economic competitiveness.

The second concept, "Technological City," was championed by Miquel Barceló (founder and future president of the municipal company 22 Arroba BCN SA) and Antoni Oliva (future secretary of the 22@ Network Association), arguing that new technologies enabled industrial development initiatives to become part of the urban ecosystem.

By the late 1990s, Poblenou faced significant challenges. The area had been progressively abandoned by traditional industries, leaving behind contaminated land, obsolete infrastructure, and declining employment. However, its strategic location—close to Barcelona's city centre, the Mediterranean coast, and well-connected to transportation networks—presented enormous potential for regeneration.

The catalyst for change came with Barcelona's successful bid to host the Universal Forum of Cultures in 2004, which required significant urban infrastructure development. This event provided the political momentum and initial financing mechanism necessary to launch the ambitious 22@ transformation.

Building the Ecosystem: 2000-2008

The 22@Barcelona project officially commenced in 2000 with the approval of the Special Infrastructure Plan, which established the regulatory framework for the area's transformation. The plan was revolutionary in its comprehensive approach, addressing not just zoning changes but also infrastructure development, heritage preservation, and social integration.

Regulatory innovation

The new 22@ zoning designation introduced unprecedented flexibility in urban planning. The plan required land developers to dedicate specific percentages of their projects to different uses: 10% for incubators, 10% for subsidised housing, and 10% for green areas. This innovative approach ensured that the district would maintain mixed-use characteristics while generating public benefits from private development.

Infrastructure development

The transformation required massive infrastructure investment, including 14,800 linear metres of new streets with high-level services, 46 new utility conduits, advanced sewage systems, selective pneumatic waste collection, district heating and cooling systems, and 47,000 metres of dark fibre optic cables. This state-of-the-art infrastructure positioned the district to attract technology companies requiring advanced connectivity and sustainability features.

Cluster strategy

The 22@Barcelona model adopted a cluster-based approach, focusing initially on five strategic sectors: Information and Communication Technologies (ICT), media, biomedical technologies, energy, and design. This cluster-based urban development approach aimed to create competitive international positions for each sector, supported by infrastructure such as the Barcelona Growth Centre and Barcelona Activa.

Early corporate anchors

The arrival of major corporations served as crucial anchors for the emerging ecosystem. Telefónica established a significant presence, bringing not only employment but also credibility that attracted other technology companies. Media companies like Mediapro chose the district for their headquarters, leveraging Barcelona's growing reputation in audiovisual production.

During this foundational period, approximately 1,500 companies established operations in the district, with nearly half being newly created start-ups. The diverse mix of established corporations and emerging ventures created a unique environment for knowledge transfer and innovation.

Institutional Development and Ecosystem Maturation: 2008-2015

The global financial crisis of 2008 paradoxically strengthened Barcelona's innovation ecosystem by forcing a focus on higher-value activities and entrepreneurship. During this period, the 22@ district evolved from a development project into a comprehensive innovation ecosystem.

Educational Infrastructure

Universities played a pivotal role in the district's maturation. The Pompeu Fabra University established its Communication Campus in the renovated Ca l'Aranyó textile mill, while the Bau Design School occupied the former Can Munné fabric warehouse. These model restorations, such as Palo Alto, an old industrial site that currently houses studios of artists and creators, including designer Javier Mariscal, demonstrate how tradition serves progress, and old factories have been redesigned to achieve greater sustainability.

Innovation support infrastructure

The establishment of specialised incubators and accelerators during this period proved crucial for nurturing start-ups. Barcelona Activa, the city's public business promotion agency, expanded its programs to include specific support for technology ventures. Barcelona Activa operated as a business incubator program with a small seed fund that provided advice and support to start-ups and entrepreneurs.

Tech Barcelona emergence

In 2013, a group of entrepreneurs led by Miguel Vicente and lawyer Ignasi Costas promoted the creation of the Barcelona E-commerce & Tech association, later renamed Barcelona Tech City to embrace a broader spectrum of technologies and companies. This organisation has become instrumental in uniting entrepreneurs, start-ups, corporations, investors, universities, and public administrations to create a cohesive ecosystem.

International recognition

During this period, Barcelona began attracting international attention as a technology hub. The Mobile World Congress, hosted annually in Barcelona since 2006, positioned the city as a global centre for mobile technology innovation. The event created significant spillover effects, attracting international investors and corporations to establish permanent presences in the city.

Expansion and Global Integration: 2015-2020

The period from 2015 to 2020 marked Barcelona's transition from a regional innovation centre to a globally recognised technology hub. The 22@ district served as the physical anchor for this transformation, while new initiatives expanded the innovation ecosystem across the entire metropolitan area.

Mobile world capital Barcelona

Mobile World Capital Barcelona was established to amplify the impact of MWC and bring it closer to citizens and the business sector, developing multiple activities aimed at highlighting the ecosystem on the world stage and attracting investment through various channels and formats. MWCapital founded *4 Years From Now* (4YFN), the start-up business platform present at MWC events around the world, creating a unique space for start-ups to connect with investors, companies, and institutions.

Venture capital ecosystem development

The maturation of Barcelona's start-up ecosystem attracted increasing venture capital investment. Notable VC firms such as Nauta Capital, K Fund, and Inveready became instrumental in providing early-stage funding to promising start-ups, while international investors increasingly recognised the potential of Barcelona-based ventures.

Key players in the local VC ecosystem included the following.

Caixa Capital Risc: The venture capital division of 'la Caixa' bank, launched in 2004, manages more than €180M and investing in early stages of innovative start-ups with investment sizes up to €3M in companies like CornerJob, Medlumics, Glovo App, and Geenapp.

Ysios Capital: A leading venture capital investor providing private equity financing to early- and mid-stage human healthcare and life science companies, founded in 2008 with more than €200 million in assets under management.

Active Venture Partners: A European venture capital company focused on digital start-ups, having invested in 25 companies across two funds totalling €74M, mainly focused on mobile retail, marketplaces, IoT, big data, SaaS, and digital business models.

Major success stories

This period witnessed several high-profile start-up successes that validated Barcelona's innovation ecosystem:

Glovo: The on-demand delivery platform became one of Spain's most successful start-ups, raising significant international funding and expanding across multiple markets.

TravelPerk: The business travel management platform emerged as a leading European SaaS company, attracting major venture capital investment.

Wallapop: The peer-to-peer marketplace demonstrated Barcelona's capability to produce consumer-focused platforms with global appeal.

Corporate innovation programs

According to the 2024 report from MWC, the number of hubs arriving in Catalonia and Barcelona has reached a stable trend of between 10 and

20 new ones each year. Almost 50% of hubs have been set up in the last six years, and 2021 stands out in particular as the year with the highest growth at a record 22 new hubs.

In 2024, 13 new hubs were established, backed by investments from Germany, the United States, China, Brazil, the United Kingdom, the Netherlands, and Japan. In 2024, there are a total of 160 active tech hubs.[8]

The report indicates that highly qualified talent is still the key factor in making the decision to set up in Catalonia (39%), followed by the previous presence of the parent company in the territory (38%) and quality of life (10%). The report noted that the availability of talent has become more relevant in recent years as opposed to previous physical presence in the territory which has lost ground.

Digital Renaissance and COVID Adaptation: 2020-Present

The COVID-19 pandemic initially posed significant challenges to Barcelona's innovation ecosystem, particularly affecting the Mobile World Congress and other major events. However, the crisis also accelerated digital transformation and strengthened the city's position as a hub for remote-first technology companies.

*Ecosystem Resilience and Growth***.** According to the Barcelona & Catalonia Start-up Hub's 2021 latest analysis[9], the region is home to more than 2,280 start-ups. The report claims that Barcelona is the 2nd most popular city among start-up founders in Europe and ranks 5th as an EU hub.

Real Estate and Infrastructure Evolution. In 2024, 32% of the total office space leased in Barcelona (291,000 m^2) was located in the 22@ district, with the year ending with a total take-up of 93,000 m^2 in the area, double the volume recorded the previous year. Currently, 21% of the city's total office stock is located in 22@, and prime rents in the district stood at €22.00/m^2/month in Q4 2024.

[8] Mobile Word Capital Barcelona (2025). Tech Hubs Overview. https://mobileworldcapital.com/app/uploads/2025/03/TECH-HUBS-OVERVIEW_DIG-ENG-DEF-1-1.pdf

[9] Catalonia Trade and Investment, Barcelona & Catalonia Startup Hub, https://startupshub.catalonia.com/

Tech Talent and Innovation Programs. Mobile World Capital Barcelona launched Talent Arena, an event dedicated to digital talent that brings together developers, leading tech companies, and educational institutions to foster learning, networking, and professional development. This initiative reflects the ecosystem's focus on talent development and retention.

Urban Technology Campus Initiative. The urban technology campus project began to take shape with the creation of Pier03, a transversal space housing companies such as Tiendeo, Viko, Novartis, and The Valley, along with the historic 7 Portes restaurant—the first place in Barcelona where Picasso lived.

Sustainability and Smart City Integration. The creation of the Center of Urban Innovation in a former industrial factory in the northern part of the 22@ is a public-private alliance with Cisco for the promotion of IoT (Internet of Things) and Smart City technologies, contributing to the growth of the existing innovation ecosystem while involving citizens and universities in decision-making regarding urban habitat improvement.

Current Ecosystem Metrics and Performance

The 22@Barcelona district presents impressive performance indicators that demonstrate the success of the urban transformation model.

Urban and infrastructures	***Companies and economic impact***	***Talent and society***
Area	*Jobs*	*Citizens (Sant Martí)*
200 hectares – 115 blocks	93.000	232.629
Potential floor	*Companies*	*Students*
4,000.000 square metres	8,223	25,000
Streets	*Turnover*	*University centres*
37 km	10,300 million Euros	10

Source: Pique, J. M., Miralles, F., & Berbegal-Mirabent, J. (2019). Areas of innovation in cities: The evolution of 22@ Barcelona. International Journal of Knowledge-Based Development, 10(1), 3-25, p 8.

Selection of interventions and their impact in the 22@Barcelona district

Dimension	***Action***	***Aim***	***Result***
Governance	*22@Network*	*Creation of a horizontal association involving companies, universities and institutions.*	*Creation of an association with 112 members and more than 15,000 professionals**

Dimension	*Action*	*Aim*	*Result*
	Cluster associations policy	*Creation of a governance structure for each cluster*	*5 clusters organisations created: Barcelona Media, Barcelona Digital, BCD, B-TECH, BIOCAT*
Talent and social development	*Porta 22*	*Promotion of the vocations and job opportunities*	*28.247* Persons have used porta 22 and Barcelona activa services*
	22@Breakfasts	*Developing a community of professionals of the district*	*100 monthly 22@Update breakfasts**
Companies and economic development	*22@Plus OAE (business one-stop service)*	*Settle of new companies in the district*	*8,223 companies*** *93,000 jobs**** *10,300 million Euros***
	Clusterisation of district	*Smart specialisation of the district*	*5 clusters promoted: ICT, MEDIA, HEALTH, ENERGY, DESIGN (40,4% companies of the district**)*
Infrastructures and urban development	*22@Law* *PEI – 22@Infrastructure plan*	*Urban transformation* *Deployment of new infrastructures*	*70% of 200 hectares*** *2.581 M€ of investment in real estate*** *14 km of streets*** *191,3 M€ of investment in infrastructures***

Note: Data refers to 2016 (*) and 2015 (**).
Source: Pique, J. M., Miralles, F., & Berbegal-Mirabent, J. (2019). Areas of innovation in cities: The evolution of 22@ Barcelona. International Journal of Knowledge-Based Development, 10(1), 3-25, p 9.

Challenges and Strategic Opportunities

Despite its remarkable success, the 22@Barcelona ecosystem faces several challenges that require strategic attention:

Talent Retention and Development. While Barcelona attracts significant international talent, retaining high-skilled professionals remains challenging due to:

- Salary disparities compared to other major European tech hubs
- Limited stock option culture in Spanish companies
- Competition from established hubs like London, Berlin, and Amsterdam

Access to Growth Capital. An estimated 70-80% of the funds raised by Spanish start-ups come from international investors, mainly in growth funding rounds, highlighting the strong appeal of the entrepreneurial ecosystem but also indicating limited local growth capital availability.

Scale-up Support Infrastructure. While the ecosystem excels at early-stage company creation, supporting scale-ups requires enhanced infrastructure, including:

- Larger office spaces for growing companies
- Specialised advisory services for international expansion
- Enhanced access to international markets and partnerships

Sector Diversification. Although the 22@ district has successfully diversified beyond its initial ICT focus, continued expansion into emerging sectors such as:

- Quantum technologies
- Artificial intelligence and machine learning
- Sustainable technology and circular economy
- Digital health and biotechnology represent significant growth opportunities.

Strategic Future Directions

Innovation District Expansion. The success of 22@Barcelona provides a model for extending innovation-focused development to other areas of the metropolitan region. Potential expansion areas include:

- Zona Franca industrial area transformation
- Barcelona Airport corridor development
- Integration with regional innovation hubs in Catalonia

International Connectivity Enhancement. Strengthening Barcelona's position as a gateway between Europe and Latin America is a significant strategic opportunity. This includes:

- Enhanced start-up exchange programs with Latin American markets
- Increased venture capital flows between regions
- Development of specialised programs for companies seeking European market entry

Sustainability Leadership. Building on Barcelona's smart city initiatives, the 22@ district can serve as a testbed for sustainable urban technologies:

- Circular economy business model innovation
- Energy efficiency and renewable energy integration

- Sustainable mobility solutions

Deep Tech Development. MWCapital's Tech Transfer projects and partnership with the European Innovation Council for deep tech venture building represent opportunities to strengthen Barcelona's position in advanced technology sectors requiring longer development cycles and higher capital requirements.

Lessons for Global Innovation Districts

The 22@Barcelona experience offers valuable insights for cities worldwide seeking to develop innovation districts:

Adaptive governance framework. The project's success demonstrates the importance of adaptive governance that can respond to internal and external changes of political, economic, and technological nature, emphasising the need to adapt to new technologies and their economic and social influence in urban regeneration projects.

Public-private collaboration model. The balanced approach to public-private partnership in 22@Barcelona provides a replicable model:

- Strategic public investment in infrastructure and regulation
- Private sector leadership in commercial development
- Shared risk and reward structures that align incentives

Heritage preservation and innovation integration. The Industrial Heritage Protection Plan, which consolidates protective measures for 114 elements, demonstrates how innovation districts can respect historical identity while fostering technological advancement.

Mixed-use development strategy. The requirement for integrated residential, commercial, and productive activities has created a vibrant 24/7 environment that supports both economic activity and community development.

Cluster-based development approach. The cluster-based approach recognises that physical proximity between firms, research centres, and universities can increase the potential for innovation through a high density of formal and informal channels of learning and exchange.

Global Benchmark and Replication

The 22@Barcelona model has already influenced urban development projects worldwide. The project's success has made it a model for other cities, including Boston, Medellín, and Montréal. The transferable elements include:

Planning and regulatory innovation

- Flexible zoning that promotes mixed-use development
- Public benefit requirements integrated into private development
- Streamlined approval processes for target industries

Infrastructure investment strategy

- Advanced digital connectivity as a foundation requirement
- Sustainable urban systems integration
- Transportation connectivity optimisation

Ecosystem development programs

- University-industry collaboration mechanisms
- Start-up incubation and acceleration programs
- International talent attraction initiatives

Conclusion

The 22@Barcelona innovation district is one of Europe's most successful examples of urban transformation for the knowledge economy. Over 25 years, the project has converted a declining industrial area into a thriving innovation hub that generates billions in economic value while preserving cultural heritage and promoting social integration.

The district's success stems from its comprehensive approach that addresses infrastructure, governance, talent development, and international connectivity as integrated elements of ecosystem development. The adaptive governance framework has enabled the project to evolve with changing technological and economic conditions while maintaining its core mission.

The 22@Barcelona case study exemplifies how strategic urban planning, productive clustering, collaborative initiatives, and creative regeneration can transform an underperforming industrial district into a thriving innovation hub, positioning Barcelona as a leader in technological innovation and providing a framework for other cities to follow.

As Barcelona looks to the future, the 22@ district provides a solid foundation for continued growth in emerging technologies, serving as a model for sustainable urban development. The lessons learned from this transformation offer valuable guidance for cities worldwide seeking to build innovation driven economies in the postindustrial era.

The ongoing evolution of 22@Barcelona demonstrates that successful innovation districts require continuous adaptation and investment. By maintaining its focus on collaboration, sustainability, and international connectivity, Barcelona is well-positioned to remain a leading innovation hub in the global knowledge economy.

Notes

Pique, J. M., Miralles, F., & Berbegal-Mirabent, J. (2019). Areas of innovation in cities: The evolution of 22@ Barcelona. *International Journal of Knowledge-Based Development*, *10*(1), 3-25.

14. Kendall Square

Kendall Square in Cambridge, Massachusetts, has earned recognition as "the most innovative square mile on the planet," representing one of the world's most successful examples of purposeful innovation district development. Located adjacent to the Massachusetts Institute of Technology (MIT) and spanning approximately one square mile along the Charles River, this district has transformed from an industrial wasteland into a global epicentre of biotechnology, artificial intelligence, and cutting-edge research commercialisation.

Context

The area encompasses approximately 750,000 square feet of laboratory space and 2.5 million square feet of office space, housing over 150 companies that employ more than 50,000 knowledge workers. With a unique density of start-ups, established corporations, research institutions, and venture capital firms, Kendall Square generates an estimated $5 billion in annual economic activity and has spawned companies with combined market capitalisations exceeding $1 trillion.

The ecosystem's foundation rests on MIT's world-class research capabilities, complemented by Harvard University, Harvard Medical School, and numerous other institutions in the Boston area. This concentration of intellectual capital, combined with strategic urban planning and sustained public-private collaboration, has created a self-reinforcing innovation environment that continues to attract global talent and investment.

Unlike many innovation districts that develop organically, Kendall Square is a deliberate transformation guided by visionary planning, substantial infrastructure investment, and policies designed to foster Collaboration between academia, industry, and government.

The district's evolution offers valuable insights for cities worldwide seeking to build knowledge-based economies and innovation driven growth.

Historical Foundation and Early Development: 1960s-1990s

Kendall Square's transformation began in the 1960s when the area was characterised by declining industrial Infrastructure, including abandoned warehouses and underutilised manufacturing facilities. The industrial decline had actually begun in the mid-1940s as companies sought cheaper labour elsewhere, culminating with the closure of Lever Brothers, Cambridge's biggest employer, in 1959.

Much of the area became what locals called "Nowhere Square," featuring vacant lots, chain-link fences, and abandoned buildings where gas works, vulcanised-rubber plants, and rendering facilities had once operated. Recognising the opportunity, MIT President James Killian announced in 1960 that the Institute would purchase the former Lever Brothers site to develop office buildings facilitating industry collaboration.

However, despite this early initiative and the proximity to MIT, the physical environment remained largely disconnected from the university's research activities, with much of Kendall Square functioning as little more than a giant parking lot through the 1960s and early 1970s.

Major developments over the period included the following.

The NASA Electronics Research Centre (1964-1970) served as an early catalyst, bringing federal research investment to the area and demonstrating the potential for high-tech industries. Although the centre closed in 1970, it established precedents for government-university-industry Collaboration and highlighted the area's potential as a technology hub.

MIT's Real Estate Development Strategy emerged in the 1970s as university leadership recognised opportunities to leverage institutional assets for economic development. The Technology Square complex, developed in partnership with Cabot, Cabot & Forbes, provided early models for combining research facilities with commercial space, creating environments where academic research could more easily transition to commercial applications.

Biotech Pioneers began establishing operations in the 1980s, with companies like Biogen and Genzyme choosing locations near MIT and Harvard Medical School. These early biotechnology firms benefited from access to university talent and research, while their presence began attracting additional life sciences companies and specialised service providers.

The *Central Transportation Planning Staff* and the *MIT Technology Licensing Office*, established during this period, created institutional frameworks that supported technology transfer and regional economic coordination. These organisations facilitated connections between university research and commercial applications, while providing Infrastructure for intellectual property management and start-up formation.

Venture Capital Emergence accelerated in the 1980s as Boston area firms began specialising in early-stage technology investments. Firms like Atlas Venture, Charles River Ventures, and General Catalyst established offices near Kendall Square, creating geographic clusters that facilitated deal flow and knowledge sharing among investors.

The *Whitehead Institute for Biomedical Research*, founded in 1982 in partnership with MIT, exemplified the institutional Innovation characterising this period. By combining university-affiliated research with independent governance and substantial private funding, the Whitehead Institute developed new models for translational research that would shape global biotechnology development.

The Biotech Boom and Infrastructure Development: 1990s-2000s

The 1990s marked Kendall Square's emergence as a global biotechnology centre, driven by breakthrough scientific discoveries, increased venture capital availability, and strategic infrastructure investments. This period established the district's reputation as a premier location for life sciences innovation.

Human Genome Project Impact created unprecedented opportunities for biotechnology commercialisation. MIT and Harvard researchers played pivotal roles in genomics research, with their discoveries giving rise to dozens of companies focused on drug discovery, diagnostics, and therapeutic development. Companies like Millennium Pharmaceuticals and Cubist Pharmaceuticals emerged directly from university research programs.

Forest City Enterprises became a major developer in Kendall Square during this period, creating mixed-use developments that combined laboratory space, office facilities, and amenities designed to support knowledge workers. Their developments, including One Kendall Square and Two

Cambridge Centre, established new standards for innovation-focused real estate.

The Kendall Square Association, formed in the mid-1990s, served as a coordinating mechanism for businesses, property owners, and institutional stakeholders. The association facilitated infrastructure improvements, advocated for supportive zoning policies, and created networking opportunities that strengthened connections within the ecosystem.

MIT's $1 Billion Capital Campaign, launched in the late 1990s, included substantial investments in biotechnology and computer science facilities. The new research buildings created additional opportunities for industry collaboration while expanding the university's capacity for cutting-edge research in emerging fields.

Akamai Technologies, founded by MIT researchers in 1998, demonstrated the expanding capabilities of Kendall Square beyond biotechnology. The company's success in internet infrastructure and content delivery networks attracted additional technology companies, highlighting the area's potential for diverse innovation sectors.

The Cambridge Innovation Centre (CIC) was established in 1999, creating flexible office space specifically designed for start-ups and early-stage companies. CIC's model of providing furnished offices, shared services, and networking opportunities became widely replicated, helping to establish Kendall Square as a start-up-friendly environment.

Transportation Infrastructure Improvements included enhancements to the MBTA Red Line Kendall/MIT station and improved pedestrian connections between the square and the MIT campus. These investments reduced barriers to Collaboration and made the area more accessible to talent from across the Boston metropolitan region.

Intellectual Property Policies evolved during this period as MIT and other institutions refined technology transfer practices. The establishment of clear intellectual property guidelines and start-up-friendly licensing terms reduced barriers to commercialisation and encouraged faculty entrepreneurship.

Ecosystem Maturation and Global Recognition: 2000s-2010s

The 2000s witnessed a surge in global recognition of Kendall Square as a leading innovation district, accompanied by substantial expansions in research capacity, corporate presence, and venture capital activity. During this period, the ecosystem matured from a promising biotechnology cluster into a comprehensive innovation environment supporting multiple high-tech sectors.

The Broad Institute, established in 2004 as a collaboration between MIT, Harvard, and Harvard-affiliated hospitals, represented a landmark development in genomics research. With over 500,000 square feet of laboratory space and hundreds of researchers, the Broad Institute created new models for large scale collaborative research while spinning out numerous biotechnology companies.

Google's Cambridge Office, opened in 2006, marked the beginning of major technology companies establishing significant presences in Kendall Square. Google's decision to locate artificial intelligence and machine learning research teams in Cambridge validated the area's capabilities beyond biotechnology and attracted additional technology firms.

MIT Computer Science and Artificial Intelligence Laboratory (CSAIL), formed through the merger of existing laboratories, consolidated the university's computer science research in state-of-the-art facilities. CSAIL's research in robotics, artificial intelligence, and computer systems generated numerous start-up companies and attracted corporate research partnerships.

Novartis Institutes for BioMedical Research established a major research campus in Kendall Square in 2002, representing one of the largest corporate research investments in the district. The facility employed over 1,000 researchers, demonstrating thew confidence of multinational corporations in the ecosystem's capabilities.

Venture Capital Concentration reached new levels during this period, with over 50 venture capital firms maintaining offices within walking distance of Kendall Square. This geographic concentration facilitated rapid deal-making and created informal networks that supported entrepreneurial development.

MIT $5 Million Entrepreneurship Competition and similar programs institutionalised start-up support within the university. These competitions, combined with expanded entrepreneurship coursework, created systematic pathways for students and faculty to engage in venture creation.

Cambridge City Planning implemented zoning changes aimed at encouraging mixed-use development and preserving affordable housing options. The Kendall Square Urban Renewal Plan balanced growth pressures with community needs while preserving the area's character as an innovation district.

International Partnerships expanded significantly, with research collaborations and corporate partnerships connecting Kendall Square institutions with universities and companies worldwide. These relationships facilitated talent exchange and created global networks supporting innovation activities.

Start-up Success Stories multiplied during this period, with companies like Akamai going public and numerous biotechnology firms achieving significant valuations through acquisitions or IPO. These successes created role models for aspiring entrepreneurs while generating wealth that funded subsequent venture activities.

Digital Transformation and Corporate Innovation: 2010s-Present

The 2010s marked Kendall Square's evolution into a comprehensive innovation ecosystem encompassing artificial intelligence, biotechnology, clean technology, and digital platforms. This period brought unprecedented corporate investment, international expansion, and recognition as a global model for innovation district development.

Amazon's Kendall Square Expansion began in 2018 with the establishment of major offices housing Alexa and other artificial intelligence teams. Amazon's presence, employing over 4,000 people in the district, attracted additional technology companies and validated Kendall Square's position as a hub of digital Innovation.

Facebook (Meta) AI Research established operations in Kendall Square in 2018, focusing on artificial intelligence R&D. The facility houses hundreds of researchers working on machine learning, computer vision, and natural language processing, contributing to the area's growing reputation in AI innovation.

MIT Stephen A. Schwarzman College of Computing, launched in 2019 with a $1 billion investment, is the largest academic venture in MIT's history. The college integrates computing across all disciplines while creating new opportunities for industry collaboration and start-up formation.

Moderna Therapeutics emerged as a Kendall Square success story, developing mRNA vaccine technology that proved crucial during the COVID-19 pandemic. The company's rapid vaccine development and subsequent public offering demonstrated the ecosystem's ability to address global challenges.

Microsoft Research New England expanded its presence in Kendall Square, focusing on machine learning, computational biology, and social computing. The facility's research collaborations with MIT and Harvard have created new models for university-industry partnerships.

Venture Capital Evolution witnessed the emergence of corporate venture arms and international investors actively investing in Kendall Square companies. Firms such as GV (Google Ventures), Intel Capital, and Johnson & Johnson Innovation established local presences to access deal flow and support their portfolio companies.

CIC (Cambridge Innovation Centre) Global Expansion demonstrated Kendall Square's influence on innovation district development worldwide. CIC's model of flexible start-up space and community building expanded to cities across North America, Europe, and Asia.

The MIT.nano Facility, opened in 2018, provides state-of-the-art nanofabrication capabilities that support research in electronics, photonics, and materials science. The facility enables new forms of university-industry Collaboration while supporting companies that develop advanced technologies.

Alexandria Real Estate Equities became a major developer of life sciences real estate in Kendall Square, creating specialised laboratory and office space designed for biotechnology and pharmaceutical companies. Their developments set new standards for flexible, technology-enabled research facilities.

International Corporate Research Centres proliferated, with companies like Takeda, Sanofi, and Shire establishing research operations in Kendall Square. These investments reflected the area's global reputation while creating opportunities for collaboration and talent exchange.

COVID-19 Response demonstrated Kendall Square's capacity for rapid Innovation, with multiple companies developing vaccines, therapeutics, and diagnostic technologies. The ecosystem's response highlighted the value of concentrated expertise and collaborative networks during crises.

Current Ecosystem Characteristics and Strengths

Today's Kendall Square is a mature innovation ecosystem with distinctive characteristics that differentiate it from other technology hubs worldwide. The district's success stems from a unique combination of institutional assets, physical Infrastructure, and a collaborative culture.

Institutional Density creates unparalleled opportunities for knowledge transfer and Collaboration. With MIT, Harvard Medical School, and numerous research hospitals within a few miles, the ecosystem benefits from continuous flows of breakthrough research and emerging talent. The proximity enables informal Collaboration, joint research projects, and rapid technology transfer.

Venture Capital Concentration provides start-ups with unprecedented access to funding and expertise. Over 100 venture capital firms maintain offices or active presences in the greater Boston area, with many focusing specifically on the life sciences and technology sectors, which are prominent in Kendall Square.

Corporate R&D creates multiple pathways for innovation commercialisation. Major corporations maintain research facilities that collaborate with local universities while also acquiring promising start-ups. This creates exit opportunities for entrepreneurs while providing corporations with access to emerging technologies.

Talent Pipeline draws from world-class universities while attracting experienced professionals from established companies. The ecosystem benefits from continuous talent circulation, with individuals moving between universities, start-ups, and established companies throughout their careers.

Physical Infrastructure supports Innovation through purpose-built laboratory and office space designed for Collaboration. Modern facilities provide flexible space that can accommodate changing research needs while fostering chance encounters and informal Collaboration.

Transportation Connectivity offers access to global markets and talent, thanks to its proximity to Logan International Airport and excellent public transportation. The MBTA Red Line connects Kendall Square directly to downtown Boston and other innovation clusters throughout the metropolitan area.

Regulatory Environment in Massachusetts supports biotechnology and technology innovation through favourable intellectual property laws, research incentives, and business-friendly policies. State-level initiatives provide additional resources for early-stage companies and research collaborations.

International Networks connect Kendall Square companies and institutions with partners worldwide. These relationships facilitate talent exchange, research collaboration, and market access while attracting international investment to the ecosystem.

Economic Impact and Performance Metrics

Kendall Square's economic performance demonstrates the potential for innovation districts to generate substantial economic returns while creating high-quality employment opportunities. The district's metrics provide benchmarks for evaluating the success of its innovation ecosystem.

Employment Concentration includes approximately 50,000 people who work in the area on a daily basis, with more than 29,000 individuals working in innovation-related employment that has also generated nearly 147,000 additional jobs. The average annual income for innovation-related employment exceeds $195,000, reflecting the high-value work being conducted in the district.

Company Formation rates demonstrate the ecosystem's entrepreneurial strength. Kendall Square is now home to more than 2,000 businesses, including over 600 start-ups. Historical data show that MIT is a pioneer in this endeavour, with an average of 25 companies spun off from MIT each year in the 1980s, and recent reports indicate continued strong formation rates.

Venture Capital Investment shows significant capital flows. The CIC innovation campus in Cambridge is home to half of the more than $14 billion in venture capital investment in Kendall Square. Regional data indicate that between Q3 2017 and Q2 2023, the value of venture deals was $99.2 billion.

Patent Generation reflects intense research activity. MIT secured 365 utility patents issued by the United States Patent and Trademark Office in 2023, marking the 10th consecutive year that MIT has been recognised as the top single-campus university for patents granted. Each year, the TLO receives over 600 invention disclosures, resulting in a high volume of issued patents.

Market Capitalisation demonstrates substantial value creation. Kendall Square is home to public companies with a market value of approximately $170 billion. MIT's broader alumni impact reveals that MIT alumni have launched approximately 30,200 active companies, employing roughly 4.6 million people and generating approximately $1.9 trillion in annual revenues as of 2014.

Real Estate and Infrastructure reflect continued investment. As of 2023, Kendall Square had approximately 56 million square feet of laboratory and R&D space, with a planned expansion of an additional 13.4 million square feet over the next two years. MIT reported its real estate investments at $4.9 billion in fiscal 2024, up 29.5 per cent since 2020.

Technology Transfer Revenue shows commercialisation success. Between 1991 and 2015, licensing MIT's patents has generated $1 billion in revenue, with 4,000 issued patents having value to licensees and 437 companies licensed MIT-owned IP.

Research Funding provides substantial foundation support. MIT received $1.7 billion in federal research funding during the 2022-23 school year, while research sponsored directly by industry totalled $175 million from over 300 companies in fiscal year 2024.

Challenges and Ongoing Issues

Despite its remarkable success, Kendall Square faces significant challenges that could affect its long-term competitiveness and sustainability. Addressing these issues requires continued collaboration between public and private sector stakeholders.

Housing Affordability is the most pressing challenge, with high real estate costs making it difficult for young professionals and researchers to live near their work. The lack of affordable housing could limit talent attraction and retention over time.

Transportation Infrastructure faces capacity constraints as the district continues to grow. The MBTA Red Line experiences overcrowding during peak hours, while limited parking creates additional challenges for some workers and visitors.

Space Constraints limit the ecosystem's physical expansion, with most developable land already utilised. Future growth may necessitate higher-density development or expansion into adjacent areas, potentially altering the district's character and social interaction patterns.

Workforce Diversity remains limited despite efforts to broaden participation in innovation activities. The ecosystem's high entry barriers and specialised skill requirements may exclude some populations from participation in high-value employment opportunities.

The cost of Business Operations has increased substantially as the district's success has driven up real estate and labour costs. Some companies, particularly early-stage start-ups, may struggle to afford Kendall Square locations.

Competition from Other Innovation Hubs has intensified as cities worldwide develop their own innovation districts. Kendall Square must continue evolving to maintain its competitive advantages in an increasingly crowded global marketplace.

Research Funding Uncertainty affects long-term planning, as federal research budgets and priorities can change with political cycles. Universities and companies must navigate uncertain funding environments while maintaining momentum in their research.

Regulatory Compliance requirements in biotechnology and other sectors can slow Innovation and increase costs. Balancing safety and efficacy requirements with innovation speed remains an ongoing challenge.

Lessons for Other Innovation Ecosystems

Kendall Square's development offers valuable insights for cities and regions seeking to establish their own innovation ecosystems. The district's experience highlights key factors that contribute to the success of innovation ecosystems.

Anchor Institution Investment is essential for creating the foundational research capabilities that attract private investment and talent. MIT's sustained commitment to excellence and Innovation provides the bedrock upon which the broader ecosystem develops.

Public-Private Collaboration enables coordinated development that balances market forces with community needs. Successful partnerships between Cambridge city government, property developers, and private companies created Infrastructure supporting Innovation while maintaining livability.

Long-term Vision and Patience characterise successful ecosystem development, as meaningful results require decades of sustained effort. Kendall Square's transformation occurred over more than 50 years, with each phase building upon previous investments and achievements.

Flexible Zoning and Planning allow ecosystems to evolve with changing technology and market conditions. Cambridge's willingness to adapt zoning regulations and development policies enabled the district to respond to new opportunities and challenges.

Transportation and Infrastructure Investment create foundations for ecosystem growth by reducing barriers to collaboration and talent access. Strategic investments in public transportation and digital Infrastructure provide essential ecosystem support.

Talent Development and Circulation require systematic attention to education, training, and career pathways. Universities, companies, and governments must collaborate to ensure adequate talent supplies while providing advancement opportunities.

International Connectivity enhances ecosystem capabilities by providing access to global markets, talent, and capital. Policies supporting international collaboration and exchange strengthen innovation ecosystems.

Risk Capital Availability enables entrepreneurship and company scaling through all development stages. Successful ecosystems require diverse funding sources, from seed funding through growth capital and public markets.

Quality of Life Amenities attract and retain talent by providing attractive living and working environments. Housing, recreation, culture, and education opportunities have a significant impact on ecosystem competitiveness.

Intellectual Property Policies must strike a balance between protection and accessibility to encourage both research investment and technology transfer. Clear, consistent policies reduce transaction costs while protecting legitimate interests.

Future Directions and Strategic Opportunities

Kendall Square's continued success requires strategic responses to emerging challenges and opportunities. The ecosystem must evolve to maintain its global leadership while addressing sustainability and inclusion concerns.

Climate Technology Innovation is a major growth opportunity, building upon the district's strengths in materials science, energy systems, and environmental engineering. MIT's recent climate commitments and research investments position Kendall Square to lead in sustainability innovation.

Artificial Intelligence and Machine Learning applications across biotechnology, healthcare, and other sectors create new opportunities for cross-disciplinary collaboration. The convergence of computing and life sciences could generate breakthrough innovations and new company formations.

Digital Health and Biotechnology convergence enables new approaches to drug discovery, personalised medicine, and healthcare delivery. Kendall Square's strengths in both sectors position it well for leadership in digital health innovation.

International Expansion through partnerships and satellite operations could extend Kendall Square's influence while providing local companies with global market access. Strategic international relationships could enhance the ecosystem's competitive position.

Workforce Development initiatives must address diversity and inclusion while meeting the growing demand for skilled workers. Partnerships with regional educational institutions could create pathways for broader participation in innovation activities.

Transportation Infrastructure investments, including a potential Green Line extension and improved bus service, could alleviate capacity constraints while connecting the district with other innovation clusters.

Housing Development strategies must balance affordability with growth, potentially including subsidised housing for essential workers and innovative financing mechanisms for workforce housing.

Regional Collaboration with other Boston area innovation clusters could create metropolitan-scale advantages while reducing pressure on Kendall Square infrastructure. Coordinated regional development could strengthen the overall ecosystem.

Regulatory Innovation in biotechnology and other sectors could accelerate Innovation while maintaining safety standards. Kendall Square could serve as a testbed for new regulatory approaches that other jurisdictions could adopt.

Corporate Venture Capital growth provides new funding sources for start-ups while creating strategic partnerships. The ecosystem could benefit from expanded corporate investment and collaboration programs.

Conclusion

Kendall Square's evolution from an industrial wasteland to a global innovation leader demonstrates the potential for purposeful ecosystem development to generate extraordinary economic and social returns. The district's success stems from a unique combination of world-class research institutions, strategic public investment, private sector collaboration, and a sustained commitment to innovation excellence.

The ecosystem's $5 billion in annual economic activity, over 50,000 knowledge workers, and hundreds of innovative companies provide compelling evidence of innovation districts' potential to drive economic growth while addressing global challenges. Kendall Square's companies have developed life-saving medicines, breakthrough technologies, and innovative business models that improve lives worldwide.

However, the district's experience also highlights challenges facing successful innovation ecosystems, including housing affordability, infrastructure capacity, and the need for continued evolution. Addressing these challenges requires continued collaboration between public and private stakeholders, along with sustained investment in the foundations supporting innovation activities.

For other cities and regions seeking to develop innovation ecosystems, Kendall Square offers valuable lessons about the importance of anchor institutions, long-term vision, public-private collaboration, and strategic infrastructure investment.

While each ecosystem must reflect local strengths and conditions, Kendall Square's principles and practices provide proven frameworks for innovation-led economic development.

As global challenges like climate change, healthcare access, and technological disruption intensify, innovation ecosystems like Kendall Square become increasingly important for developing solutions and driving economic progress. The district's continued evolution will influence innovation ecosystem development worldwide while contributing to solutions for humanity's most pressing challenges.

The Kendall Square experience proves that with appropriate vision, investment, and collaboration, cities can create environments where Innovation flourishes, generating economic prosperity while improving the quality of life for residents and contributing to global progress. This blueprint remains relevant for innovation ecosystem development in an era of rapid technological change and global competition.

15. The London Knowledge Quarter

Knowledge Quarter London (KQ) is one of the world's most concentrated innovation ecosystems, encompassing a one-mile radius around King's Cross railway station in north-central London. The area contains one of the highest densities of knowledge-based businesses and science organisations in the world, with the KQ's contribution to the wider UK economy equivalent to the City of London and competing with leading innovation districts such as Kendall Square in Boston, USA.

Context

In a one-mile radius, KQ has 15 universities, 18 Cultural institutions, 15 Higher Education Institutes, 22 Museums and Galleries, 72 Libraries and archives, 400 Research Centres, 4,500 Scientists, 3,200 Published journals, 18,000 members of academic staff, 30,000 Graduates annually, 70,000 members of staff, 100,000 students, 200 million catalogued items, and eight million annual visitors a year.

The quarter spans the historic neighbourhoods of King's Cross, St Pancras, Bloomsbury, and parts of Euston, bringing together an unprecedented concentration of universities, research institutes, cultural institutions, and innovative companies across two London boroughs (Camden and Islington). A significant recent development within the quarter is the Argent site, a major regeneration of 67 acres in King's Cross led by property developer Argent since the late 1990s.

Partners in the development include The British Library, The British Museum, Google, The Alan Turing Institute, Francis Crick Institute, University College London (UCL), and organisations ranging from global brands to small businesses, all drawn to an area exceptionally rich in knowledge resources. KQ is known as world leader in precision medicine and personalised care.

The ecosystem emerged through a unique combination of historic academic excellence, strategic urban regeneration, and deliberate policy coordination to create what many consider a blueprint for 21st-century innovation districts. The KQ builds on an institutional landscape that has existed for generations. UCL, The British Museum, The Wellcome Trust, Great Ormond Street Hospital and numerous other world-leading universities, research centres and cultural institutions have occupied this part of London for decades or even centuries.

What has changed over the past two decades is the addition of major new facilities through the King's Cross regeneration, particularly the 67-acre Argent development on former railway lands. This strategic investment has brought institutions like the Francis Crick Institute, Central Saint Martins, and The Alan Turing Institute to the area, complementing the established knowledge base.

The formal KQ partnership emerged to coordinate these historic and new assets, creating an interconnected ecosystem that generates scientific breakthroughs, spawns successful start-ups, and attracts billions in international investment. The model demonstrates how deliberate collaboration between long-standing institutions and targeted urban regeneration can catalyse innovation driven economic growth.

Historical Foundations: Building on Centuries of Knowledge

The Knowledge Quarter's success builds upon London's deep historical foundation as a centre of learning and discovery. The Royal Society, established in 1660 following a lecture at Gresham College in London by Christopher Wren, set the precedent for London's role in advancing natural philosophy and scientific inquiry. This tradition of institutional excellence created the cultural and intellectual bedrock upon which the modern Knowledge Quarter would later flourish.

The British Library's relocation to its current St. Pancras site in 1997 marked an early catalyst for the area's transformation into a knowledge hub. As one of the world's largest libraries, housing over 200 million items, the British Library brought immediate scholarly gravitas to the district and established a foundation for the knowledge-intensive ecosystem that would follow.

University College London, founded in 1826 as England's first institution to admit students regardless of race, class or religion, has maintained a central London presence for nearly two centuries. UCL's central position in the emerging KQ, combined with its research excellence across multiple disciplines, positioned it as a natural anchor tenant for the developing innovation district.

The area's cultural institutions also provided essential foundations. The British Museum in nearby Bloomsbury, established in 1753, and various smaller cultural organisations created an environment where knowledge preservation, creation, and dissemination were valued and supported.

This cultural density would prove crucial in attracting both researchers and knowledge-intensive businesses to the district.

Urban Regeneration as Innovation Catalyst: 2000-2010

The transformation of King's Cross from industrial wasteland to innovation hub began with one of London's largest urban regeneration projects. King's Cross had long been a strategic location, connected to industrial cities in the north through the completion of Regent's Canal in 1820, but by the 1980s it had become notorious for drugs, sex work and homelessness.

The catalyst for change came with the 1996 decision to move the Channel Tunnel Rail Link (High Speed 1) from Waterloo to St Pancras, providing the impetus for the comprehensive redevelopment of the 67-acre site between King's Cross and St Pancras stations. London & Continental Railways Limited and DHL, as landowners, selected Argent as the development partner in 2001, beginning intensive community consultation and strategic planning.

The masterplan, prepared by Allies and Morrison, Porphyrios Associates and Townshend Landscape Architects, received outline planning permission in 2006, allowing for approximately 50 new buildings, 20 new streets, 10 new major public spaces, restoration of 20 historic buildings, and up to 2,000 homes.

The key to success was consolidated land ownership combined with long-term capital, allowing for an integrated and patient approach to development with emphasis on placemaking rather than simply maximising short-term returns. This approach enabled the creation of a high-quality public realm, mixed-use environments, and spaces specifically designed to attract knowledge-intensive businesses competing for talent in tight labour markets.

The redevelopment coincided with broader trends in London's economy, particularly the growth of the technology and creative sectors. King's Cross was designated an 'opportunity area' in the 2000 Camden Council Unitary Development Plan and again in the 2004 London Plan to provide substantial new office space to meet growing demand in the capital. The Euston development/HS2 is the latest large-scale development in the KQ which is of equivalent size to the Argent site.

Institutional Anchors and Scientific Excellence: 2010-2016

The period from 2010 to 2016 witnessed the establishment of several world-class research institutions that would become the scientific anchors of the KQ. The most significant was the creation of the Francis Crick Institute, Europe's largest biomedical research facility.

The Francis Crick Institute was established in 2010 as a partnership between Cancer Research UK, Imperial College London, King's College London, the Medical Research Council, University College London and Wellcome Trust. The institute began operating in April 2015 when the MRC's National Institute for Medical Research and CRUK's London Research Institute merged and was officially opened by The Queen in November 2016.

Built at a cost of nearly £700 million and directed by Nobel Laureate Sir Paul Nurse, the institute houses 1,500 staff, including 1,250 scientists, with an annual budget of over £100 million. The Francis Crick Institute's location next to St Pancras railway station created an immediate focal point for biomedical research excellence within the emerging Knowledge Quarter.

The Crick represented not simply a merger of existing institutes but a new entity with a different ethos – a multidisciplinary institute with substantial investment in biomedically related physical sciences, emphasising research in human biology and disease through clinical sciences connections.

University College London strengthened its presence in the area during this period, recognising the strategic value of proximity to other knowledge institutions. UCL's contribution to the Knowledge Quarter has been significant, with the university playing a central role in various collaborative research initiatives and spin-off companies.

The presence of the British Library also evolved during this period, with plans emerging for major expansions that would house additional research institutes. A major extension to the British Library was planned to open in 2025, providing a new headquarters for the Alan Turing Institute, the UK's National Institute for Data Science.

Corporate Investment and Ecosystem Development: 2016-2020

The period from 2016 to 2020 marked the Knowledge Quarter's emergence as a globally recognised innovation district, attracting major corporate investments and spawning numerous start-ups. The announcement of Google's massive headquarters development became a watershed moment for the district's reputation.

Since 2007, Google, Facebook/Meta, Samsung and Toyota have established major hubs in the area. The number of firms roughly doubled to 800 between 2010 and 2021, while jobs increased from 8,000 to 27,000 between 2011 and 2019.

Google's King's Cross development, designed by Heatherwick Studio and Bjarke Ingels Group, is the first building owned and designed by Google outside the United States. Combined with existing buildings, the development will create capacity for 7,000 Google employees and contribute significantly to the Knowledge Quarter's growing knowledge-based economy.

The corporate investment created a demonstration effect, attracting additional knowledge-intensive businesses. Office rents in King's Cross more than doubled in value, going from 48 per cent below the London city centre average in 2010 to 19 per cent above in 2022, indicating strong demand for space in the district.

During this period, the formal Knowledge Quarter organisation was established. Knowledge Quarter London Limited was incorporated on 14 September 2016 as a private company limited by guarantee without share capital. The consortium is governed by a board of ten organisations supported by a steering group, with founding members being the British Library and the University of the Arts London.

The organisation developed a structured approach to fostering collaboration and innovation. Partners pay annual membership fees based on organisation size, ranging from £500 for organisations under 10 employees to £20,000 for those with 5,001-10,000 employees, with special discounts for charities and community associations.

Start-up Ecosystem and Innovation Networks: 2018-Present

The mature phase of the Knowledge Quarter's development has been characterised by a thriving start-up ecosystem and sophisticated innovation networks. The area has become particularly attractive to data-driven start-ups, with machine learning being especially popular sectors.

Notable start-ups that emerged or relocated to the Knowledge Quarter include:

BenevolentAI, which uses AI to identify drug compounds that have been abandoned for specific diseases and suggest more effective applications, with algorithms capable of designing new molecules based on knowledge of over a billion relationships between drugs and the body.

Altmetric, which tracks the attention scholarly work receives outside traditional academic literature, providing Attention Scores that reflect research popularity and impact.

Brainpool is an AI consultancy that bridges the gap between academia and industry with a network of over 300 AI and machine learning specialists from prestigious universities worldwide.

The *Sanger Institute's Mosaic,* a cancer genomics venture, exemplifies successful research commercialisation, having attracted over £22 million in investment through access to extensive research resources, business development support, and Innovate UK funding.

The Knowledge Quarter has also attracted significant infrastructure investment to support growing companies. BlackRock, British Airways, and GIC are developing the *Tribeca Life Sciences Hub*, providing 600,000 square feet of lab space for emerging enterprises. *MAYDE*, located at 5-10 Brandon Road, offers 113,000 sq ft of combined GMP manufacturing, laboratory, and office space, while *KILN* will provide 7,000 sqm of purpose-built laboratory and office facilities.

Research Excellence and Global Impact

The Knowledge Quarter has established itself as a global leader in several key research domains, generating world-class scientific output and technological innovation. Publications from the Knowledge Quarter are frequently aligned with the United Nations Sustainable Development Goals and have been influential in policy-related discussions, underscoring the area's commitment to addressing pressing global challenges through impactful research.

When compared to other international innovation districts such as Kendall Square in Boston and Nanyang Technological University in Singapore, the Knowledge Quarter consistently ranks highly across various indicators of research collaboration and impact, particularly in co-publication rates with local organisations, indicating a robust collaborative culture.

The Francis Crick Institute has been particularly productive in generating high impact research. The partnership between King's College London and the Crick has led to notable spin-outs, including GammaDelta Therapeutics, a pioneering biotechnology company focused on exploiting Gamma Delta T cells for immunotherapy, and the MAGIC consortium, bringing 15 international partners together to transform treatment for muscular dystrophies.

The *UK Dementia Research Institute* employs more than 800 researchers developing treatments and technologies for individuals affected by dementia, while the London BioScience Innovation Centre spans 2,500m2 as one of the UK's largest hubs of life science and biotechnology companies.

University College London has strengthened its research profile significantly within the Knowledge Quarter. UCL boasts one of the world's largest and highest-impact neuroscience centres, including the development of a £300m landmark facility bringing together over 500 research scientists, clinicians and patients to tackle neurological diseases like dementia.

Economic Impact and Investment Flows

The Knowledge Quarter has become a significant driver of economic activity, attracting substantial domestic and international investment while generating high-value employment. A Socioeconomic Impact Assessment found that the KQ's contribution to the wider UK economy is equivalent to that of the City of London and competes with leading innovation districts such as Kendall Square in Boston.

The area has demonstrated remarkable growth in property values and business activity. Office rents more than doubled in value between 2010 and 2022, while the number of businesses doubled to 800 between 2010 and 2021, creating approximately 19,000 additional jobs.

London-based start-ups attracted large investments in 2024, with companies like *BMLL Technologies* raising approximately $20.3 million in Series B funding and *Zepz* securing substantial funding, highlighting strong investor confidence in the city's start-ups.

The system also benefited from major infrastructure investments. At the UK's first International Investment Summit in October 2024, investors pledged $80 billion, including a $25 billion commitment from Macquarie to EV infrastructure, while US tech firms invested $7.6 billion in data centres.

Government support has been crucial in enabling this growth. London-based SMEs can claim up to 186% in R&D tax relief, cutting innovation costs and boosting investment in new ideas, while London & Partners' free 12-month Grow London Global program helps London tech firms expand internationally.

Challenges and Strategic Responses

Despite its remarkable success, the Knowledge Quarter faces several significant challenges that threaten its continued growth and competitiveness. The most pressing issue is the availability and affordability of space for growing businesses.

The KQ's success has placed multiple pressures on its infrastructure: local housing is in short supply and extraordinarily expensive, commercial property prices have skyrocketed, and employment sites are under pressure from conversion to residential use under Permitted Development Rights.

Brexit has created additional challenges for international talent recruitment and collaboration. International recruitment is key to the Crick's success and UK biomedicine generally, but Brexit poses a threat to both through potential restrictions on EU talent mobility.

The COVID-19 pandemic highlighted vulnerabilities in the business model of some knowledge-intensive organisations. TechHub, a shared office space provider that housed hundreds of promising tech start-ups, including companies acquired by Google, Facebook and Microsoft, filed for administration in August 2020 after revenues fell by 75% due to COVID-19.

However, the ecosystem has demonstrated resilience and adaptability. Real estate experts predict a shift towards creating offices as 'destinations' that encourage staff to commute for specific purposes like collaboration and socialising, rather than simply downsizing office space.

Future Directions and KQ 2050 Vision

The Knowledge Quarter has developed ambitious plans for continued growth and development through its 'KQ 2050' vision. The long-term spatial plan aims to accelerate development of the KQ as an 'incubator for innovation' that supports the economic prosperity of the UK while ensuring future development can continue to deliver research, technology and entrepreneurialism.

Strategic expansion beyond the core King's Cross area is already underway. UCL has moved elements of civil engineering and architecture to Stratford, taking advantage of abundant grow-on space at lower build costs, while maintaining connections to the KQ through Knowledge Quarter London Ltd.'s dialogue with the London Legacy Development Corporation.

Regional collaboration is expanding the ecosystem's reach. An emerging cluster in gene therapy, stem cell therapy and tissue engineering connects the KQ with the *Cell & Gene Therapy Catapult's* £55-million facility at Stevenage in Hertfordshire, the Advanced Therapies Division at Potters Bar, and the RVC's campus at Hawkshead Lane.

The Knowledge Quarter is also strengthening its connections with other London innovation districts. Knowledge Quarter London Ltd. is engaging with Imperial College at White City in west London, which emphasises multidisciplinary research and innovation with application to the Life Sciences.

Emerging technologies are creating new opportunities for growth. The KQ's focus on thematic areas like artificial intelligence and quantum technologies positions it as a competitive player in the global innovation landscape.

Educational and Community Engagement

The Knowledge Quarter has developed comprehensive programs to engage local communities and develop future talent pipelines. The Francis Crick Institute's outreach program, developed alongside teachers and leading science education organisations, engages over 20,000 young people in Camden's state schools, along with their teachers and families.

Camden STEAM envisions Camden as a hub of inclusive opportunity, inspiring young people to pursue STEAM careers by connecting them with local employers and educators, with 60 STEAM employers and over 200 STEAM Ambassadors. Camden Learning and Camden Council's STEAM program, with support from Google, are developing an AI Campus for borough pupils to enhance students' knowledge of AI and address its ethical use.

The LIFT program, a collaboration between Camden, Hackney, Islington and Tower Hamlets, provides residents with opportunities in local technology, science and digital sectors, supporting businesses in recruiting skilled, diverse local talent and facilitating comprehensive apprenticeship support.

These educational initiatives ensure that the benefits of the Knowledge Quarter extend beyond the immediate research and business community to create pathways for local residents to participate in the knowledge economy.

Global Implications and Lessons

The Knowledge Quarter model offers valuable insights for other cities and regions seeking to develop innovation ecosystems. Several key factors have contributed to its success.

Historical Foundation: The presence of established world-class institutions provided credibility and critical mass from the outset, demonstrating the importance of building on existing strengths rather than creating innovation districts from scratch.

Strategic Urban Regeneration: The comprehensive redevelopment of King's Cross created purpose-built spaces designed specifically for knowledge-intensive activities, while maintaining connections to London's transport networks and cultural amenities.

Patient Capital and Coordinated Development: Long-term investment approaches and coordinated landowner strategies enabled placemaking and quality public realm development that attracted knowledge workers and innovative companies.

Institutional Collaboration: The formal Knowledge Quarter organisation provides a mechanism for coordination and collaboration between diverse institutions, while maintaining their independence and distinct missions.

Government Support: Targeted policy interventions, including R&D tax incentives, infrastructure investment, and planning permission processes, have facilitated growth while addressing market failures.

International Connectivity: London's position as a global financial centre and international transport hub provides access to global talent, capital, and markets that smaller cities cannot easily replicate.

Conclusion

London's Knowledge Quarter is one of the world's most successful examples of deliberate innovation ecosystem development. Through a strategic combination of historic institutional excellence, comprehensive urban regeneration, and coordinated policy intervention, the district has created a globally competitive hub for research, innovation, and knowledge-intensive business activity.

The transformation from postindustrial wasteland to world-leading innovation district demonstrates the potential for urban regeneration to catalyse economic transformation. The Knowledge Quarter's success in attracting major corporate investments, generating scientific breakthroughs, and fostering start-ups provides a model for other cities seeking to develop knowledge-based economies.

However, the challenges facing the Knowledge Quarter – particularly around space availability, infrastructure capacity, and international talent access – highlight the ongoing policy attention required to maintain ecosystem vitality. The KQ 2050 vision is an ambitious attempt to address these challenges while continuing to build on the district's fundamental strengths.

As global competition for knowledge-intensive economic activity intensifies, the Knowledge Quarter's experience offers valuable lessons about the importance of institutional excellence, strategic placemaking, and coordinated governance in creating sustainable innovation ecosystems. The district's continued evolution will provide important insights into how established innovation clusters adapt to changing economic and technological conditions while maintaining their competitive advantages.

Other cities and regions can learn from the Knowledge Quarter's approach to leveraging existing institutional strengths, creating enabling physical and policy environments, and fostering collaboration between diverse stakeholders. However, successful replication will require careful attention to local context, patient capital formation, and long-term commitment from both public and private sector partners.

The Knowledge Quarter is a testament to the potential for strategic urban development to create environments where knowledge creation, innovation, and economic prosperity can flourish together, providing a model for sustainable growth in an increasingly knowledge-driven global economy.

16. MaRS Discovery District, Toronto

Toronto, Canada's largest city with over 2.7 million residents in the city proper and 6.2 million in the Greater Toronto Area, has emerged as one of North America's most dynamic innovation hubs. At the heart of this transformation stands MaRS Discovery District, a pioneering urban innovation ecosystem that has redefined how cities can foster technological advancement and entrepreneurial growth.

Context

The name "MaRS" was originally derived from a computer file name and later attributed to "Medical and Related Sciences." However, the organisation has since expanded far beyond medical research to encompass information and communications technology, cleantech, fintech, and social innovation. This evolution reflects the dynamic nature of innovation ecosystems and their ability to adapt to emerging opportunities.

Located in Toronto's Discovery District at the intersection of College Street and University Avenue, MaRS occupies 1.5 million square feet and houses more than 120 tenants, ranging from early-stage start-ups to global technology companies. Adjacent to the University of Toronto and its affiliated research hospitals, MaRS has become North America's largest urban innovation hub and a registered charity dedicated to commercialising research and fostering high growth companies.

Since its founding in 2000, MaRS has transformed from a visionary concept into a globally recognised model for urban innovation districts. The organisation has supported over 1,400 Canadian science and technology companies, helping them navigate the complex journey from research to commercialisation. With a mission to "create successful global businesses from Canada's science, technology and social innovation," MaRS has become instrumental in positioning Toronto as a leading global technology centre.

Between 2010 and 2022, MaRS-supported ventures achieved remarkable milestones, including the creation of over 33,000 jobs, $11.5 billion in cumulative revenue, $19 billion in capital raised, and a contribution of $29.6 billion to Canada's GDP. These companies demonstrate a 20.7% annualised revenue growth rate, exceeding Ontario's GDP growth tenfold, while generating $7.5 billion in tax revenue for the province.

Early Vision and Foundation: 2000-2005

The story of the MaRS Discovery District cannot be told without acknowledging the role of Dr. John Robert Evans (1929-2015). Evans was a visionary Canadian leader whose role in developing the MaRS Discovery District is the culmination of a remarkable career spanning medicine, academia, business, and innovation policy. His story reveals fascinating insights into how transformative innovation hubs emerge from the intersection of individual vision and institutional expertise.

Evans was a cardiologist, academic, and businessman who served as founding dean of McMaster University Medical School, President of the University of Toronto (1972-1978), and later as the founding Director of the Population, Health and Nutrition Department of the World Bank. What makes his trajectory particularly compelling is how each role built towards his ultimate contribution to Canadian innovation.

In 2000, Evans co-founded MaRS Discovery District along with 12 other civic leaders, donating $14 million and gathering additional support from government, corporations, and the University of Toronto. The location choice was deeply symbolic, being built on the site where insulin was first administered to a diabetic patient, connecting past medical breakthroughs to future innovation potential.

Evans developed the concept behind MaRS as "a bold idea for a Toronto convergence hub that would help innovation flourish, bring important research discoveries to patients and the marketplace, and foster the growth and success of promising new companies". In this way MaRS got underway to address what he saw as a critical gap in Canada's innovation ecosystem.

The early vision centred on creating a dense urban innovation district that would co-locate universities, start-ups, corporations, and investors in a single geographic area. This concept, which seemed experimental at the time, drew inspiration from successful innovation clusters while adapting to Toronto's unique urban context and institutional strengths.

Key to the early development was the strategic location within Toronto's Discovery District, which already housed major research institutions including the University of Toronto, several affiliated teaching hospitals, and research facilities. The proximity to the Hospital for Sick Children, University Health Network, and Princess Margaret Cancer Centre provided immediate access to world-class medical research and clinical expertise.

The founding team understood that successful innovation ecosystems require more than physical infrastructure. They envisioned a comprehensive support system that would provide start-ups with access to capital, mentorship, specialised facilities, and professional services. This holistic approach distinguished MaRS from traditional business incubators, positioning it as an ecosystem orchestrator rather than merely a landlord.

The project secured early support from the Government of Ontario and the Federal Government of Canada, establishing the public-private partnership model that continues to underpin MaRS operations today. This funding structure recognised innovation as a public good while leveraging private sector expertise and market mechanisms to drive commercialisation.

Evans' approach to MaRS reflected his broader philosophy about innovation. As he put it: "One of my interests was to address the discrepancy between the amount of investment in research and the relatively small portion of that investment that was related to research output that would lead to commercialisation". This insight drove the fundamental design of MaRS as a convergence space.

Evans served as the first Chair of the Board of Directors for 10 years overseeing the development from concept to reality. The MaRS Centre opened in 2005 and has since become North America's largest urban innovation hub, supporting over 1,200 start-ups and hosting roughly 6,000 innovation workers.

The Evans family's continued commitment to MaRS speaks to the depth of his vision. In 2012, the Evans family donated $10 million to fund the MaRS Solutions Lab, designed to help young Canadian leaders approach complex challenges from an integrated, multidisciplinary perspective. The Heritage Building was also designated as the Evans Wing in recognition of his contributions.

What is particularly striking about Evans' role is how it exemplified his broader approach to institutional building. As he described his method: "I look for the collection of people – the kinds of individuals who bring their own networks to be part of programs going forward. This network-centric thinking became fundamental to MaRS's design as a convergence hub.

Evans represented a remarkable synthesis of academic rigour, business acumen, and civic vision. His development of MaRS was not just about creating another innovation space—it was about addressing a systemic challenge in how Canada translates research excellence into economic and social impact. The fact that MaRS has become a model studied globally suggests Evans successfully created something that transcended its original Toronto context to become a blueprint for innovation driven economic development.

Building the Infrastructure: 2005-2010

The MaRS Centre officially opened in May 2005, marking the beginning of what would become a revolutionary approach to urban innovation. The facility incorporated *the Heritage Building*, a beautifully restored four-story brick structure that formerly served as part of Toronto General Hospital and was the site of the discovery of insulin in 1921-1922. This symbolic connection to one of medical history's most important breakthroughs reinforced MaRS's mission to translate research into world-changing innovations.

The Heritage Building received the Heritage Toronto Award of Excellence for Architectural Conservation and Craftsmanship in 2006, demonstrating how innovation districts can successfully integrate historical preservation with cutting-edge functionality. The building houses professional services, industry associations, pharmaceutical companies, and offices of Canadian universities and the Government of Ontario.

Adjacent to the Heritage Building, *the South Tower* provides 200,000 square feet of labs and offices, including the MaRS Incubator on the second and third floors. This dedicated space houses approximately two dozen life science and technology firms, offering furnished offices, laboratory space, meeting rooms, and kitchen facilities specifically designed for emerging start-ups.

The complex also includes the *Toronto Medical Discovery Tower*, a 15-story structure with 400,000 square feet of state-of-the-art wet labs. This facility accommodates advanced scientific equipment and houses basic research activities for the *University Health Network*, one of Canada's premier research hospitals. The tower was designed with typical R&D lab floors configured for maximum flexibility, with 80% wet lab and 20% dry lab capacity.

The MaRS atrium serves as a glass-roofed public thoroughfare, providing walkway access to all buildings while featuring retail vendors, conference areas, video conferencing rooms, and a public food court. This design creates natural opportunities for the "creative collisions" that drive innovation, allowing researchers, entrepreneurs, investors, and corporate partners to interact informally.

During this period, MaRS established its core programming model, focusing on four key areas: *providing physical infrastructure, offering advisory services, facilitating access to capital, and building communities of practice.* The organisation developed specialised programs for different stages of company development, from early-stage incubation through growth-stage acceleration.

The early tenant mix was carefully curated to create a diverse ecosystem. Beyond start-ups, MaRS attracted established pharmaceutical companies, professional service firms, government agencies, and university research groups. This diversity ensured that start-ups had immediate access to potential customers, partners, and service providers within the same building complex.

Ecosystem Expansion: 2010-2018

The period from 2010 to 2018 marked significant expansion and maturity of the MaRS ecosystem. The organisation launched several flagship programs that would become models for innovation hubs worldwide.

Construction of the *West Tower (Phase 2)* began during this period, with the support of loans from the Ontario Government. Originally scheduled for completion in 2010, construction was halted in 2008 due to the global financial crisis and resumed in 2011. The project faced public criticism due to leasing difficulties and the need for government financial support. However, by November 2016, 93 per cent of the building had been leased, validating the long-term demand for innovation-focused real estate in Toronto's core.

MaRS repaid the province's loan in 2017 three years earlier than it was due. The financial viability of the building had always been the long-term funding model.

During this expansion phase, MaRS attracted major corporate tenants and partners. In 2016, IBM announced plans to join the fintech hub at MaRS, joining financial services companies including *CIBC, Manulife, and Moneris.* This corporate presence provided start-ups with direct access to industry partners and potential customers while giving established companies exposure to emerging technologies and business models.

The organisation's programming became increasingly refined, with the development of sector-specific expertise and tailored support services. The health sector programs leveraged Toronto's concentration of world-class medical research institutions and teaching hospitals. The cleantech initiatives aligned with growing global concern about climate change and Ontario's commitment to clean energy.

MaRS's fintech programming emerged as the sector gained prominence, positioning Toronto as one of North America's leading financial technology centres. The proximity to Bay Street, Canada's financial district, provided start-ups with unprecedented access to banking partners and regulatory expertise.

The period also saw the emergence of notable success stories that validated the MaRS model. Companies like *Wealthsimple,* which would become one of Canada's fintech unicorns, benefited from MaRS's programming and network. The success of these early graduates helped attract subsequent cohorts of high-quality entrepreneurs and investors.

International recognition grew significantly during this period. MaRS began hosting delegations from innovation hubs worldwide and sharing its model with cities seeking to develop their own innovation districts. The organisation's leadership became sought-after speakers at global conferences on innovation and economic development.

Technology Leadership and Global Recognition: 2018-Present

Since 2018, MaRS has achieved unprecedented scale and global recognition as a model for urban innovation ecosystems. The organisation has supported over 1,400 Canadian science and technology companies, with portfolio companies achieving remarkable financial milestones that demonstrate the maturation of Toronto's innovation ecosystem.

New ventures and infrastructure

MaRS Momentum program for high growth potential ventures and specialised sector-focused initiatives in health, cleantech, and financial technology, was officially launched in 2020, but soft launched in 2019. This was the key program created by MaRS as part of instrumental financial support from the federal government and it was three major hubs working together

Recent years have witnessed several landmark exits that validate the long-term potential of MaRS-supported companies. Notable acquisitions include *Wattpad,* a storytelling platform that was acquired, and ecobee, a smart thermostat company that became a major player in the IoT market. In December 2021, *Opus One Solutions*, a grid modernisation company supported by MaRS, was acquired by GE Digital to further the corporation's global energy transition goals.

The organisation's portfolio has grown to include two unicorn companies: *Wealthsimple* and *Trulioo,* both of which have achieved valuations exceeding $1 billion. This success has helped establish Toronto as one of the world's leading fintech centres, with MaRS playing a central role in connecting start-ups with the financial services sector.

MaRS has expanded its physical footprint with the opening of *MaRS Waterfront*, a 50,000-square-foot facility developed in partnership with the University of Toronto and Waterfront Toronto. This expansion provides purpose-built workspace for professionals in technology and creative industries while offering spectacular meeting and event spaces near Lake Ontario.

The organisation's programming has evolved to address emerging technological frontiers, including artificial intelligence, quantum computing, digital health, and advanced manufacturing. MaRS has become particularly prominent in climate technology, hosting annual Climate Impact conferences that attract global leaders and investors in the cleantech sector.

The COVID-19 pandemic demonstrated the resilience and adaptability of the MaRS ecosystem. Many portfolio companies successfully pivoted to address pandemic-related challenges, with companies like Maple leading Canada's telemedicine revolution by securing $75 million in investment from Shoppers Drug Mart while doubling in size.

Recent partnerships have strengthened MaRS's global reach. The collaboration with the Toyota Mobility Foundation, through the Mobility Unlimited Hub, focuses on accessibility and assistive technology, while partnerships with leading Japanese consortia aim to accelerate Canada's carbon removal industry.

Sector Focus and Specialisation

MaRS has developed deep expertise across four primary sectors that generate the greatest social and financial impact: health, cleantech, fintech, and enterprise software. This strategic focus allows the organisation to provide specialised support while building critical mass in each sector.

Health Innovation: Leveraging Toronto's concentration of world-class medical research institutions, MaRS supports companies in digital health, biotechnology, medical devices, and health services. The proximity to major teaching hospitals provides start-ups with access to clinical expertise, patient populations for testing, and regulatory guidance. Recent success stories include companies developing AI-powered diagnostic tools, novel therapeutics, and digital health platforms.

Cleantech and Climate Innovation: As the urgency of climate action has increased, MaRS has positioned itself as a leader in supporting clean technology ventures. The annual MaRS Climate Impact conference has become a premier gathering for cleantech innovators, featuring panel discussions, company demonstrations, and investor meetings. Portfolio companies work on solutions ranging from renewable energy and energy storage to carbon removal and sustainable agriculture.

Financial Technology: Toronto's fintech ecosystem has flourished with the support of MaRS, benefiting from the city's position as Canada's financial capital. The fintech hub at MaRS features major financial institutions, including CIBC and Manulife, alongside innovative start-ups that are developing everything from robo-advisers to blockchain applications. The success of companies like Wealthsimple has attracted global attention to Toronto's fintech potential. MaRS was critical in fintech in the early days but stepped back once the sector really took off to focus in areas that really needed support.

Enterprise Software and AI: With the rise of artificial intelligence and machine learning, MaRS has supported numerous companies developing enterprise software solutions. These ventures focus on enhancing business efficiency, automating processes, and facilitating digital transformation across various industries. The organisation's AI programming includes specialised workshops on implementation strategies and ethical considerations.

Investment and Capital Formation

MaRS has played a crucial role in developing Toronto's venture capital ecosystem, both through direct investment activities and by connecting start-ups with external investors. The organisation operates *the MaRS*

Investment Accelerator Fund (MaRS IAF), described as one of Canada's most active early-stage investors.

The investment landscape has evolved dramatically since MaRS's founding. In the early years, access to venture capital was limited, and many promising companies struggled to secure growth funding. Today, Toronto ranks among the top global start-up ecosystems, with the Toronto-Waterloo corridor placing 14th in Start-up Genome's annual Global Start-up Ecosystem Report.

MaRS spun out *TIAP* (previously known as MaRS Innovation) which was and is a global leader in bringing together IP from member research organisations and providing funding and commercialisation support from the lab to market phase. MaRS was instrumental in the creation of *Triphase Accelerator* through a collaboration and option agreement with a leading innovator of cancer therapies, *Celgene Corporation.*

MaRS created *Graphite Ventures* from its IAF fund and co-developed *ArcTern Ventures* for cleantech investments. It also spun-out *StandUp Ventures* for women-led start-ups. Key early-stage investors include Tom Rand (ArcTern's managing partner and MaRS senior adviser) and *Radical Ventures* (which recruited former MaRS executive Salim Teja).

MaRS' success led, in part, to the setup of the Vector Institute (literally over the road) and the two organisations now have a 1+1=100 effect.

MaRS has developed high-level approaches to investor relations, including detailed market intelligence services powered by partnerships with data providers like *PitchBook*. This enables the organisation to provide start-ups with comprehensive competitive analysis, investor targeting, and market sizing support.

The organisation's track record demonstrates the effectiveness of its investment strategy. MaRS-supported ventures have collectively raised $19 billion in capital, with individual companies achieving funding rounds that would have been unimaginable in Canada's earlier venture capital environment.

MaRS has been an important talent incubator and has also spun out critical projects that would never have seen the light of day without the convening power of a neutral organisation that brought public, private and government together.

In this phase of MaRS Procurement has been a major focus. MaRS learned that helping start-ups is a critical and necessary element, but it is not sufficient for success. Working with governments and corporations to procure these innovations is essential.

The Broader Community Role of MaRS

MaRS has played a pivotal role as an event and storytelling organisation, deliberately architecting its physical space to foster community building and narrative sharing.

The MaRS Centre complex design includes comprehensive event infrastructure, including a glass-roofed atrium with sub-dividable conference areas for public and private events, a 4,500-square-foot auditorium accommodating up to 300 people, and the CIBC Live Lounge serving as the go-to venue for tenant community events and meetups. These spaces enable MaRS to convene key stakeholders and amplify the stories of supported ventures and their societal impact.

Beyond physical convening, MaRS became instrumental in Canada's social innovation movement through institutional innovation. A $10 million Evans family donation funded *the MaRS Solutions Lab*, designed to help young Canadian leaders approach complex 21st-century challenges from an integrated, multidisciplinary perspective.

MaRS established the *Centre for Impact Investing* in 2011, which spun out SVX (Social Venture Connexion) in 2013 as "the first local impact investment platform of its kind in North America". The organisation also incubated Canada's B-Corporation movement, celebrating the rise of 41 *Canadian B-Corps* spanning 25 industries, from renewable energy companies like Bullfrog Power to impact investing firms like *Renewal2.* This combination of physical infrastructure and institutional innovation established MaRS as a central node in Canada's broader social innovation ecosystem.

Challenges and Future Directions

Despite its remarkable success, MaRS faces several challenges common to mature innovation ecosystems. Competition for talent has intensified as global technology companies establish significant operations in Toronto. Companies like Google, Microsoft, and Amazon have built major engineering centres in the city, creating both opportunities and challenges for the start-up ecosystem.

Access to growth-stage capital remains a challenge, particularly for companies seeking to scale internationally. While early-stage funding has improved dramatically, many Canadian companies still struggle to access the large funding rounds required for global expansion. MaRS has responded by strengthening relationships with international investors and facilitating cross-border partnerships.

The organisation faces ongoing pressure to demonstrate a return on public investment. As a recipient of government funding, MaRS must strike a balance between its social mission and financial sustainability. The organisation has addressed this through diversified revenue streams, including real estate income, corporate partnerships, and fee-for-service programs.

MaRS is currently focused on several strategic priorities. It is aiming to help more scale-ups reach $100 million in revenue, with close to 70 portfolio companies already on track to achieve this milestone. Continued investment in emerging technologies like quantum computing, synthetic biology, and advanced materials positions MaRS at the forefront of next-generation innovation.

The expansion to MaRS Waterfront is a new model for innovation hubs, integrating workspace with residential and recreational amenities. This approach recognises that attracting and retaining talent requires comprehensive quality of life considerations beyond traditional office environments.

Global Impact and Model Replication

MaRS has become a globally recognised model for urban innovation districts, influencing the development of similar initiatives worldwide. When MaRS opened in 2005, the concept of urban innovation was largely untested. Today, cities from Melbourne to Amsterdam are creating dense innovation districts that co-locate universities, start-ups, corporations, and investors.

The organisation's influence extends through formal partnerships, advisory relationships, and knowledge sharing initiatives. MaRS leadership regularly consults with other cities and regions seeking to develop innovation ecosystems, sharing lessons learned from over two decades of experience.

Key principles that have made MaRS successful include the following.

Strategic Location: Proximity to research universities and medical institutions provides access to talent, research, and partnership opportunities.

Mixed-Use Development: Combining start-ups, established companies, research institutions, and service providers creates natural opportunities for collaboration and knowledge transfer.

Comprehensive Programming: Beyond physical space, successful innovation districts require well-developed support services, including access to capital, mentorship, and market intelligence.

Public-Private Partnership: Combining public mission with private sector efficiency enables sustainable operations while maintaining focus on social impact.

Sector Specialisation: Developing deep expertise in specific industries allows for more targeted and effective support services.

Community Building: Regular events, networking opportunities, and shared spaces foster the relationships that drive innovation ecosystems.

Economic Impact and Measurement

The economic impact of MaRS extends far beyond its immediate portfolio of companies. The organisation has contributed to positioning Toronto as one of the world's leading technology centres, attracting international investment and talent while retaining Canadian innovations that might otherwise migrate to larger markets.

Quantitative measures demonstrate the ecosystem's success. Since 2010, MaRS-supported ventures have contributed $29.6 billion to Canada's GDP and generated $7.5 billion in tax revenue. These companies employ 33,000 individuals across diverse sectors, with an average revenue growth rate significantly exceeding broader economic indicators.

The multiplier effects extend throughout the regional economy. Success stories like Wealthsimple and other unicorn companies have created wealth that funds subsequent generations of entrepreneurs and investors. The concentration of technology talent has attracted global companies to establish operations in Toronto, further strengthening the ecosystem.

Beyond financial metrics, MaRS has contributed to Canada's innovation capacity and global competitiveness. The organisation has helped retain intellectual property and talent that might otherwise migrate to larger technology centres, while building Indigenous capabilities in critical sectors like artificial intelligence and clean technology.

Conclusion

MaRS Discovery District is one of the most successful examples of purposeful innovation ecosystem development in the 21st century. From its founding vision in 2000 to its status as North America's largest urban innovation hub, MaRS has demonstrated how cities can systematically build innovation capacity through strategic planning, sustained investment, and adaptive management.

The organisation's success stems from its comprehensive approach to ecosystem development, combining physical infrastructure, programming excellence, capital formation, and community building. By focusing on sectors with significant social and economic impact, MaRS has created value that extends far beyond traditional business metrics.

Lessons for Other Regions

The MaRS model offers valuable lessons for other cities and regions seeking to build innovation ecosystems. The importance of anchor institutions, the value of sector specialisation, the necessity of public-private partnerships, and the critical role of community building all emerge as key success factors that can be adapted to different contexts.

As MaRS looks towards the future, continued evolution will be essential. The organisation must navigate increasing competition for talent, changing investor expectations, and emerging technological frontiers while maintaining its core mission of translating research into social and economic benefit.

The transformation of Toronto into a globally recognised technology centre, with MaRS as its anchor institution, demonstrates the potential for cities to systematically build innovation capacity. This success provides both inspiration and practical guidance for the next generation of innovation districts worldwide.

Other cities seeking to replicate this success should focus on leveraging existing institutional strengths, building diverse and inclusive communities, maintaining a long-term perspective, and adapting continuously to changing market conditions. The MaRS experience proves that with vision, persistence, and strategic execution, cities can build thriving innovation ecosystems that drive economic prosperity and social progress.

17. Dublin Docklands

Dublin Docklands, stretching along both sides of the River Liffey from the Talbot Memorial Bridge to the 3Arena, has experienced one of the most remarkable urban transformations in Europe. What was once a neglected postindustrial wasteland has transformed into the "Silicon Docks", now home to the European headquarters of technology giants such as Google, Facebook, LinkedIn, Twitter, and Amazon. Technology companies in the area employ around 7,000 people, and the wider ecosystem supports thousands more throughout the region.

Context

The area primarily falls within Dublin's D01 and D02 postal districts, and by 2008, it was home to more than 599 businesses. The transformation reflects a carefully planned policy strategy that has developed over more than thirty years. It started with financial services and has evolved into a comprehensive innovation district that draws in global technology investment.

Dublin's success can be attributed to strategic government intervention, favourable tax policies, educational excellence, and its unique position as an English-speaking gateway to European markets. The city has capitalised on its status as the only English-speaking EU member state after Brexit, along with its 12.5% corporate tax rate and strong partnerships with universities, to establish a globally competitive innovation ecosystem.

However, this success has also brought about significant challenges. The average salary in Silicon Docks has increased to €127,000, while rents are 20% higher than the Dublin average, reaching €2,479 per month. Housing affordability has become a pressing issue, as many workers are unable to afford to live in the areas where they work. This situation raises concerns about social displacement and the sustainability of community development.

Early Foundations: The IFSC and Urban Regeneration (1987-2000)

The story of Dublin Docklands started with the establishment of the International Financial Services Centre (IFSC) in 1987. The IFSC was established as an urban regeneration area and special economic zone on 11 hectares of abandoned state-owned land that once belonged to the port authority, located in the North Wall and George's Dock areas. The Custom House Docks Development Authority was established to oversee the transformation of these lands, with construction beginning in 1988 and Allied Irish Banks becoming the first tenant in 1990.

The IFSC represented an innovative approach to urban regeneration by utilising targeted economic incentives. The EU has approved a special corporate tax rate of 10% for "designated financial services activities", which is significantly lower than Ireland's standard rate of 32%. This tax differential acted as a strong incentive for international financial institutions looking to establish operations in Europe.

By 1997, the initial phase had produced significant results. The area included 114,000 square metres of office space, more than 300 apartments, a hotel, and numerous retail outlets. By 2010, the International Financial Services Centre IFSC) accounted for 16% of Ireland's corporation tax revenues, while the wider international financial services industry contributed more than one-fifth of total corporate tax receipts.

The success of IFSC 1 led to further expansion. The Dublin Docklands Development Authority (DDDA) was established in 1997 to oversee the broader regeneration of the area, including the development of an additional 4.8 hectares at the IFSC along North Wall Quay and Lower Mayor Street. The Irish Taxes and Consolidated Act of 1997 modernised legal and tax structures, establishing Irish section 110 special purpose vehicles SPVs) and foundations for advanced financial instruments.

However, the model encountered several challenges. The tax differential approved by the EU faced pressure and was set to expire in 2005. In response, the Irish Government made a significant policy change by lowering the national corporate tax rate to 12.5% for all companies by 2003. This move was intended to maintain competitiveness while also addressing EU concerns regarding state aid.

Building the Tech Ecosystem: Digital Convergence (2000-2010)

The groundwork for Dublin's technology transformation was established in the early 2000s through strategic government initiatives and infrastructure development. The Irish Government established The Digital Hub in the Liberties area in 2003, creating a cluster of digital content and technology companies on a state-of-the-art campus. The

Hub was established with Media Lab Europe, a collaboration with MIT, as its main tenant.

The Digital Hub was created to embody the Irish Government's goal of establishing a world-class digital media industry while also fostering the urban regeneration of the historically working-class Liberties neighbourhood. Since its inception, over 200 companies have emerged from the enterprise cluster, creating thousands of skilled jobs. Notable alumni include Amazon, Stripe, Slack, Havok, and Etsy.

The turning point occurred when Google decided to set up its European headquarters in Dublin. Following the dotcom collapse from 1999 to 2001, Dermot Tuohy, the director of operations for IDA Ireland in California, actively sought out emerging technology companies such as PayPal, eBay, and Google. In 2002, Google executives explored various locations in Dublin, initially considering the Digital Hub before ultimately choosing a site at Grand Canal Dock on Barrow Street, where they moved into Gordon House in 2004.

The IDA viewed Google's decision as a major turning point for investment in Dublin, which prompted other Silicon Valley companies such as Twitter and Facebook to set up operations in the area. On October 2, 2008, Facebook announced the opening of its Dublin office, establishing its international headquarters at a time when the platform had just 100 million users. The move initially created 70 jobs, supported by IDA Ireland.

The tech companies were drawn to various factors in addition to tax incentives. Dublin provided easy access to the city centre, a campus-like atmosphere in the Grand Canal Dock area, and a pool of highly educated talent. By 2012, Citibank had ranked Dublin as the city with the best "human capital" in the world, home to numerous colleges and universities, including Trinity College Dublin, University College Dublin, and Dublin City University.

The area experienced significant infrastructural development, highlighted by the opening of the Grand Canal Dock railway station in 2001 and the decontamination of the former gasworks site, which was carried out by the DDDA between 2002 and 2006 at a cost of €52 million. By the time the decontamination was complete, increased demand—partly fuelled by Google's presence—enabled the authorities to sell the land for €300 million.

Expansion and Ecosystem Development (2010-2020)

Between 2010 and 2020, Dublin Docklands emerged as a globally recognised technology Hub. The nickname "Silicon Docks" was adopted around 2011 to refer to the concentration of European headquarters for high-tech companies in the Grand Canal Dock area. During this period, Facebook experienced significant growth, with engineer Colm Doyle joining the company in 2010. He later moved to Silicon Valley in 2012 to help establish Facebook's North American developer support presence.

Major infrastructure investments enhanced connectivity and urban appeal. The new infrastructure features the Samuel Beckett Bridge and the LUAS Docklands extension, along with venues such as the Bord Gáis Energy Theatre, the renovated 3Arena, and the Convention Centre Dublin. Thirteen Dublin Bike stations opened in the Docklands between 2013 and 2014, enhancing sustainable transportation options.

The area's development was sped up due to planning reforms. In May 2014, An Bord Pleanála approved a fast-track planning process for the Docklands Strategic Development Zone (SDZ), which will allow for the development of 366,000 square metres of office space and 2,600 homes across 22 hectares in the North Lotts and Grand Canal Dock areas.

Dublin City Council, which took over the responsibilities of the DDDA, has identified five specific development hubs: Spencer Dock, Point Village, Grand Canal Dock, Britain Quay, and Boland's Mill. One-third of the entire Docklands area has been designated for development. Buildings that had remained unfinished since the 2008 financial crisis were completed, most notably the former Anglo Irish Bank building, which was finished in 2017 by its new owner, the Central Bank of Ireland.

Technology companies continued to expand their presence significantly. Google has completed the acquisition of the iconic Bolands Quay site in a deal valued at €170 million. Facebook grew to 2,500 employees, while Google exceeded 7,000 employees in Dublin and is approaching 8,000.

The development of venture capital supported the maturation of the ecosystem. Enterprise Ireland's Competitive Start Fund invests in 15 seed-stage start-ups each quarter, while other accelerators, such as Launchpad, provide essential seed funding. Professor Vinny Cahill of Trinity College Dublin noted in 2012 that although Dublin had a growing venture capital community, it had not yet reached the level of networks and experienced investors found in Silicon Valley.

Innovation District Development and Academic Integration (2018-Present)

The latest phase has focused on developing a comprehensive innovation district that integrates academic institutions with commercial enterprises. In July 2018, Trinity College Dublin announced plans to develop a €1 billion Grand Canal Innovation District campus, inspired by innovation districts in cities such as Boston, Toronto, Rotterdam, and Barcelona. The 5.5-acre campus at Grand Canal Quay and Macken Street is designed to host start-ups, attract foreign direct investment companies, and accommodate venture capital firms, all while providing public and civic spaces.

Trinity submitted planning applications in November 2020 for the first phase of a 5,500 square metre Innovation Hub. This facility will provide space for early-stage start-ups, research-active corporations, meeting areas, and community programs, with full operations scheduled to begin in the first quarter of 2022. The development is funded by government grants from Enterprise Ireland's Regional Enterprise Development Fund and the Urban Regeneration Development Fund, as part of Project Ireland 2040.

The broader Smart Docklands initiative, established in 2018 and funded by Dublin City Council and CONNECT (the Science Foundation Ireland Research Centre for Future Networks at Trinity College), has created a two-square-kilometre smart city testbed. The program is supported by corporate partners such as Google, Deloitte, Accenture, Microsoft, Mastercard, Intel, IBM, and SoftBank.

Smart Docklands serves as an "honest broker" between government, industry, universities, and citizens, facilitating trials of smart city technologies, including connected bike lights engaging 500 participants, flood sensors using LoRaWan and SIGFOX technologies, and partnerships on NB-IoT and 5G connectivity.

Academic excellence continues to be a key factor in Dublin's appeal as an Innovation Hub. Trinity College Dublin, founded in 1592, opened a new Business School building in 2019, which cost €80 million. The building includes an Innovation and Entrepreneurial Hub, smart classrooms equipped with digital technology, and facilities for executive education. Trinity Innovation & Enterprise promotes a culture of innovation by supporting 37 campus companies that have emerged from the university. A state-of-the-art Innovation Hub, Portal, is set to open in September 2025.

Economic Impact and Global Recognition

Dublin's transformation has led to significant economic benefits. Ireland's tech sector is home to nine of the top ten US technology companies, eight of the top ten global software companies, and the top three global enterprise software companies. The ICT sector employs 37,000 people and generates €35 billion in annual exports.

The current workforce in the Docklands is 40,000, and it is projected to double in the next two years. Approximately 25% of the Docklands population was born outside of Ireland, making it one of Dublin's most multicultural areas. Only 8% of tenants in the Docklands are Irish, while Europeans make up 52%, and 32% are categorised as "international".

The area's success has drawn ongoing investment. Microsoft has been present in Ireland since 1985. The company recently began construction on a €380 million data centre in Clondalkin, covering an area of 35,000 square metres. This project will add 150 full-time employees to its existing workforce of over 1,000. Tech companies are selecting Ireland for their data centres, partly because the cool climate helps to lower air-conditioning costs. For instance, Microsoft's facility could save around €17 million each year.

Challenges and Sustainability Concerns

Dublin's success has led to major challenges that could jeopardise its long-term sustainability. Housing affordability has become the most pressing issue. Dublin City Council Chief Executive Owen Keegan acknowledged that there are concerns about the low levels of social housing and homes for middle-income workers in the Docklands. Many workers cannot afford to rent or buy-in the Docklands, as the council can acquire two social homes outside the area for the price of one apartment in the Docklands.

In 2017, the average rent for a two-bedroom apartment in the Docklands was €2,310 per month for rent-controlled properties and €2,406 for noncontrolled units. The average price of a two-bedroom apartment rose by 15.1% to €485,129, with 66% of buyers being investors rather than owner-occupiers.

Concerns about social displacement have grown. Resident Michael Ingle described how the area has transformed from a place that few wanted to live to one that few can afford to live in. Long-term resident Declan Byrne remarked that if his father were brought to the area today, he would not recognise it. This could be Brazil for all he knows.

The planning framework has been criticised for not providing adequate social services. The social housing quota has been reduced from 20% to 10% of new developments, contributing to rising prices and limited job opportunities for local residents, which have increased social tensions. Councillor Éilis Ryan of the Workers' Party warned that the residential units being constructed are only suited for individuals with high-paying jobs in the area.

There are also risks associated with economic dependency. The housing crisis and rising cost of living have prompted tech firms to retain employees, leading many companies to consider locations in other parts of Ireland where expenses are more manageable. Technology commentator Karlin Lillington cautioned that Dublin risks repeating the issues faced by San Francisco, where the rapid expansion of the tech industry has led to significant displacement and a lack of affordable housing.

COVID-19 exposed weaknesses in the system. The pandemic temporarily drained life from the neighbourhood, as work-from-home policies and reduced public transport usage significantly affected the area's vibrancy. Local Green Party councillor Claire Byrne described the area's reduced activity as "crushing", expressing concern about the long-term impacts on the community.

Policy Interventions and Responses

Both government and private sector responses to these challenges have been diverse and multifaceted. Irish companies have started offering employee accommodations, with Ryanair purchasing 40 houses near Dublin Airport for its cabin crew. Meanwhile, Musgrave manages 50 rental properties for staff, and Windmill Healthcare owns 28 properties, with plans for further expansion. These initiatives come at a substantial cost, with Supermacs reportedly spending between €6 and €7 million on staff housing.

Planning authorities have recognised the need for reform. Fine Gael Councillor Ray McAdam stated that "the planning department within the city council needs to take a more critical look at planning applications and how they comply with the Docklands SDZ". Some councillors advocate for increasing housing density; however, this approach addresses supply issues rather than affordability for lower-income workers.

Investment in transportation infrastructure aims to improve accessibility. Projects like MetroLink aim to connect high-employment areas, such as the Docklands, with commuters from outer Dublin, potentially alleviating pressure on local housing markets.

The Digital Hub has encountered its own set of challenges. In April 2021, the government announced the dissolution of the Digital Hub Development Agency, transferring its assets to the Land Development Agency. However, operations will continue until at least 2025, pending redevelopment plans. Digital Hub CEO Fiach Mac Conghail anticipates operations continuing "until at least the end of this decade".

Future Directions and Strategic Positioning

Dublin's future as a technology Hub depends on addressing sustainability challenges while also preserving its competitive advantages. Brexit has strengthened Dublin's strategic position. Postbrexit, Ireland has solidified its position as a Hub for foreign direct investment, becoming the only English-speaking country in the EU with a young, educated workforce and a stable, business-friendly political environment.

The concept of an innovation district offers promising solutions. Dublin City Council's planning framework proposes the development of an innovation district at Grand Canal Quay by Trinity College. This initiative aims to promote local employment by leveraging available skills and to support sustainable tourism. This involves implementing a Water Animation Strategy to keep the River Liffey navigable while also providing diverse, high-quality architecture in public spaces.

Trinity's expanded role could serve as a vital academic foundation. The Grand Canal Innovation District aims to create "an internationally visible focal point for excellence in research and innovation in Ireland", providing scale, critical mass, and density to resonate with global companies while realising ambitions of Indigenous start-ups.

Smart city initiatives provide opportunities for inclusive development. Smart Docklands emphasises community engagement by collaborating with diverse stakeholder groups to identify challenges that can be addressed through smart city technologies. Additionally, it seeks to raise awareness of ongoing projects and activities.

However, fundamental questions about development models still remain. As technology commentator Karlin Lillington argued, "we really need to talk" about what kind of Dublin society wants before accommodating more tech companies, requiring broader societal consultation on planning decisions that seem to prioritise adding ever more tech employers to confined geographic areas.

International Implications and Lessons

Dublin's experience provides valuable insights for other cities that are developing innovation districts. The success illustrates the effectiveness of consistent policy commitment, strategic tax incentives, and high-quality education in attracting global technology investment. The combination of urban regeneration and economic development demonstrates how former industrial areas can be transformed into globally competitive innovation hubs.

The Irish strategy of integrating national tax policy with local development planning has created significant competitive advantages. The decision to nationalise corporate tax benefits before the expiration of EU approval, rather than trying to maintain special zones, demonstrated adaptability in policy while still appealing to international investors.

However, the challenges facing Dublin emphasise the risks associated with concentrated technology development, particularly when there is insufficient focus on housing provision and social inclusion. The experience indicates that successful innovation districts need pro-active planning for affordable housing, community services, and social integration from the beginning, rather than waiting to respond reactively to displacement pressures.

The significance of transportation connectivity in managing growth pressures is clearly evident. Dublin's investment in LUAS extensions, DART connectivity, and the planned MetroLink infrastructure demonstrates how transportation policy can more evenly distribute the benefits of development while still preserving the advantages of urban clusters.

Cities that are considering similar strategies should recognise the important role of anchor institutions. Google's early decision to establish its location sparked the development of a broader ecosystem, while Trinity College's plans for an innovation district aim to provide academic support for long-term sustainability. The collaboration between multinational corporations and leading universities fosters resilient innovation ecosystems.

The Smart Docklands initiative shows how cities can use a concentrated presence of technology to drive broader innovation. By positioning government as an "honest broker" between different stakeholders, cities can capture innovation benefits while addressing community needs.

Conclusion

The Dublin Docklands is one of the most successful examples of innovation driven urban regeneration in Europe. The transformation of derelict port lands into a global technology Hub illustrates how strategic policy intervention, educational excellence, and ongoing investment can foster internationally competitive innovation districts.

The foundation established by the IFSC's financial services cluster provided crucial infrastructure and policy frameworks that facilitated the growth of the technology sector. The decision to establish the Digital Hub and actively attract technology companies through the IDA created a critical mass that drew successive waves of global investment.

However, Dublin's experience also highlights the complexities involved in managing successful innovation districts. The housing crisis, concerns about social displacement, and risks of economic dependency emphasise the necessity for comprehensive planning that promotes both community development and economic growth.

Lessons for Other Regions

As cities around the world strive to develop innovative economies, Dublin's experience highlights several key principles:

- *A sustained commitment to policy across different political cycles* promotes long-term ecosystem development.
- *Educational institutions play a crucial role* in supporting sustainable innovation districts.
- *A pro-active planning for housing and social inclusion is essential* for fostering community acceptance and ensuring long-term viability.
- *Improved transport connectivity can help distribute the benefits* of development while preserving the advantages of clustering.

- *Government can play a valuable role as a facilitator and an honest broker* among various stakeholder interests. In conclusion, global connectivity and access to markets are essential for the success of innovation districts.

Dublin is facing ongoing challenges that require continuous innovation in policy. The Trinity College Innovation District, the Smart Docklands initiatives, and investments in transportation infrastructure all represent efforts to tackle sustainability issues while preserving a competitive advantage.

Success hinges on balancing growth management with inclusive development, ensuring that innovation districts not only meet the broader needs of the community but also maintain global competitiveness.

Other cities can learn from the successes and challenges that Dublin has experienced. The experience demonstrates that innovation driven regeneration can revitalise urban areas and create substantial economic value. However, it also emphasises that achieving success requires careful planning for sustainable development that benefits everyone in the community, not just high-income technology workers.

Dublin's story is constantly evolving, as Brexit, COVID-19, and housing pressures create both new challenges and opportunities. The city's response to these challenges will determine if it can maintain its status as a leading global Innovation Hub while also tackling the social and economic sustainability issues that have emerged from its success.

APPENDIX: SHORT PROFILES OF ALL CASE STUDIES COVERED IN THE PROJECT

Australia

1. *Canberra Innovation Ecosystem.* National capital's distributed innovation ecosystem leveraging government agencies, universities, and research institutions across multiple specialised precincts. Development Pathway: Research organisation and corporate anchors Sectoral Focus: Defence and aerospace technology

2. *Tech Central, Sydney.* Central business district transformation creating Australia's largest tech precinct through adaptive reuse of railway corridor infrastructure. Development Pathway: Adaptive reuse and urban transformation Sectoral Focus: Financial technology and services

3. *Macquarie Park Innovation District, Sydney.* Suburban business park evolution into integrated innovation district anchored by Macquarie University and major corporate headquarters. Development Pathway: Research organisation and corporate anchors Sectoral Focus: Advanced manufacturing and deep technology

4. *Bradfield City and Aerotropolis*, Western Sydney. Greenfield smart city development around new international airport creating advanced manufacturing and technology ecosystem from ground up. Development Pathway: Property led development and government partnership Sectoral Focus: Advanced manufacturing and deep technology

5. *Fishermans Bend Innovation District*, Melbourne. Industrial brownfield transformation into mixed-use innovation precinct integrating research, residential, and commercial development near city centre. Development Pathway: Adaptive reuse and urban transformation Sectoral Focus: Advanced manufacturing and deep technology

6. *Monash Technology Precinct*, Melbourne. University-anchored R&D hub specialising in advanced manufacturing, materials science, and engineering innovation in Clayton. Development Pathway: Research organisation and corporate anchors Sectoral Focus: Advanced manufacturing and deep technology

7. *The Precinct (Fortitude Valley)*, Queensland. Creative industries quarter leveraging heritage buildings and urban density to foster digital media,

gaming, and start-up ecosystem. Development Pathway: Adaptive reuse and urban transformation Sectoral Focus: Creative and digital industries

8. *Innovation Districts (City North)*, Melbourne. Northern Melbourne manufacturing corridor transformation into integrated innovation ecosystem leveraging existing industrial heritage and university partnerships. Development Pathway: Response to industrial decline, visionary leadership and institution building Sectoral Focus: Advanced manufacturing and deep technology

9. *Future Economy Precinct*, Geelong. Regional manufacturing city's transition to advanced manufacturing and technology hub leveraging Deakin University and automotive industry transformation. Development Pathway: Response to industrial decline, visionary leadership and institution building Sectoral Focus: Advanced manufacturing and deep technology

10. *Boggo Road Innovation Precinct*, Brisbane. Former prison site redevelopment into health and technology research precinct anchored by University of Queensland and Princess Alexandra Hospital. Development Pathway: Adaptive reuse and urban transformation Sectoral Focus: Health and biomedical

11. *The Australian Automation and Robotics Precinct*, Perth. Mining industry transformation driving automation and robotics innovation hub leveraging Western Australia's resource sector expertise and requirements. Development Pathway: Research organisation and corporate anchors Sectoral Focus: Advanced manufacturing and deep technology

12. *Bentley Technology Park and Curtin Exchange*, Perth. University-industry collaboration precinct focusing on mining technology, engineering innovation, and resource sector digital transformation near Curtin University. Development Pathway: Research organisation and corporate anchors Sectoral Focus: Advanced manufacturing and deep technology

13. *Henderson Defence Precinct,* Perth. Naval shipbuilding and defence industry cluster leveraging strategic infrastructure investment and proximity to Australian Marine Complex facilities. Development Pathway: Property led development and government partnership Sectoral Focus: Defence and aerospace technology

14. *Tonsley Innovation District*, Adelaide. Former Mitsubishi automotive manufacturing site transformed into integrated innovation ecosystem for advanced manufacturing, clean technology, and education.

Development Pathway: Response to industrial decline, visionary leadership and institution building Sectoral Focus: Advanced manufacturing and deep technology

15. *Lot 14*, Adelaide. Former Royal Adelaide Hospital site redevelopment creating technology and entrepreneurship hub in central Adelaide with government agency anchors. Development Pathway: Adaptive reuse and urban transformation Sectoral Focus: Advanced manufacturing and deep technology

16. *The Australian Technology Park*, Sydney, A Missed Opportunity. Sydney's ambitious technology park development that struggled to achieve critical mass and integration with broader innovation ecosystem. Development Pathway: Property led development and government partnership Sectoral Focus: Advanced manufacturing and deep technology

17. *Adelaide's Multifunction Polis*: A Vision Unrealised. 1980s Japan-Australia joint venture concept for futuristic city combining technology, education, and residential development that never materialised. Development Pathway: Property led development and government partnership Sectoral Focus: Advanced manufacturing and deep technology

18. *UNE Smart Farm and Incubator*. University of New England's agricultural innovation hub combining precision agriculture research, technology development, and start-up incubation programs. Development Pathway: Research organisation and corporate anchors Sectoral Focus: Agricultural technology and food innovation

19. *Bendigo Innovation District*. Regional Victorian city's knowledge economy development leveraging La Trobe University campus and health sector to drive technology innovation. Development Pathway: Research organisation and corporate anchors Sectoral Focus: Health and biomedical

20. *Hobart Marine and Antarctic Precinct*. Southern Ocean research excellence cluster leveraging University of Tasmania, CSIRO, and Australian Antarctic Division for marine science innovation. Development Pathway: Research organisation and corporate anchors Sectoral Focus: Clean technology and sustainability

21. *Silicon Highlands*: An Aspirational Innovation Ecosystem. Regional New South Wales technology development concept aimed at leveraging natural amenities and digital connectivity for knowledge workers.

Development Pathway: Opportunistic connections and serendipitous development Sectoral Focus: Creative and digital industries

22. *Sunshine Coast Innovation Ecosystem.* Queensland coastal region's knowledge economy development centred on University of Sunshine Coast and health technology innovation clusters. Development Pathway: Research organisation and corporate anchors Sectoral Focus: Health and biomedical

23. *Toowoomba Ag Tech and Logistics Hub.* Strategic agricultural technology development leveraging regional Queensland's farming expertise, university research, and transport infrastructure connectivity. Development Pathway: Research organisation and corporate anchors Sectoral Focus: Agricultural technology and food innovation

24. *TropiQ Health & Knowledge Precinct*, Townsville. Tropical medicine and health research cluster anchored by James Cook University, focusing on Indigenous health and tropical disease innovation. Development Pathway: Research organisation and corporate anchors Sectoral Focus: Health and biomedical

25. *Newcastle Innovation Ecosystem.* Postindustrial city transformation leveraging the University of Newcastle and the health sector to create a technology and start-up ecosystem in the Hunter region. Development Pathway: Response to industrial decline, visionary leadership and institution building Sectoral Focus: Clean technology and sustainability

26. *Wagga Wagga AgriPark.* Regional agricultural innovation hub combining research, education, and commercial development to advance farming technology and food production systems. Development Pathway: Research organisation and corporate anchors Sectoral Focus: Agricultural technology and food innovation

27. *Albury-Wodonga Innovation District:* Australia's Cross-Border Innovation Hub. Bi-state regional innovation development leveraging dual university presence and strategic location for advanced manufacturing and technology development. Development Pathway: Research organisation and corporate anchors Sectoral Focus: Advanced manufacturing and deep technology

28. *Gold Coast Health and Knowledge Precinct.* Integrated health, education, and research development anchored by Griffith University and Gold Coast Health creating biomedical innovation cluster. Development Pathway: Research organisation and corporate anchors Sectoral Focus: Health and biomedical

29. *Parkville Biomedical Precinct*, Melbourne. World-leading biomedical research cluster integrating University of Melbourne, major hospitals, and research institutes in concentrated urban precinct. Development Pathway: Opportunistic connections and serendipitous development Sectoral Focus: Health and biomedical

30. *Adelaide BioMed City*. Integrated biomedical research and healthcare delivery ecosystem anchored by major hospitals, universities, and research institutes in central Adelaide. Development Pathway: Research organisation and corporate anchors Sectoral Focus: Health and biomedical

31. *The Floriat Biomedical Precinct*: Perth's Emerging Innovation Hub. Western Australian biomedical research cluster development leveraging hospital infrastructure and university partnerships for health technology innovation. Development Pathway: Research organisation and corporate anchors Sectoral Focus: Health and biomedical

32. *Liverpool*, Western Sydney. Emerging health and education precinct anchored by Western Sydney University and Liverpool Hospital creating integrated knowledge and healthcare cluster. Development Pathway: Research organisation and corporate anchors Sectoral Focus: Health and biomedical

33. *Monash-Clayton Health*, Medicine, and Biotech Precinct. University-hospital integration creating comprehensive biomedical R&D ecosystem with strong industry collaboration and commercial translation focus. Development Pathway: Research organisation and corporate anchors Sectoral Focus: Health and biomedical

34. *Herston Health Precinct*, Brisbane. Major quaternary health precinct integrating Royal Brisbane Hospital, Queensland University of Technology, and biomedical research institutes. Development Pathway: Research organisation and corporate anchors Sectoral Focus: Health and biomedical

35. *Randwick Health and Innovation Precinct*, Sydney. Comprehensive health and biomedical research cluster anchored by major teaching hospitals and University of New South Wales integration. Development Pathway: Research organisation and corporate anchors Sectoral Focus: Health and biomedical

36. *Westmead Health District*, Western Sydney. Integrated health, research, and education precinct combining major hospital, medical research institute, and university campus development. Development Pathway:

Research organisation and corporate anchors Sectoral Focus: Health and biomedical

37. *Wagga Wagga Health and Knowledge Precinct.* Regional health and education cluster leveraging Charles Sturt University and major regional hospital to create rural health innovation focus. Development Pathway: Research organisation and corporate anchors Sectoral Focus: Health and biomedical

Europe

38. *22@Barcelona.* Former industrial district transformation into innovation ecosystem through strategic urban planning integrating technology companies, research centres, and residential development. Development Pathway: Response to industrial decline, visionary leadership and institution building Sectoral Focus: Creative and digital industries

39. *The London Knowledge Quarter.* Central London research and education cluster leveraging proximity of major universities, museums, and research institutions for knowledge economy development. Development Pathway: Opportunistic connections and serendipitous development Sectoral Focus: Health and biomedical

40. *Amsterdam Science Park.* University-anchored research and technology development cluster integrating academic research, corporate R&D, and start-up ecosystem in planned environment. Development Pathway: Research organisation and corporate anchors Sectoral Focus: Advanced manufacturing and deep technology

41. *Aspern Seestadt, Vienna.* New urban development project creating sustainable smart city with integrated research, residential, and commercial development on former airfield site. Development Pathway: Property led development and government partnership Sectoral Focus: Clean technology and sustainability

42. *Birmingham Innovation Precinct.* Postindustrial city centre regeneration leveraging university partnerships and digital technology focus to create innovation ecosystem in Midlands. Development Pathway: Response to industrial decline, visionary leadership and institution building Sectoral Focus: Advanced manufacturing and deep technology

43. *Bristol: The SetSquared Incubator.* University-industry collaboration creating technology incubation and entrepreneurship ecosystem leveraging aerospace and engineering expertise in Southwest England.

Development Pathway: Research organisation and corporate anchors Sectoral Focus: Defence and aerospace technology

44. The *Cambridge Cluster*. World-renowned technology and biotech ecosystem radiating from university creating dense network of research, industry, and venture capital. Development Pathway: Opportunistic connections and serendipitous development Sectoral Focus: Health and biomedical

45. *Copenhagen's Ørestad District*. New urban development on former common land creating modern business and residential district with strong sustainability focus and architectural innovation. Development Pathway: Property led development and government partnership Sectoral Focus: Clean technology and sustainability

46. *Dublin Docklands*. Former port area transformation into international financial services and technology hub leveraging European headquarters location and skilled workforce. Development Pathway: Adaptive reuse and urban transformation Sectoral Focus: Financial technology and services

47. *Eindhoven Brainport*. Technology region built around Philips heritage creating high-tech ecosystem specialising in electronics, design, and advanced manufacturing innovation. Development Pathway: Response to industrial decline, visionary leadership and institution building Sectoral Focus: Advanced manufacturing and deep technology

48. *Edinburgh BioQuarter*. Integrated life sciences cluster combining university research, hospital facilities, and commercial development creating comprehensive biomedical innovation ecosystem. Development Pathway: Research organisation and corporate anchors Sectoral Focus: Health and biomedical

49. *Hammarby Sjöstad*, Sweden. Sustainable urban development demonstration project integrating environmental technology, circular economy principles, and innovative infrastructure systems design. Development Pathway: Property led development and government partnership Sectoral Focus: Clean technology and sustainability

50. *Manchester: The Sister Innovation District*. Urban regeneration project leveraging university partnerships and digital technology focus to create innovation ecosystem in former industrial areas. Development Pathway: Response to industrial decline, visionary leadership and institution building Sectoral Focus: Creative and digital industries

51. *Medicon Valley, Denmark-Sweden.* Cross-border life sciences cluster connecting Copenhagen and Malmö regions creating integrated pharmaceutical and biotech ecosystem spanning national boundaries. Development Pathway: Opportunistic connections and serendipitous development Sectoral Focus: Health and biomedical

52. *The* Mitte Tech Corridor, Berlin. Central Berlin technology ecosystem leveraging start-up culture, university partnerships, and creative industries to create integrated innovation environment. Development Pathway: Opportunistic connections and serendipitous development Sectoral Focus: Creative and digital industries

53. *The Munich Innovation Corridor.* Technology corridor connecting research institutions, major corporations, and start-up ecosystem creating comprehensive innovation environment in Bavaria. Development Pathway: Research organisation and corporate anchors Sectoral Focus: Advanced manufacturing and deep technology

54. Cyber Valley, Baden-Württemberg. Europe's largest artificial intelligence research consortium in Stuttgart-Tübingen region combining universities, Max Planck Institute and major industry partners to create internationally competitive AI ecosystem. Development Pathway: Public-private research partnership and institutional collaboration. Sectoral Focus: Artificial intelligence, machine learning and intelligent robotics.

55. *Newcastle University (UK) Science Precinct.* University-centred research and innovation development focusing on engineering, digital technology, and energy systems in Northeast England. Development Pathway: Research organisation and corporate anchors Sectoral Focus: Clean technology and sustainability

56. *Oslo Science City, Norway.* University and research institute cluster creating comprehensive knowledge economy development with strong focus on sustainability and energy technology. Development Pathway: Research organisation and corporate anchors Sectoral Focus: Clean technology and sustainability

57. *Paris-Saclay Innovation Ecosystem.* Planned research and education cluster integrating major universities, research institutes, and corporate R&D to create European innovation powerhouse. Development Pathway: Property led development and government partnership Sectoral Focus: Advanced manufacturing and deep technology

58. *The Ruhr Innovation District.* Industrial heritage region transformation leveraging universities and research institutions to create post-coal economy focused on technology innovation. Development Pathway: Response to industrial decline, visionary leadership and institution building Sectoral Focus: Clean technology and sustainability

59. *Station F, Paris.* World's largest start-up campus created in former railway freight station providing entrepreneurship support, corporate partnerships, and innovation programming. Development Pathway: Adaptive reuse and urban transformation Sectoral Focus: Creative and digital industries

60. *Strathclyde Innovation and Manufacturing Ecosystem.* University-industry collaboration creating advanced manufacturing and engineering innovation cluster leveraging Scottish industrial heritage and research excellence. Development Pathway: Research organisation and corporate anchors Sectoral Focus: Advanced manufacturing and deep technology

61. *Stockholm Innovation Districts.* Distributed innovation ecosystem across multiple districts leveraging strong technology sector, university partnerships, and sustainable development focus. Development Pathway: Opportunistic connections and serendipitous development Sectoral Focus: Clean technology and sustainability

62. *Skolkovo Innovation Centre*, Moscow. Government-led technology development project creating comprehensive innovation ecosystem with strong focus on technology transfer and commercialisation. Development Pathway: Property led development and government partnership Sectoral Focus: Advanced manufacturing and deep technology

63. *Zurich, Switzerland Innovation Park.* Research and technology development cluster leveraging proximity to ETH Zurich and strong financial sector to create comprehensive innovation ecosystem. Development Pathway: Research organisation and corporate anchors Sectoral Focus: Financial technology and services

North America

64. *MaRS Discovery District*, Toronto. Biomedical and technology commercialisation hub integrating research institutions, hospitals, and start-up ecosystem in downtown Toronto innovation cluster. Development Pathway: Research organisation and corporate anchors Sectoral Focus: Health and biomedical

65. *Kendall Square*. World-renowned biotech and technology cluster anchored by MIT creating dense ecosystem of research, industry, and venture capital. Development Pathway: Opportunistic connections and serendipitous development Sectoral Focus: Health and biomedical

66. *Boston's Seaport District*. Waterfront redevelopment creating modern innovation district with strong technology and life sciences focus complementing established Cambridge ecosystem. Development Pathway: Property led development and government partnership Sectoral Focus: Advanced manufacturing and deep technology

67. *Cummings Research Park, Alabama*: Huntsville research park development leveraging NASA presence and defence contractors to create aerospace and technology innovation cluster. Development Pathway: Research organisation and corporate anchors Sectoral Focus: Defence and aerospace technology

68. *Cortex St. Louis*. Innovation district development integrating universities, hospitals, and technology companies to create biomedical and technology commercialisation ecosystem. Development Pathway: Research organisation and corporate anchors Sectoral Focus: Health and biomedical

69. *Ion District Houston*. Energy industry transformation creating innovation district focused on energy technology, sustainability, and start-up ecosystem development in Texas. Development Pathway: Response to industrial decline, visionary leadership and institution building Sectoral Focus: Clean technology and sustainability

70. *Research Triangle Park*, North Carolina. Planned research park integrating major universities and corporate R&D facilities creating comprehensive technology and biomedical innovation ecosystem. Development Pathway: Property led development and government partnership Sectoral Focus: Health and biomedical

71. *San Diego Innovation District*. Biotech and technology cluster leveraging university research, defence industry presence, and venture capital to create comprehensive innovation environment. Development Pathway: Research organisation and corporate anchors Sectoral Focus: Health and biomedical

72. *Silicon Valley*. World's premier technology ecosystem combining university research, venture capital, and entrepreneurial culture creating global innovation benchmark. Development Pathway: Opportunistic

connections and serendipitous development Sectoral Focus: Advanced manufacturing and deep technology

73. *South Lake Union, Seattle.* Former industrial area transformation into technology hub anchored by Amazon and other major tech companies creating urban innovation district. Development Pathway: Adaptive reuse and urban transformation Sectoral Focus: Advanced manufacturing and deep technology.

74. Bellevue Innovation District, Washington. Suburban city transformation into a technology hub, anchored by Amazon, T-Mobile, and Meta headquarters alongside emerging startup ecosystem. Development Pathway: Transit-oriented development and mixed-use urban transformation Sectoral Focus: Enterprise software, telecommunications, cloud infrastructure, and technology platforms

Asia-Pacific

75. *Cyberjaya: Malaysia's Digital Innovation Ecosystem.* Planned smart city development creating comprehensive digital technology ecosystem with government, education, and industry integration in Selangor. Development Pathway: Property led development and government partnership Sectoral Focus: Creative and digital industries

76. *Dubai Technology Entrepreneur Campus.* Government-led innovation hub development creating start-up ecosystem and technology commercialisation focus within broader economic diversification strategy. Development Pathway: Property led development and government partnership Sectoral Focus: Financial technology and services

77. *HITEC City Innovation Ecosystem,* Hyderabad. Information technology and biotech cluster development leveraging government support and skilled workforce to create major Indian innovation centre. Development Pathway: Property led development and government partnership Sectoral Focus: Creative and digital industries

78. *King Abdullah Economic City.* New planned city development integrating industrial, residential, and innovation functions creating comprehensive economic zone on Saudi Arabian coast. Development Pathway: Property led development and government partnership Sectoral Focus: Financial technology and services

79. *One-North, Singapore.* Integrated biomedical and technology development cluster combining research institutes, universities, and

commercial development in planned urban environment. Development Pathway: Property led development and government partnership Sectoral Focus: Health and biomedical

80. *Taiwan Tech Arena.* Start-up and technology commercialisation ecosystem leveraging Taiwan's manufacturing expertise and creating bridge between Asia-Pacific innovation networks. Development Pathway: Property led development and government partnership Sectoral Focus: Advanced manufacturing and deep technology

81. *Tel Aviv Innovation Corridor.* High-tech ecosystem leveraging military technology transfer, university research, and entrepreneurial culture creating 'Start-up Nation' innovation environment. Development Pathway: Opportunistic connections and serendipitous development Sectoral Focus: Advanced manufacturing and deep technology

82. *Tsukuba Science City, Japan.* Planned research city integrating universities, national research institutes, and corporate R&D facilities creating comprehensive science and technology ecosystem. Development Pathway: Property led development and government partnership Sectoral Focus: Advanced manufacturing and deep technology

83. *Shenzhen, China Special Zone.* Manufacturing centre transformation into innovation hub leveraging electronics industry, proximity to Hong Kong, and entrepreneurial manufacturing culture. Development Pathway: Response to industrial decline, visionary leadership and institution *building Sectoral Focus: Advanced manufacturing and deep technology*

84. Yokohama Minato Mirai 21. Waterfront redevelopment is creating a business and innovation district with a strong international focus and modern urban planning principles. Development Pathway: Property-led development and government partnership Sectoral Focus: Advanced manufacturing and deep technology

85. *Yongsan, South Korea.* Former military base redevelopment is creating a technology and innovation district in central Seoul with a strong focus on digital industries. Development Pathway: Adaptive reuse and urban transformation Sectoral Focus: Creative and digital industries

86. *Zhongguancun, China.* Beijing's technology corridor is leveraging university presence and government support to create a comprehensive high-tech industry and research ecosystem. Development Pathway: Research organisation and corporate anchors Sectoral Focus: Advanced manufacturing and deep technology

Latin America

87. *DistritoTec Innovation District, Monterrey.* University-anchored urban innovation development integrating education, research, and commercial activity to create comprehensive technology ecosystem. Development Pathway: Research organisation and corporate anchors Sectoral Focus: Advanced manufacturing and deep technology

88. *Medellín Innovation District, Colombia.* Urban transformation project leveraging education and innovation to create a post-conflict economic development model in a Colombian city. Development Pathway: Response to industrial decline, visionary leadership and institution building. Sectoral Focus: Creative and digital industries

89. *The Polo Tecnológico Rosario UNR Medical Precinct, Argentina.* University-hospital integration is creating a biomedical R&D cluster focusing on health technology innovation in a regional Argentine city. Development Pathway: Research organisation and corporate anchors Sectoral Focus: Health and biomedical

90. *Porto Digital, Recife Brazil.* Software and digital services cluster development leveraging university partnerships and government support to create a technology innovation ecosystem. Development Pathway: Research organisation and corporate anchors Sectoral Focus: Creative and digital industries

91. *UFRJ Science Park and Health-Science Cluster, Brazil.* University-anchored R&D cluster integrating health sciences, engineering, and technology commercialisation in Rio de Janeiro. Development Pathway: Research organisation and corporate anchors Sectoral Focus: Health and biomedical

92. *University of Chile Health & Science Precinct, Santiago.* University-hospital integration is creating a comprehensive health research and technology development cluster in the Chilean capital city. Development Pathway: Research organisation and corporate anchors Sectoral Focus: Health and biomedical

Africa

93. *The Innovation Hub, Pretoria, South Africa.* Technology incubation and commercialisation cluster leveraging university partnerships and government support to create an African innovation ecosystem.

Development Pathway: Research organisation and corporate anchors Sectoral Focus: Advanced manufacturing and deep technology

94. *Kigali Innovation City, Rwanda.* Planned smart city development creating a comprehensive technology and education ecosystem as part of Rwanda's economic transformation strategy. Development Pathway: Property-led development and government partnership Sectoral Focus: Creative and digital industries

95. *Konza Technopolis, Kenya.* Planned technology city development, creating a comprehensive ICT and innovation ecosystem, leveraging strategic location and government vision. Development Pathway: Property-led development and government partnership Sectoral Focus: Creative and digital industries

96. *Lagos Technology Ecosystem, Nigeria.* Organic technology ecosystem development leveraging entrepreneurial culture, large market, and growing digital economy to create 'Silicon Lagoon'. Development Pathway: Opportunistic connections and serendipitous development Sectoral Focus: Financial technology and services

ADDENDUM: NINE FORCES OF ECOSYSTEM CONVERGENCE—AN EXPLORATION

Innovation districts and precincts are critical urban and regional platforms for knowledge creation, economic growth, and societal resilience. However, their success depends on more than physical co-location or investment. It hinges on managing nine fundamental forces: knowledge dynamics, economic and policy incentives, spatial and urban planning, organisational cultures, social networks, governance structures, technological infrastructure, societal inclusivity, and historical legacies.

This Part of the book offers a structured framework to analyse these forces, drawing insights from leading international and Australian case studies. The findings challenge assumptions that innovation districts emerge automatically from clustering or market forces. Instead, they require deliberate stewardship, adaptive governance, inclusivity, and investment in trust and connectivity. For Australia's innovation cities to compete globally, attention must shift from infrastructure delivery to ecosystem cultivation—building environments where diverse actors collaborate meaningfully to address shared challenges and create sustainable futures.

Method

The analysis draws on an examination of a sample of 44 selected urban (metropolitan) innovation ecosystems and innovation districts, listed in Table 1. This diverse sample provided a robust comparative base, spanning different development models (university-led, government-planned, corporate-initiated, and mixed approaches), various cultural and governance contexts, and different stages of maturity

Table 1: Innovation Ecosystems and Distracts Studied

Urban (Metropolitan) Innovation Ecosystems	Innovation Districts - Australia	Innovation Districts - International
1. Amsterdam (Netherlands) 2. Austin (Texas, USA) 3. Bangalore (Bengaluru), India 4. Barcelona (Spain) 5. Berlin (Germany) 6. Boston (Massachusetts, USA)	24. Canberra Innovation District (ACT) 25. Macquarie Park Innovation District (Sydney, New South Wales) 26. Monash-Clayton Innovation District (Melbourne, Vic.)	33. 22@Barcelona (Barcelona, Spain) 34. Berlin Adlershof Science City (Germany) 35. Cambridge Innovation Cluster (United Kingdom) 36. Eindhoven High-Tech Campus (Netherlands)

Urban (Metropolitan) Innovation Ecosystems	Innovation Districts - Australia	Innovation Districts - International
7. Dubai, United Arab Emirates 8. Manchester (United Kingdom) 9. Melbourne (Australia) 10. Montréal (Canada) 11. Munich (Germany) 12. New York City, USA 13. Paris (France) 14. San Diego (USA) 15. San Francisco (USA) 16. Santiago, Chile 17. Seoul, South Korea 18. St. Louis, Missouri (USA) 19. Stockholm, Sweden 20. Tel Aviv, Israel 21. Sydney (Australia) 22. Toronto (Canada) 23. Vienna, Austria	27. Parkville-Carlton Innovation District (Melbourne, Vic.) 28. Paramatta-Westmead Innovation District 29. Queensland University of Technology Precinct (Brisbane, Queensland) 30. Tech Central (Sydney, New South Wales) 31. Tonsley Innovation District (Adelaide, SA) 32. Western Sydney Aerotropolis (New South Wales)	37. Kendall Square (Cambridge, Massachusetts, USA) 38. Kista Science City (Stockholm, Sweden) 39. London Knowledge Quarter (United Kingdom) 40. MaRS Discovery District (Toronto, Canada) 41. One-North (Singapore) 42. Porto Digital (Recife, Brazil) 43. Station F (Paris, France) 44. Zhongguancun Science Park (Beijing, China)

Through studying these districts, along with a comprehensive review of scholarly research spanning urban studies, innovation systems, economic geography, and organisational science, as well as a systematic analysis of innovation district literature, it was possible to identify *nine recurring factors* that consistently influence innovation district dynamics.

These were organised into a coherent analytical framework, grouping related concepts and identifying overarching categories. This framework was refined through detailed examination of well-documented innovation districts, including Kendall Square, 22@Barcelona, and several Australian examples.

The resulting nine forces naturally organised into complementary groupings:

- *Structural elements*: knowledge, economic, and spatial dimensions
- *Social and institutional factors*: cultures, networks, governance

During our research process, the framework naturally organised into two tiers:

- *Primary structural elements* (the first five forces): knowledge dynamics, economic incentives, spatial planning, organisational cultures, and social networks—present in all successful innovation districts

- *Critical enabling factors* (the additional four forces): governance, technological infrastructure, inclusivity, and historical influences, shape how the primary forces manifest in specific contexts

This tiered structure balances analytical comprehensiveness with practical utility, providing both theoretical rigour and actionable insights for practitioners. The nine forces are described and discussed in the pages that follow.

1. *Knowledge Dynamics: Beyond Spatial Proximity*

Effective knowledge flow underpins vibrant innovation ecosystems. It is not achieved through proximity alone but through designed mechanisms that encourage cross-institutional learning.

Knowledge dynamics fundamentally determine innovation district performance, encompassing the mechanisms, processes, and pathways through which intellectual capital flows across institutional boundaries. The effectiveness of these dynamics influences whether an innovation district functions as an integrated ecosystem or merely a spatial agglomeration of disconnected entities.

Facilitators of Knowledge Convergence

High-functioning innovation districts enable knowledge flows across institutional boundaries through deliberately designed mechanisms. Kendall Square in Boston exemplifies this phenomenon, where the interface between MIT and the surrounding commercial ecosystem has been systematically developed to enhance permeability. It has involved:

- Strategic interventions facilitating knowledge convergence include:
- Purpose-built collaborative infrastructure
- Intermediary organisations functioning as knowledge brokers
- Shared physical research facilities co-locating diverse expertise
- Cross-sectoral appointments enabling individuals to span boundaries
- Open innovation platforms reducing transaction costs for knowledge exchange
- Collaborative research programs aligning incentives around common challenges

Barriers to Knowledge Flow

Spatial proximity, while necessary, is insufficient to ensure effective knowledge dynamics. Innovation districts frequently exhibit significant barriers despite physical co-location. These barriers often stem from:

- Misaligned incentive structures across the institutional spectrum
- Restrictive intellectual property regimes impeding open exchange
- Disciplinary structures creating internal boundaries within institutions

Addressing Institutional Tensions

A fundamental reality complicates knowledge dynamics: people are not automatically motivated to share knowledge, and behavioural norms differ significantly between institutional forms.

- Academics typically operate where knowledge dissemination brings rewards—attending conferences to present papers, building reputations through publication
- Business professionals approach knowledge as a strategic asset with commercial value—attending conferences primarily to sell ideas, products and services
- Consultants carefully manage what proprietary knowledge they reveal

These tensions manifest physically even in celebrated environments. During our field research in Kendall Square, we observed buildings with armed guards and security systems—visible security measures revealing an underlying reality: many organisations actively protect knowledge rather than freely sharing it.

These divides create significant friction points:

- Business professionals sign confidentiality agreements facing dismissal if they divulge IP
- Academic researchers face institutional pressure to publish findings quickly
- Companies withhold information about strategic directions from potential competitors—including those co-located within innovation districts

Our work on "From Transactions to Partnerships" has demonstrated that shifting from transactional mindsets to genuine collaboration requires developing interpersonal and inter-institutional trust. People do business with people they trust.

Without this foundation of trust, innovation districts risk becoming archipelagos of isolated knowledge islands—physically proximate but intellectually disconnected.

Strategic Approaches to Knowledge Integration

Advanced innovation districts implement effective strategies to overcome these barriers while respecting legitimate institutional differences:

- Differentiated intellectual property frameworks provide appropriate protection while enabling sufficient knowledge circulation
- Structured intermediary spaces bridging institutional logics
- Investment in boundary-spanning capabilities—individuals and organisations skilled in navigating diverse institutional environments

The patterns of knowledge flow within an innovation district provide critical diagnostic insights into its functional dynamics, significantly determining whether it will develop as an integrated innovation ecosystem.

2. *Economic and Policy Incentives*

Incentive structures shape behaviours more powerfully than spatial configurations by aligning institutional logics.

Economic and policy incentives function as powerful structural determinants shaping collaborative behaviour within innovation districts. The architecture of funding mechanisms, taxation frameworks, and policy instruments significantly influences whether institutions converge productively or remain fragmented.

Facilitators of Economic Convergence

Exemplary innovation systems deploy incentive structures that deliberately align diverse institutional interests. Germany's Fraunhofer Institutes represent a paradigmatic case, operating on a tripartite funding model:

- Approximately one-third core government funding
- One-third competitive public grants
- One-third industry contracts

This balanced framework creates structural alignment between fundamental research excellence, public mission fulfilment, and industrial relevance.

Similar approaches have proven effective internationally:

- The United Kingdom's Catapult Centres employ matched funding requirements necessitating industry-university collaboration

- Singapore's Agency for Science, Technology and Research (A*STAR) utilises innovation vouchers to reduce barriers between research capabilities and small to medium enterprises

Economic Drivers of Divergence

Many conventional funding mechanisms inadvertently reinforce institutional separation:

- Academic research funding rewards scholarly publication rather than application
- Corporate investment prioritises near-term product development over fundamental research
- Government programs align with Ministerial portfolio boundaries rather than cross-cutting innovation challenges

These divergent incentives are reinforced by incongruent performance metrics across sectors:

- Universities evaluate success through citation indices and academic reputation
- Commercial entities focus on market returns and competitive positioning
- Government initiatives track sectoral growth indicators or employment outcomes

Even well-intentioned collaborative funding programs can produce superficial rather than substantive convergence when emphasising formal compliance over authentic engagement.

Strategic Incentive Architecture

Advanced innovation districts develop more nuanced economic frameworks that acknowledge institutional diversity while creating structural imperatives for meaningful collaboration.

Australia's Cooperative Research Centres exemplify this approach, combining sustained public investment with substantial industry commitment while preserving academic participation. They establish funding stability for long-term investigation while maintaining commercial discipline for relevance and application.

Democratising participation requires additional mechanisms, such as innovation vouchers, tiered matching requirements, and capability development programs, that reduce barriers for smaller enterprises and non-traditional innovation actors.

Fundamentally, assessment frameworks must evolve beyond narrowly defined institutional metrics to capture collaborative outcomes and system-level impacts, evaluating:

- Universities not merely on scholarly output but on engagement effectiveness and societal contribution
- Commercial entities not exclusively on financial returns but on ecosystem development
- Public investments through multidimensional impact frameworks rather than simplistic output indicators

3. *Spatial and Urban Planning*

The physical environment either encourages or inhibits relational dynamics essential for innovation.

The spatial configuration of innovation districts significantly influences patterns of interaction, access, and integration among diverse stakeholders. Effective spatial planning creates environments conducive to knowledge exchange and collaborative engagement, while inadequate planning can inadvertently reinforce fragmentation and exclusion.

Spatial Facilitators of Convergence

Strategic urban planning catalyses convergence through deliberate design decisions that enhance proximity, interaction, and accessibility. Barcelona's 22@ innovation district exemplifies this approach, employing integrated spatial planning to deliberately intersperse educational, research, commercial, and residential functions.

High-performing innovation districts incorporate multiple spatial elements facilitating convergence:

- Permeable interfaces between institutional campuses and surrounding urban fabric
- Diverse building typologies accommodating different organisational scales and stages
- Public spaces designed for both programmed and spontaneous interaction
- Integrated mobility systems prioritising pedestrian movement and sustainable transit
- These elements collectively enhance "collision density"—the frequency of meaningful interactions across institutional and social boundaries.

Strategic spatial planning can democratise innovation participation through deliberate inclusion mechanisms:

- Affordable space provision for emerging enterprises
- Flexible facilities for community-based innovation
- Accessible public realms ensuring districts function as genuinely inclusive ecosystems

Spatial Drivers of Divergence

Inadequate spatial planning frequently reinforces divergence dynamics. Commercial real estate imperatives often lead to:

- Monofunctional developments
- Institutional enclaves
- Prohibitive occupancy costs excluding smaller innovation actors

Property market dynamics, when insufficiently moderated, generate spatial exclusion through escalating costs. This market-driven gentrification displaces precisely the diversity of occupants essential for resilient innovation ecosystems.

Planning approaches that segregate functions through rigid zoning or impose standardised building typologies further constrain collaborative potential, physically embodying and reinforcing institutional silos.

Integrated Spatial Strategies

Advanced innovation districts employ spatial strategies balancing market dynamics with ecosystem imperatives:

- Adaptive reuse of heritage structures maintaining affordability and character
- Careful curation of ground floor activities enhancing public engagement
- Integrated mobility systems prioritising sustainable connectivity
- Deliberate preservation of spatial diversity accommodating various scales of innovation

MaRS Discovery District in Toronto and Station F in Paris demonstrate effective spatial approaches through their transformation of historic structures into innovation environments preserving character while accommodating contemporary needs.

Effective spatial strategies require governance mechanisms moderating market pressures and maintaining long-term ecosystem priorities through planning frameworks incorporating affordability provisions, public realm requirements, and mixed-use mandates.

4. Organisational Cultures and Institutional Logics

Organisational mindsets and reward systems often clash, impeding collaboration despite co-location.

Organisational cultures and institutional logics represent powerful influences on convergence and divergence within innovation districts. These intangible factors—encompassing fundamental values, operational assumptions, decision-making processes, and reward systems—shape collaborative capacity at a profound level.

Cultural Foundations of Convergence

Innovation districts achieving sustained convergence develop "interinstitutional competence"—the capacity for organisations to maintain distinctive identities while developing mutual recognition and operational compatibility with diverse partners.

Stanford University's relationship with Silicon Valley exemplifies this balance, maintaining robust academic integrity while engaging with industry through structured mechanisms including:

- The Stanford Research Park
- Joint faculty-industry appointments
- Deliberately designed innovation environments like the d.school bridging academic and commercial approaches

Successful innovation districts cultivate particular cultural attributes enabling effective cross-sectoral collaboration:

- Institutional ambidexterity—simultaneously pursuing both foundational research and practical application
- Value complementarity—recognising different institutional missions as mutually reinforcing
- Temporal flexibility—aligning different organisational timeframes

Cultural Barriers to Convergence

Innovation districts frequently encounter challenges stemming from cultural and institutional misalignment. Fundamental differences in organisational purpose, planning horizons, performance metrics, and risk tolerance create persistent friction despite formal collaborative structures.

These tensions reflect deep-seated institutional logics—the organising principles shaping how institutions perceive their roles, evaluate success, and engage with external partners. When these logics remain rigid and unexamined, they constitute invisible yet formidable barriers to effective collaboration.

The challenge extends beyond surface-level differences to fundamental questions of institutional purpose and value.

- Universities must balance knowledge advancement with knowledge application
- Commercial entities must reconcile competitive positioning with ecosystem contribution
- Government agencies must address public accountability while enabling innovation risk-taking

Strategic Cultural Integration

Advanced innovation districts actively address cultural dimensions through a range of integration mechanisms:

- Boundary-spanning roles—positions specifically designed to bridge institutional divides
- Deliberately designed hybrid spaces enabling cultural exchange and adaptation
- Professional development and mobility programs enhancing cross-sectoral understanding

Over time, these interventions stimulate the emergence of genuinely hybrid organisational forms:

- Technology transfer offices balancing academic excellence with market relevance
- Social enterprises integrating community mission with business discipline
- Corporate research laboratories combining proprietary development with ecosystem contribution

While cultural and institutional dimensions may be less immediately apparent than physical infrastructure or funding flows, they frequently exert greater influence on collaborative outcomes.

5. *Social Networks and Human Capital*

Networks, not just institutions, move knowledge and opportunity.

Social networks and human capital constitute essential determinants of convergence and divergence within innovation districts. The configuration, density, and diversity of relationships fundamentally shape knowledge flows, resource access, and collaborative potential.

Network Structures Enabling Convergence

High-performing innovation districts develop network architectures characterised by both density and diversity. Kendall Square in Boston exemplifies this complexity, featuring overlapping professional communities spanning academic, entrepreneurial, corporate, and investment domains.

The qualitative characteristics of these networks prove as significant as their structural configuration. Successful innovation ecosystems cultivate "bridging social capital"—connections spanning diverse communities rather than merely reinforcing existing homogeneous groups.

Trust functions as a critical lubricant within these networks, reducing transaction costs for collaboration and enabling knowledge exchange that formal contracts alone cannot facilitate. Innovation districts developing high trust environments demonstrate enhanced capacity for:

- Risk-sharing
- Exploratory collaboration
- Knowledge exchange across institutional boundaries

Network Fragmentation and Human Capital Constraints

More commonly, innovation districts exhibit significant network fragmentation along institutional, disciplinary, and demographic boundaries. This network homophily—the tendency to associate primarily with similar others—creates invisible but powerful barriers to knowledge flow and collaborative opportunity.

Such network segregation frequently reinforces existing power asymmetries and institutional hierarchies. Information, resources, and opportunities circulate within closed loops rather than throughout the ecosystem, creating inequities undermining both social inclusion and innovation potential.

Human capital mobility further influences convergence dynamics. Districts characterised by limited cross-sectoral movement demonstrate reduced capacity for institutional learning and knowledge integration, reinforcing cultural divergence.

Strategic Network Development and Human Capital Mobility

Advanced innovation districts implement deliberate strategies enhancing network connectivity and human capital integration:

- Structured opportunities for cross-sectoral engagement through curated events, mentoring programs, and collaborative projects
- Investment in dedicated "network orchestrators" identifying potential synergies and facilitating introductions
- Human capital mobility mechanisms facilitating movement between academia, industry, government, and civil society
- Inclusive approaches addressing barriers to participation and enhancing both equity outcomes and innovation performance

The configuration and quality of social networks provide critical diagnostic indicators of a district's convergence potential, significantly determining whether it will function as an integrated ecosystem or remain a collection of proximate but disconnected entities.

6. *Governance and Leadership*

Governance structures either facilitate or impede ecosystem dynamism.

Governance arrangements and leadership approaches constitute fundamental determinants of convergence and divergence within innovation districts. The mechanisms through which decisions are made, resources allocated, and conflicts resolved significantly influence collaborative potential across diverse institutional stakeholders.

Governance Structures Enabling Convergence

Effective innovation districts develop governance models balancing structure with adaptability, strategic direction with distributed agency. Toronto's MaRS Discovery District illustrates this balance through its governance architecture—an independent, mission-driven organisation with a diverse board representing academic, commercial, governmental, and community interests.

Successful governance frequently involves intermediary organisations positioned at the interstices of traditional sectors, providing neutral institutional space for negotiating priorities and developing shared strategic frameworks.

Collaborative governance mechanisms transcend conventional hierarchical structures through deliberative approaches engaging diverse perspectives in decision-making processes, establishing forums where stakeholders with different institutional logics can negotiate mutual interests.

Governance Fragmentation and Leadership Limitations

More frequently, innovation districts suffer from fragmented or misaligned governance. Multiple authorities with overlapping jurisdictions pursue separate agendas with limited coordination, producing incoherent strategic direction, resource misalignment, and contradictory incentives.

Power asymmetries further complicate governance challenges. When decision-making processes privilege established institutions while marginalising smaller organisations or community stakeholders, they generate resistance, disengagement, or performative rather than substantive collaboration.

Conventional leadership approaches often prove inadequate for complex, multi-institutional environments. Command-and-control leadership styles typically generate resistance in networked innovation environments, while transactional leadership focused exclusively on balancing competing interests frequently lacks transformative capacity.

Strategic Governance Design and Collaborative Leadership

Advanced innovation districts implement governance approaches that acknowledge complexity while providing sufficient direction for collective action. They establish multi-level governance frameworks incorporating:

- Strategic boards with diverse representation from key stakeholder groups
- Advisory councils engaging broader constituencies beyond formal decision-makers
- Working groups addressing specific innovation challenges
- Regular stakeholder forums enabling transparent communication
- Clear frameworks for decision rights, resource allocation, and accountability

Effective leadership in innovation districts requires distinctive capabilities:

- Systems thinking comprehending interconnections across institutional boundaries
- Facilitative capacity enabling diverse stakeholders to find common ground

- Translational ability bridging different institutional languages and perspectives
- Transformative vision articulating compelling shared purposes while respecting institutional diversity

Distributed leadership approaches prove particularly effective, developing leadership capacity throughout the network rather than concentrating authority in single individuals or organisations.

7. Technological Infrastructure

Digital platforms complement physical co-location, shaping access, reach, and collaboration.

Technological infrastructure has emerged as a critical determinant of convergence and divergence within contemporary innovation districts. Digital connectivity, collaborative platforms, data ecosystems, and computational capacity increasingly function as essential complements to physical proximity in enabling productive interactions across institutional boundaries.

Digital Enablers of Convergence

Advanced technological infrastructure, designed with openness and accessibility as guiding principles, substantially enhances convergence by reducing transactional barriers to collaboration. High performance computing facilities, shared data repositories, and digital collaboration environments allow diverse stakeholders to engage in joint problem-solving despite physical or organisational separation.

Open digital platforms demonstrate particular value in supporting convergence, enabling participants from diverse institutional contexts to engage without prohibitive technical barriers or proprietary constraints. Innovation districts such as Station F in Paris and MaRS Discovery District in Toronto have leveraged such digital infrastructure to extend collaborative reach beyond physical boundaries.

Data infrastructure is an increasingly significant convergence mechanism:

- Open data initiatives
- Federated data platforms
- Shared digital twins (used in urban planning, life sciences, and advanced manufacturing)

In health innovation particularly, technologies enabling privacy-preserving collaboration—such as federated learning approaches used in

Europe's GAIA-X project—allow multiple institutions to develop collective intelligence while respecting regulatory requirements.

Digital Drivers of Divergence

Despite its convergence potential, technological infrastructure frequently reinforces or creates new divisions when inadequately designed or governed. Digital divides manifest across multiple dimensions:

- Between large organisations with high-level technical capabilities and smaller entities with limited digital resources
- Between technically specialised knowledge workers and broader community participants
- Between well-resourced districts in developed economies and under-resourced environments in developing regions

Proprietary technological ecosystems present particular challenges when large technology providers implement closed systems with limited interoperability, effectively creating digital enclaves fragmenting rather than integrating the innovation ecosystem.

Technological path dependence further complicates convergence dynamics. Systems adopted by influential actors can become entrenched, creating technological lock-in limiting ecosystem adaptability and excluding alternative approaches or emerging participants.

Strategic Technology Governance for Inclusive Innovation

Forward-thinking innovation districts implement deliberate strategies ensuring technological infrastructure enhances rather than undermines convergence, prioritising digital inclusivity through:

- Investments in shared infrastructure
- Technical capacity building
- Governance frameworks enabling broad participation

Effective approaches include development of open digital commons—shared technological resources governed by stakeholder communities rather than single institutions. Barcelona has pioneered this approach through its technological sovereignty initiatives, creating digital infrastructure explicitly designed to serve public purposes while enabling private innovation.

Interoperability emerges as a central concern. Rather than imposing monolithic systems forcing all participants into identical technological environments, successful districts establish interoperability standards enabling diverse systems to communicate effectively.

8. *Societal and Cultural Inclusivity*

Inclusivity is a social good and a key driver of innovation.

Societal and cultural inclusivity constitutes a critical dimension of convergence and divergence within innovation districts, though it has frequently received insufficient attention. This dimension encompasses the extent to which innovation ecosystems incorporate diverse demographic, socioeconomic, and cultural perspectives, and the degree to which benefits are equitably distributed.

Inclusive Approaches Enabling Convergence

Innovation districts fostering inclusivity recognise diversity as a fundamental driver of creative capacity rather than a peripheral concern. Substantial empirical research demonstrates that cognitive diversity—derived from varied backgrounds—enhances problem-solving capability, expands potential solutions, and improves decision-making processes.

Effective inclusive practices include participatory planning and co-design processes engaging diverse stakeholders as active contributors. Barcelona's 22@ district exemplifies this approach through its integration of social housing, community facilities, and educational infrastructure, ensuring socioeconomic diversity while creating multiple pathways for local community participation.

Inclusive governance structures further strengthen convergence by ensuring strategic direction and resource allocation reflect diverse priorities. When innovation district leadership incorporates representation from varied demographic, sectoral, and community groups, it enhances both legitimacy and effectiveness.

Exclusionary Patterns Driving Divergence

Many innovation districts exhibit patterns of exclusion undermining both social equity and innovative potential through multiple mechanisms:

- Economic exclusion: Prohibitive costs creating insurmountable barriers for smaller organisations or emerging enterprises
- Educational exclusion: Innovation participation requiring specialised credentials accessible only through elite educational pathways

- Cultural exclusion: Innovation environments reflecting dominant cultural norms implicitly marginalising alternative perspectives
- Spatial exclusion: Gentrification processes displacing existing communities

These exclusionary patterns not only raise ethical concerns but fundamentally constrain innovation ecosystems by limiting the diversity of perspectives, experiences, and knowledge resources available for creative recombination.

Strategic Approaches to Inclusive Innovation

Forward-thinking innovation districts implement comprehensive inclusion strategies addressing structural barriers while creating multiple pathways for diverse participation, recognising inclusion as fundamental to ecosystem resilience, adaptability, and creative capacity.

Effective strategies include:

- Affordable space policies ensuring economic diversity through mixed-cost facilities or subsidised innovation spaces
- Inclusive procurement initiatives directing purchasing power towards underrepresented businesses
- Talent development programs creating alternative pathways into innovation careers
- Cultural programming celebrating diverse traditions and approaches to innovation
- Community benefit agreements formalising commitments to local economic inclusion

Advanced innovation districts embed inclusion through systematic monitoring of participation patterns and outcome distributions across demographic dimensions, establishing clear metrics and accountability mechanisms creating feedback loops driving continuous improvement.

9. *Historical and Path-dependent Influences*

Understanding Embedded Contexts. Every district inherits opportunities and constraints from its past.

Historical legacies and path-dependent trajectories exert profound influence on the dynamics within innovation districts. Institutional memories, prior collaborative experiences, established relationships, and physical infrastructure inherited from earlier periods significantly shape current possibilities and constraints.

Historical Foundations for Convergence

In certain contexts, historical legacies create robust foundations for contemporary convergence. Regions with established traditions of cross-institutional collaboration often possess substantial "relational capital"—accumulated trust, shared norms, mutual understanding—facilitating current and future collaborative initiatives.

The evolution of Eindhoven's High-Tech Campus exemplifies how historical relationships shape contemporary innovation ecosystems. The long-standing cooperation between Philips, regional technical universities, and governmental authorities created institutional frameworks enabling the district's transformation from a single company campus to a multi-organisational innovation hub.

Physical infrastructure legacies similarly influence convergence potential. Industrial heritage sites repurposed as innovation districts—such as Barcelona's 22@ district or Toronto's Distillery District—often possess distinctive architectural character, spatial configurations conducive to interaction, and established connections to surrounding urban systems.

Historical Barriers to Convergence

Historical legacies frequently present significant barriers to contemporary convergence:

- Previous competitive conflicts or failed collaborative initiatives generating persistent mistrust
- Established organisational routines exhibiting substantial inertia resisting innovation-oriented reconfiguration
- Academic institutions with long-established disciplinary structures struggling to develop interdisciplinary frameworks
- Physical infrastructure developed for earlier paradigms constraining contemporary convergence

Strategic Approaches to Historical Legacies

Advanced innovation districts develop approaches that acknowledge historical context while creating capacity for strategic renewal:

- Systematic institutional memory work documenting prior collaborative initiatives and identifying informal networks spanning organisational boundaries
- Adaptive reuse strategies for physical infrastructure creating productive balance between historical continuity and contemporary requirements

- Narrative reconciliation addressing historical impacts, persistent inequities, and developing shared narratives supporting inclusive participation

The historical dimensions provide critical context for understanding current dynamics and future potential, significantly influencing whether a district can achieve the contextual integration necessary for sustained innovation.

Summary

Our analysis reveals that innovation district performance is fundamentally shaped by the interplay between forces of convergence and divergence. Convergence facilitates collaborative alignment across institutional boundaries, while divergence reflects the centrifugal forces that push stakeholders towards institutional self-interest and siloed activity.

The framework identifies nine critical forces that determine whether an innovation district functions as an integrated ecosystem or merely as a spatial agglomeration of disconnected entities.

1. Knowledge dynamics establish how intellectual capital flows between institutions, either enriching collective intelligence or remaining confined within organisational boundaries.
2. Economic and policy incentives either create structural imperatives for meaningful collaboration or reinforce institutional separation through misaligned funding mechanisms and performance metrics.
3. Spatial and urban planning decisions physically embody convergence or divergence through their impact on interaction patterns, accessibility, and inclusion.
4. Organisational cultures and institutional logics—encompassing values, operational assumptions, and reward systems—can either find productive alignment or remain rigidly separated despite physical proximity.
5. Social networks and human capital mobility similarly determine whether knowledge and capabilities flow across institutional boundaries or remain locked within homogeneous communities.
6. Governance structures and leadership approaches significantly influence collaborative potential, either enabling genuine partnerships or perpetuating fragmentation and misalignment.

7. Technological infrastructure can democratise innovation participation or create new digital divides.
8. Societal and cultural inclusivity determines whether diverse perspectives are meaningfully integrated or excluded, affecting both equity outcomes and innovation performance.
9. Historical and path-dependent influences shape both opportunities and constraints based on inherited relationships, physical infrastructure, and institutional memories.

Successful innovation districts achieve productive balance between convergence and divergence rather than maximising either dimension. Convergence without divergence risks homogenisation that undermines innovation; divergence without convergence produces fragmentation that prevents effective collaboration.

The key challenge lies in creating frameworks that enable sufficient alignment for collective action while preserving the productive differences that fuel creativity and adaptation.

An overview summary of convergence and divergence forces is provided below.

Summary of Convergence and Divergence Forces in Innovation Districts

Driving Forces	Facilitators of Convergence	Barriers Promoting Divergence	Approaches to Achieving Balance
1. Knowledge Dynamics	Open knowledge exchange, co-location, open innovation platforms, collaborative research centres	IP protectionism, institutional silos, disciplinary boundaries	Flexible IP frameworks, shared labs, knowledge intermediaries
2. Economic and Policy Incentives	Joint funding schemes, R&D tax incentives, PPPs, mission-driven programs	Short-termism, competitive grants, sectoral bias, rigid metrics	Blended funding, inclusive incentives, cross-sector co-investment
3. Spatial and Urban Planning	Mixed-use planning, walkability, integrated transport, inclusive zoning	Speculative development, zoning segregation, poor transit connectivity	Participatory planning, adaptive reuse, integrated mobility design

Driving Forces	Facilitators of Convergence	Barriers Promoting Divergence	Approaches to Achieving Balance
4. Organisational Cultures and Institutional Logics	Shared values, hybrid models, joint appointments, boundary-spanning roles	Institutional rigidity, cultural misalignment, conflicting incentives	Cultural training, incentives for collaboration, flat structures
5. Social Networks and Human Capital	Dense, diverse networks, talent mobility, inclusive programs and mentoring	Network homophily, social exclusion, limited mobility pathways	Inclusive governance, social infrastructure, talent bridges
6. Governance and Leadership	Collaborative governance, facilitative leadership, shared decision-making bodies	Hierarchical structures, fragmented mandates, leadership churn	Dedicated system integrators, inclusive advisory boards
7. Technological Infrastructure	Open platforms, shared digital infrastructure, data interoperability	Digital divide, proprietary platforms, data monopolies	Public investment in digital inclusion, open-source tools
8. Societal and Cultural Inclusivity	Community co-design, inclusive governance, social innovation frameworks	Gentrification, demographic exclusion, elite decision-making	Equity targets, diverse leadership, co-created innovation strategies
9. Historical and Path-dependent Influences	Legacy partnerships, institutional memory, cross-sector trust	Institutional mistrust, legacy silos, outdated infrastructure	Narrative reconciliation, institutional renewal, legacy infrastructure adaptation

Insights and Implications

The analytical framework presented in this Paper offers both conceptual clarity and practical guidance for understanding and steering the complex dynamics of innovation districts.

By systematically examining the nine forces that shape convergence and divergence—knowledge dynamics, economic incentives, spatial planning, organisational cultures, social networks, governance structures, technological infrastructure, inclusivity, and historical influences—we can develop approaches to innovation ecosystem development that balance excellence with inclusion, specialisation with integration, and institutional diversity with collective purpose.

Key Insights

Several critical insights emerge from this integrative analysis:

- Innovation districts function as complex adaptive systems rather than merely spatial agglomerations or policy instruments. Their performance depends on the dynamic interaction of multiple subsystems operating across different temporal and organisational scales. This complexity necessitates systems thinking rather than reductionist approaches focused on individual factors or linear causal relationships.
- The most successful innovation districts achieve productive balance between convergence and divergence rather than maximising either dimension. Convergence without divergence risks homogenisation and groupthink that undermines innovation; divergence without convergence produces fragmentation that prevents effective collaboration. The challenge lies in creating frameworks that enable sufficient alignment for collective action while preserving the productive differences that fuel creativity and adaptation.
- Effective innovation districts develop what might be termed "contextual integration"—they achieve alignment appropriate to their specific historical, institutional, economic, and cultural circumstances rather than implementing standardised models disconnected from local context. This contextual sensitivity applies not only to initial design but to ongoing governance and evolution as districts respond to changing conditions and emergent patterns.
- Design—both institutional and spatial—emerges as a central determinant of innovation district performance. Districts that consciously design for permeability, diversity, and adaptability demonstrate enhanced capacity to overcome historical path dependencies and institutional rigidities. Moreover, convergent outcomes prove more durable when supported by participatory leadership and governance structures that transcend electoral, funding, and planning cycles.
- Importantly, convergence does not imply uniformity or conformity. The most innovative districts maintain what might be characterised as "generative tension" between different institutional logics, disciplinary perspectives, and operational approaches. They create frameworks within which multiple ways of knowing, learning, and creating can productively interact without being forced into artificial consensus or standardisation.

Practical Implications

These insights suggest several practical implications for different stakeholder groups engaged in innovation district development:

- For *policymakers*, the framework highlights the importance of integrated approaches that span traditional departmental boundaries. Innovation policy, urban planning, higher education, infrastructure investment, and social inclusion initiatives must be conceptualised and implemented as complementary elements of coherent innovation ecosystems rather than separate policy domains. This integration requires both institutional mechanisms for cross-departmental coordination and conceptual frameworks that enable shared understanding across policy communities.
- For *university and research leaders*, the analysis emphasises the strategic importance of boundary-spanning capabilities. Academic institutions that develop effective approaches to external engagement—through dedicated intermediary organisations, permeable campus boundaries, joint appointments, and collaborative research structures—enhance both their societal impact and their core research and educational missions. This engagement requires more nuanced performance metrics that capture both excellence within disciplines and integration across institutional boundaries.
- For *business and industry partners*, the framework suggests the value of ecosystem thinking alongside conventional competitive positioning. Firms that contribute to shared innovation infrastructure, participate in knowledge commons, and engage with diverse stakeholders beyond immediate commercial relationships often derive significant long-term advantage through enhanced access to talent, ideas, and collaborative opportunities. This perspective requires longer strategic horizons than quarterly performance metrics typically encourage.
- For *urban planners and local authorities*, the analysis highlights the importance of integration between innovation strategies and broader urban systems. Effective innovation districts function as integral elements of their metropolitan contexts rather than isolated enclaves. This integration encompasses physical connectivity, housing provision, educational pathways, and cultural amenities that collectively support innovation while distributing its benefits across diverse communities.

- For *communities and civil society organisations*, the framework provides conceptual tools for substantive engagement with innovation agendas. By understanding the multiple dimensions of innovation districts, community stakeholders can advocate for inclusive approaches that address their priorities and leverage their distinctive assets, moving beyond reactive positions to pro-active participation in innovation ecosystem development.

Research agenda

This framework suggests several promising directions for further research on innovation district dynamics:

- Longitudinal studies tracking the evolution of convergence and divergence patterns over extended time periods, particularly through economic cycles and leadership transitions
- Comparative analyses examining how these forces manifest differently across diverse national, regional, and cultural contexts
- Action research methodologies that engage stakeholders in collaborative investigation and experimentation with alternative approaches to innovation district governance
- Mixed-method studies combining quantitative network analysis with qualitative investigation of institutional cultures and leadership practices
- Interdisciplinary research integrating insights from urban studies, innovation studies, institutional theory, network analysis, and design thinking

Conclusion

Understanding the interplay of convergence and divergence forces is essential for steering effective innovation districts. These districts are not static geographic zones, but dynamic socio-technical systems shaped by historical legacies, institutional logics, policy environments, and everyday practices. Their success depends on the thoughtful alignment of diverse actors and interests, which requires deliberate strategies to navigate tensions, build trust, and generate shared value.

Each force identified in this Paper carries both integrative potential and centrifugal risk. If left unmanaged, these forces can fragment ecosystems, reinforce exclusion, and inhibit innovation. However, when actively shaped through inclusive governance, shared infrastructure, and adaptive institutional arrangements, they become the foundation for vibrant, resilient, and mission-oriented innovation systems.

The framework presented here offers a starting point for more sophisticated approaches to innovation district development—approaches that transcend simplistic economic or technological determinism to engage with the complex social, institutional, and cultural dimensions of innovation ecosystems. By systematically addressing the drivers of convergence and divergence, innovation districts can become not merely economic engines but generative environments for inclusive knowledge creation and application that addresses our most pressing societal challenges.

APPROACH TO THE PROJECT

This book adopts an alternative approach compared to much of the existing literature on innovation. It is not intended to be a technical manual or a study concentrated on a single discipline. Instead, it takes a comprehensive and reflective approach, analysing the concept of "innovation ecosystem" from various perspectives and showing how these viewpoints can be effectively integrated in practice.

The methodology is built around several interlocking themes.

Conceptual framework

Innovation ecosystems are examined through four interrelated perspectives: placemaking, economics, business, and governance. Each offers a unique perspective on how to view and appreciate innovation. The book presents these perspectives in dialogue, emphasising their tensions and commonalities. This highlights the assumptions that typically go unnoticed in policy design.

Case-based inquiry

The analysis is based on case studies from both international and Australian contexts. These are written as narratives within the following headings:

- Context
- Development stages – from inception through to maturity
- Research excellence, translation and commercial outcomes
- Challenges and future directions
- Lessons for other regions
- Conclusions

Over 90 case narratives have been prepared and are included in a searchable text database. Narratives average 3, 000 words, with some longer, reflecting the amount of publicly available information. Narratives verified by ecosystem leaders are included at the end of this Book.

The criteria for inclusion of a case study in the project were:

- Concentration and co-location of universities, hospitals, institutes and firms to enable collaboration on R&D, research translation and commercialisation

- Shared infrastructure and programs, including public university capital, and shared platforms, including incubators and testbeds that reduce transaction costs for industry collaboration
- Industry development policy frameworks that link land use, transportation, and investment attraction to create employment and export-facing capacity in areas such as technology, healthcare, medical, and biotechnology, as well as regional development.

The case narratives demonstrate how various locations have integrated spatial planning, commercial strategies, institutional arrangements, and cultural influences. By comparing various cases, the book highlights what might be adapted to the Australian context and what should not be implemented.

Systems thinking and design logic

The book approaches innovation as a complex adaptive system influenced by numerous ecosystems, rather than a linear progression from research to commercialisation. Innovation is *an outcome* that arises from feedback, collaboration, and long-term interactions. This perspective underscores the significance of *systems integrators* and their ability to align institutions, strategies, and resources across diverse areas of expertise.

Reflexive and practice-oriented analysis

The approach is deliberately reflexive. It recognises the uncertainty, competing interests, and the dangers of reducing the term "innovation ecosystem" to a mere policy slogan. The goal is to equip policymakers, university leaders, and business strategists with the vocabulary and framework needed to navigate complexity. It is not intended to provide formulas, menus, or checklists.

Through this methodology, the book sets out to establish a connection between academic theory, innovation policy, and public decision-making. It aims to offer insights that are clear in concept, relevant in practice, and mindful of the unique institutional and cultural contexts in which innovation takes place.

REFERENCES

A tribute to our founder, Dr. John Evans, MaRS Discovery District. February 17, 2015, https://www.marsdd.com/our-story/a-tribute-to-our-founder-dr-john-evans/

ABC News. (2015, October 19). *Fishermans Bend development approval lacked infrastructure plans.* https://www.abc.net.au/news/2015-10-19/no-thought-to-infrastructure-at-fishermans-bend-report-finds/6868274 (ABC)

Abetti, P.A. (2004). Government-supported incubators in the Helsinki region, Finland: Infrastructure, results, and best practices. *Journal of Technology Transfer*, 29(1), 19-40.

Adner, R. (2006). Match your innovation strategy to your innovation ecosystem. *Harvard Business Review, 84*(4), 98–107.

Agtmael, A. van & Bakker, F. (2016). *The smartest places on earth: Why rustbelts are the emerging hotspots of global innovation.* PublicAffairs.

Amazon Staff. (2020). *Catalyst for growth* [Company blog]. https://www.aboutamazon.com/news/job-creation-and-investment/catalyst-for-growth (Amazon News)

Amin, A., & Thrift, N. (1995). Globalisation, institutional 'thickness' and the local economy. *Environment and Planning A, 27*(8), 1133–1154.

Ansell, C., & Gash, A. (2008). Collaborative governance in theory and practice. *Journal of Public Administration Research and Theory, 18*(4), 543–571.

Arthur, W. B. (1999). Complexity and the economy. *Science, 284*(5411), 107–109.

Audretsch, D. B., & Feldman, M. P. (2004). Knowledge spillovers and the geography of innovation. In J. V. Henderson & J. F. Thisse (Eds.), *Handbook of Regional and Urban Economics* (Vol. 4, pp. 2713–2739). Elsevier. https://doi.org/10.1016/S1574-0080(04)80017-3

Audretsch, D.B. (2001). The role of small firms in U.S. biotechnology clusters. *Small Business Economics*, 17(1-2), 3-15.

AusBiotech. (2023). Biotechnology Industry Positioning Paper. AusBiotech Ltd

Australian Bureau of Statistics. (2025, August 22). *Research and Experimental Development, Businesses, Australia: 2023–24.* https://www.abs.gov.au/statistics/industry/technology-and-innovation/research-and-experimental-development-businesses-australia

Autio, E., & Thomas, L. D. W. (2014). Innovation ecosystems: Implications for innovation management. In M. Dodgson, D. M. Gann, & N. Phillips (Eds.), *The Oxford Handbook of Innovation Management* (pp. 204–228). Oxford University Press.

Autio, E., Nambisan, S., Thomas, L. D. W., & Wright, M. (2018). Digital affordances, spatial affordances, and the genesis of entrepreneurial ecosystems. *Strategic Entrepreneurship Journal*, 12(1), 72–95. https://doi.org/10.1002/sej.1266

Azoulay, P., Jones, B. F., Kim, J. D., & Miranda, J. (2018). Age and High-Growth Entrepreneurship. *National Bureau of Economic Research Working Paper No. 24489.* https://doi.org/10.3386/w24489

Bakhshi, H., & Stathoulopoulos, K. (2016). *Dynamic mapping of the UK's creative industries.* Nesta. https://media.nesta.org.uk/documents/dynamic_mapping.pdf

Bakhshi, H., Freeman, A., & Potts, J. (2011). *State of the Arts: The UK's Creative Industries.* NESTA.

Barca, F. (2009). *An agenda for a reformed cohesion policy: A place-based approach to meeting European Union challenges and expectations.* European Commission.

Barcelona City Council. (2024). *Discover 22@, Barcelona's innovation district.* https://www.barcelona.cat/internationalwelcome/en/activites/discover-22-barcelonas-innovation-district

Barcelona City Council. (n.d.). *22@ Barcelona: 10 years of urban renewal* [Publication page]. https://www.barcelona.cat/en/conei xbcn/barcelonallibres/22-barcelona (Barcelona)

Bathelt, H., Malmberg, A., & Maskell, P. (2004). Clusters and knowledge: Local buzz, global pipelines and the process of knowledge creation. *Progress in Human Geography, 28*(1), 31–56.

Baum, A. E., & Hartzell, D. (2022). Global Property Investment Strategies, Structures, Decisions. Wiley.

Beer, A., McKenzie, F., Blakely, E. J., Panelli, R., & Maude, A. (2020). *Developing Australia's Regions: Theory and Practice.* Routledge.

Benner, M., & Sandström, U. (2000). Institutionalizing the triple helix: Research funding and norms in the academic system. *Research Policy, 29*(2), 291–301.

Bennett, T., Dodsworth, F., Noble, G., Poovey, M., & Watkins, M. (2013). *Collecting, Ordering, Governing: Anthropology, Museums and Liberal Government.* Duke University Press.

Biomedical Translation Fund. (n.d.). *Program overview* [Factsheet]. https://business.gov.au/Grants-and-Programs/Biomedical-Translation-Fund

Bloom, N., Brynjolfsson, E., Foster, L., Jarmin, R. S., Saporta-Eksten, I., & Van Reenen, J. (2019). What drives differences in management practices? *American Economic Review*, 109(5), 1648–1683. https://doi.org/10.1257/aer.20170491

Bogle, A. (2023, February 10). Atlassian's new Sydney HQ Innovation hub or corporate real estate play? The Sydney Morning Herald. Retrieved from https//www.smh.com.au/business/companies/atlassians-new-sydney-hq

Borrás, S., & Edler, J. (2020). The governance of societal challenges: New modes of innovation policy. *Research Policy, 49*(10), 104006.

Boschma, R. (2022). Proximity and Innovation: Recent Advances and Future Directions. Regional Studies, 56(1), 5-14.

Breznitz, D. (2021). *Innovation in real places strategies for prosperity in an unforgiving world.* New York, Oxford University Press.

Breznitz, D., & Feldman, M. (2012). The engaged university. *Journal of Technology Transfer, 37*(2), 139–157.

Breznitz, D., & Ornston, D. (2018). The revolutionary power of peripheral agencies: Explaining radical policy innovation in Finland and Israel. *Comparative Political Studies, 51*(10), 1219–1255.

Buderi, R. (2022). *When Futures Converge: Kendall Square and the Making of a Global Innovation Hub.* MIT Press.

Camagni, R. (1991). Local 'milieu', uncertainty and innovation networks: Towards a new dynamic theory of economic space. In R. Camagni (Ed.), *Innovation networks: Spatial perspectives* (pp. 121-144). Belhaven Press.

Carayannis, E. G., & Campbell, D. F. J. (2009). 'Mode 3' and 'Quadruple Helix': Toward a 21st century fractal innovation ecosystem. *International Journal of Technology Management, 46*(3-4), 201–234.

Carayannis, E. G., & Campbell, D. F. J. (2011). *Mode 3 knowledge production in quadruple helix innovation systems: 21st-century democracy, innovation, and entrepreneurship for development.* Springer.

Chandler, A. D. (1962). *Strategy and structure: Chapters in the history of the industrial enterprise.* MIT Press.

Chesbrough, H. W. (2003). *Open innovation: The new imperative for creating and profiting from technology.* Harvard Business School Press.

Chiesa, V. & Chiaroni, D. (2005). Industrial clusters in biotechnology: Driving forces, development processes, and management practices. Imperial College Press.

City of Seattle. (2013). *South Lake Union Urban Design Framework.* https://www.seattle.gov/Documents/Departments/OPCD/OngoingInitiatives/SouthLakeUnion/SouthLakeUnionUrbanDesignFramework.pdf

Clark, G., & Moonen, T. (2020). The Innovation Districts Handbook Translating Research into Practice. Brookings Institution.

Clark, J. (2020). *Uneven Innovation: The Work of Smart Cities.* Columbia University Press.

Cochlear. (2024). *Annual Report 2024* [PDF]. https://assets.cochlear.com/api/public/content/85ed5fe549814d91abddf0a975162644?v=771b1a7d

Colombo, M.G. & Delmastro, M. (2002). How effective are technology incubators? Evidence from Italy. *Research Policy*, 31(7), 1103-1122.

Cooke, P., Uranga, M. G., & Etxebarria, G. (1997). Regional innovation systems: Institutional and organisational dimensions. *Research Policy, 26*(4–5), 475–491.

Corrado, C., Hulten, C., & Sichel, D. (2006). *Intangible capital and economic growth.* Finance and Economics Discussion Series, 2006-24. Board of Governors of the Federal Reserve System.

Cortex Innovation District. (n.d.). *Our history.* https://www.cortexstl.org/about/our-history (Cortex STL)

Crossick, G., & Kaszynska, P. (2016). *Understanding the Value of Arts and Culture: The AHRC Cultural Value Project.* Arts and Humanities Research Council, https://www.ukri.org/wp-content/uploads/2021/11/AHRC-291121-UnderstandingTheValueOfArts-CulturalValueProjectReport.pdf

CSL. (2024). *Annual reports and results overview* [Investor page]. https://investors.csl.com/investors/annual-reports

Cushman & Wakefield. (2024). *Transparency 22@ Barcelona: The urban innovation district.* https://www.cushmanwakefield.com/en/spain/insights/transparency-22-arroba-barcelona

Department of Health and Aged Care. (2024–2025). *MRFF 3rd 10-year Investment Plan (2024–25 to 2033–34).* https://www.health.gov.au/our-work/mrff/about/10-year-investment-plan

Department of Health and Aged Care. (2025, August 11). *Medical Research Future Fund: Supporting researchers to tackle Australia's health priorities.* https://www.health.gov.au/news/mrff-medical-research-future-fund-supporting-researchers-to-tackle-australias-health-priorities

Department of Health and Aged Care. (2025, June 5). *Medical Research Future Fund (MRFF).* https://www.health.gov.au/our-work/mrff

Department of Industry, Science and Resources. (2018). *Australian automotive industry: Transition following the end of Australian motor vehicle production* [Report]. https://www.industry.gov.au/publications/australian-automotive-industry-report

Department of Industry, Science and Resources. (2024). *Industry Growth Centres Initiative.* https://www.industry.gov.au/funding-and-programs/industry-growth-centres-initiative

Department of the Prime Minister and Cabinet. (2024, April). *A Future Made in Australia* [Speech by the Prime Minister]. https://www.pmc.gov.au/news-centre/pmc/future-made-australia

Destination NSW. (2023). *The value of SXSW to the NSW visitor economy.* https://www.destinationnsw.com.au/newsroom/the-value-of-sxsw-to-the-nsw-visitor-economy

Department of Industry, Innovation and Science, Australia. (2019). *Cooperative Research Centres Program Evaluation Report.* Canberra: Commonwealth of Australia.

Dr. John Evans: Truly gifted, MaRS Discovery District, February 27, 2012, https://www.marsdd.com/news/dr-evans-101/

Edquist, C. (1997). *Systems of innovation: Technologies, institutions and organizations.* Pinter.

Eickelpasch, A., & Fritsch, M. (2005). Contests for cooperation—A new approach in German innovation policy. *Research Policy*, 34(8), 1269–1282. https://doi.org/10.1016/j.respol.2005.03.008

Engel, J. S., & del-Palacio, I. (2011). Global clusters of innovation: The case of Israel and Silicon Valley. *California Management Review*, 53(2), 27–49. https://doi.org/10.1525/cmr.2011.53.2.27

Etzkowitz, H., & Leydesdorff, L. (1995). The Triple Helix: University-industry-government relations: A laboratory for knowledge-based economic development. EASST Review, 14(1), 14-19.

Etzkowitz, H., & Leydesdorff, L. (1995). The Triple Helix: University–industry–government relations. *EASST Review, 14*(1), 14–19.

Etzkowitz, H., & Leydesdorff, L. (2000). The dynamics of innovation: from National Systems and "Mode 2" to a Triple Helix of university–industry–government relations. Research Policy, 29(2), 109-123.

European Commission. (2021). *Study on Open Innovation and Knowledge Transfer in EU Member States.* European Union.

Ezell, S., & Atkinson, R. D. (2019). *Manufacturing USA at Five: Progress and Challenges.* Information Technology and Innovation Foundation.

Fagerberg, J., Martin, B. R., & Andersen, E. S. (2020). Innovation policy: Rationales, lessons and challenges. *Journal of Economic Surveys, 34*(3), 515–548. https://doi.org/10.1111/joes.12325

Feldman, M. P., & Lowe, N. (2018). Evidence-based economic development policy. *Cambridge Journal of Regions, Economy and Society*, 11(1), 1–7. https://doi.org/10.1093/cjres/rsx028

Ferris, S., Johns, K., & Wallace, R. (2021). The rise and offshore shift of Australian technology companies Causes and consequences. Australian Economic Review, 54(2), 225-240.

Florida, R. (2002). *The rise of the creative class.* Basic Books.

Florida, R. (2017). *The new urban crisis.* Basic Books.

Florida, R., & Adler, P. (2021). The power of place The dynamics of innovation in urban clusters. Journal of Urban Affairs, 43(1), 49-67.

Florida, R., & Mellander, C. (2021). Post-pandemic Innovation Ecosystems. Journal of Economic Geography, 21(3), 311-329.

Florida, R., Adler, P., & Mellander, C. (2017). The city as innovation machine. *Regional Studies, 51*(1), 86–96.

Foray, D. (2015). *Smart specialisation: Opportunities and challenges for regional innovation policy.* Routledge.

Freeman, C. (1987). *Technology policy and economic performance: Lessons from Japan.* Pinter.

Gibbons, M., Limoges, C., Nowotny, H., Schwartzman, S., Scott, P., & Trow, M. (1994). *The new production of knowledge: The dynamics of science and research in contemporary societies.* Sage.

Global Institute on Innovation Districts. (2025). *Research and insights for impact* and *Placemaking resources.* https://giid.org/research/giid-research-amp-insights-for-impact/; https://giid.org/topics/placemaking/

Granovetter, M. (1985). Economic action and social structure: The problem of embeddedness. *American Journal of Sociology, 91*(3), 481–510.

Granovetter, M. S. (1973). The strength of weak ties. *American Journal of Sociology, 78*(6), 1360–1380.

Harvey, D. (1989). *The condition of postmodernity: An enquiry into the origins of cultural change.* Blackwell.

Haskel, J., & Westlake, S. (2017). *Capitalism without capital: The rise of the intangible economy.* Princeton University Press.

Hauser, H. (2014). *Review of the Catapult Network.* Innovate UK.

Healey, P. (1997). *Collaborative planning: Shaping places in fragmented societies.* Macmillan.

Higgs, P., Cunningham, S., & Bakhshi, H. (2008). *Beyond the creative industries: Mapping the creative economy in the United Kingdom*, NESTA.

Hirsch-Kreinsen, H., & Schwinge, I. (2022). Innovation Policy in the Age of Technological Transformation. Routledge.

Honouring Dr. John Evans: Evans family donates $10 million to MaRS Discovery District, MaRS Discovery District, February 24, 2012 https://www.marsdd.com/media-centre/honouring-dr-john-evans-evans-family-donates-10-million-mars-discovery-district/

Howard Partners, 2005. *The Emerging Business of Knowledge Transfer: Creating Value from Intellectual Products and Services*, Department of Education, Science and Training. Download Paper.

Howard Partners, 2007. *Melbourne: Australia's Knowledge Capital: The Contribution of Melbourne's Universities to the City's Economic, Cultural, and Community Development*, City of Melbourne, 2007.

Howard, J. H., 2005. *Knowledge Exchange Networks in Australia's Innovation System: Overview and Strategic Analysis*, Department of Education, Science, and Training.

Howard, J. H., Argaral R., Williams T. (2016). "Smart Specialisation as an Engagement Framework for Triple Helix Interactions". Paper presented at the XIV International Triple Helix Conference 25-27 September 2016, Heidelberg Germany, Triple Helix Models of Innovation: Addressing Ecosystem Challenges in the Era of Crises Regional Dimensions of Triple Helix? Clusters, Cities and Geographic Boundaries.

Howard, J. H., Argaral R., Williams T. (2016). Governance Models and Frameworks for Smart Specialisation" 'Paper presented at Regional Studies Association and European Commission Joint Research Centre, Changing Patterns of Territorial Policy: Smart Specialisation & Innovation in Europe, First SMARTER Conference on Smart Specialisation and Territorial Development, 28-30 September 2016, Seville, Spain.

Howard, J.H. (2008). *Between a Hard Rock and a Soft Space: Design, Creative Practice and Innovation, A Background Paper for the National Innovation Review*, Council for Humanities, Arts and Social Sciences, Occasional Paper No 5, https://www.actoninstitute.au/_files/ugd/bbb23e_a689118c7dd0411596c53c8498d2cb9f.pdf

Howard, J.H. (2008). *Innovation Creativity and Leadership: A Study of the ACT Innovation System*, ACT Government.

Howard, J.H. (2015). *Translating Research for Economic and Social Benefit: Measures that facilitate the transfer of knowledge from publicly funded research organisations to industry: country comparisons (Australia),* Securing Australia's Future project, ACOLA.

Howard, J.H. and Matthews, M., 2001. *Mapping the Nature and Extent of Business-University Interaction in Australia*, Australian Research Council.

Howard, J.H., Johnson, R., and Grigg, L., (2002). *Best Practice Processes for University Research Commercialisation*, Department of Education, Science and Training, Canberra.
https://www.actoninstitute.au/_files/ugd/bbb23e_038669588289495ba8861102f6324d70.pdf

Hunt, V., Layton, D., & Prince, S. (2015). *Why Diversity Matters.* McKinsey & Company.

Iammarino, S., & McCann, P. (2020). Multilevel Governance and Smart Specialisation: Implications for Innovation Clusters. Cambridge Journal of Regions, Economy and Society, 13(2), 267-288.

Isenberg, D. (2011). *The entrepreneurship ecosystem strategy as a new paradigm for economic policy: Principles for cultivating entrepreneurship.* Babson Global. Retrieved from https://www.babson.edu/media/babson/site-assets/content-assets/academics/centers/blank-institute/global-research/eep/The-Entrepreneurship-Ecosystem-Strategy-of-Isenberg.pdf

Jackson, D. J. (2011). What is an innovation ecosystem? *National Science Foundation, Arlington.* Retrieved from https://erc-assoc.org/sites/default/files/topics/policy_studies/DJackson-Innovation%20Ecosystem.pdf

Jackson, D. J. (2011). *What is an innovation ecosystem?* National Science Foundation White Paper, Arlington, VA.

Jacobs, J. (1961). *The death and life of great American cities.* Random House.

JLL. (2023). Australian Office Investment Review Market Update Q1 2023. Jones Lang LaSalle Australia.

John R. Evans, MaRS Discovery District (Biography), April 9, 2024, https://www.marsdd.com/bio/john-r-evans/

John Robert Evans, RCP Museum, https://history.rcp.ac.uk/inspiring-physicians/john-robert-evans

John Robert Evans, Wikipedia, Updated January 21, 2025, https://en.wikipedia.org/wiki/John_Robert_Evans

JTC. (n.d.). *Biopolis at one-north.* https://www.jtc.gov.sg/find-space/biopolis (JTC)

Kagan, S., & Kirchberg, V. (2016). *Sustainability: A New Frontier for the Arts and Cultures.* Transcript Verlag.

Katz, B., & Wagner, J. (2014). *The Rise of Innovation Districts: A New Geography of Innovation in America.* Brookings Institution.

Kendall Square Association. (2025). *The Innovation Trail.* https://kendallsquare.org/the-innovation-trail/

Kendall Square Initiative, Massachusetts Institute of Technology. (2016). *A new era set to begin in Kendall Square* [MIT News]. https://news.mit.edu/2016/new-era-kendall-square-initiative-cambridge-planning-board-0518 (news.mit.edu)

Klein, E. (2022). *Why We're Polarized.* Simon & Schuster. (for insights into narrative, identity, and political culture)

Klerkx, L., Aarts, N., & Leeuwis, C. (2010). Adaptive management in agricultural innovation systems: The interactions between innovation networks and their environment. *Agricultural Systems*, 103(6), 390–400. https://doi.org/10.1016/j.agsy.2010.03.012

Koh, F.C.C., Koh, W.T.H. & Tschang, F.T. (2005). An analytical framework for science parks and technology districts with an application to Singapore. *Journal of Business Venturing*, 20(2), 217-239.

Koschatzky, K. (1998). Firm innovation and region: The role of space in innovation processes. *International Journal of Innovation Management*, 2(4), 383-408.

KPMG. (2014). *Innovation precincts: Creating vibrant innovation economies.* KPMG Australia.

Landabaso, M., Oughton, C. & Morgan, K. (1999). Learning regions in Europe: Theory, policy and practice through the RIS experience. Paper presented at the 3rd International Conference on Technology and Innovation Policy, Austin, Texas.

Landry, C. (2000). *The creative city: A toolkit for urban innovators.* Earthscan.

Leclercq-Vandelannoitte, A., & Isaac, H. (2022). Hybrid Spaces and Innovation: New Forms of Collaboration in Co-working Environments. Organization Science, 33(6), 1978-1997.

Lefebvre, H. (1991). *The production of space* (D. Nicholson-Smith, Trans.). Blackwell. (Original work published 1974)

Lester, R.K. (2003). Universities as public spaces. MIT Industrial Performance Center Working Paper Series.

Lester, R.K. & Piore, M.J. (2004). *Innovation: The missing dimension.* Harvard University Press.

Leydesdorff, L., & Ivanova, I. (2016). Open innovation and triple helix models of innovation. *Journal of Open Innovation: Technology, Market, and Complexity, 2*(1), 11.

Lundvall, B. Å. (1992). *National systems of innovation: Towards a theory of innovation and interactive learning.* Pinter.

Markusen, A., & Gadwa, A. (2010). *Creative placemaking.* National Endowment for the Arts. https://www.arts.gov/about/publications/creative-placemaking

MaRS Discovery District founder Dr. John Evans dies at 85, The Globe and Mail February 14, 2015 https://www.theglobeandmail.com/report-on-business/mars-discovery-district-founder-dr-john-evans-dies-at-85/article23002737/

MaRS Discovery District. (2014). *A tribute to our founder, Dr. John Evans.* https://www.marsdd.com/our-story/a-tribute-to-our-founder-dr-john-evans/ (MaRS Discovery District)

MaRS Discovery District. (n.d.). *Events and venues.* https://hubs.marsdd.com/events/

MaRS is North America's Largest Innovation Hub, MaRS Disovery District, March 4, 2025, https://www.marsdd.com/about/

Marshall, A. (1920). *Principles of economics* (8th ed.). Macmillan.

Mazzucato, M. (2018). *The entrepreneurial state: Debunking public vs. private sector myths.* Penguin Books.

Metcalfe, J. S. (1995). The economic foundations of technology policy: Equilibrium and evolutionary perspectives. In P. Stoneman (Ed.), *Handbook of the Economics of Innovation and Technological Change* (pp. 409–512). Blackwell.

Miles, I., & Green, L. (2008). *Hidden innovation in the creative industries.* NESTA.

Miric, N. (2020). Diversity in innovation: The role of cultural inclusion. *Journal of Innovation Management, 8*(1), 55–71.

MIT News. (2022). *New MIT Museum in Kendall Square invites exploration and discussion.* https://news.mit.edu/2022/mit-museum-opening-0930

Moore, J. F. (1993). Predators and prey: A new ecology of competition. *Harvard Business Review, 71*(3), 75–86.

Moore, J. F. (1996). *The death of competition: Leadership and strategy in the age of business ecosystems.* Harper Business.

Morisson, A. (2020). Smart Specialisation Strategies and Sustainable Urban Innovation: Lessons from Barcelona's 22@ District. Sustainability, 12(14), 5725.

Morisson, A. (2021). Innovation districts: An investigation of global best practices and implications for urban policy. Cities, 115, 103234.

Morisson, A., & Panetti, E. (2020). Place-based innovation policies. *Regional Studies, Regional Science, 7*(1), 1–12.

MTPConnect. (2017). *Clinical trials in Australia: The economic profile and competitive position of the sector* [Report summary]. https://www.mtpconnect.org.au/Story?Action=View&Story_id=59

Mulgan, G. (2019). *Social Innovation: How Societies Find the Power to Change.* Policy Press.

Nelson, R. R. (1993). *National innovation systems: A comparative analysis.* Oxford University Press.

Nelson, R. R., & Winter, S. G. (1982). *An evolutionary theory of economic change.* Harvard University Press.

NESTA. (2007). *Hidden innovation: How innovation happens in six 'low innovation' sectors.* NESTA.

NESTA. (2014). *The impact of innovation policy: Evidence from the UK Innovation Investment Fund.* Nesta Working Paper Series. Retrieved from https://media.nesta.org.uk/documents/impact_of_innovation_policy.pdf

Nowotny, H. (2005). *The Public Nature of Science Under Assault: Politics, Markets, Science and the Law.* Springer.

NSW Government. (2023). Tech Central Precinct Strategy. Retrieved from https//www.investment.nsw.gov.au/tech-central

NSW Government. (2025). *Bradfield City Centre construction kickstart* [Project page, including Advanced Manufacturing Research Facility]. https://www.planning.nsw.gov.au/policy-and-legislation/city-shaping-projects/western-sydney/aerotropolis/construction-kickstart

OECD. (2011). *Regions and innovation policy.* OECD Publishing.

OECD. (2020). *The OECD Innovation Strategy: Getting a Head Start on Tomorrow.* OECD Publishing.

OECD. (2021). *Building Resilient Clusters for Inclusive Innovation.* OECD Publishing.

OECD. (2021). *The geography of higher education for smart specialisation.* OECD Publishing. https://doi.org/10.1787/0d9b1cb7-en

OECD. (2023). *Agile Governance in Regional Innovation Systems.* OECD Publishing.

OECD/Eurostat. (2018). *Oslo Manual 2018: Guidelines for collecting, reporting and using data on innovation* (4th ed.). OECD Publishing.
Oh, D. S., Phillips, F., Park, S., & Lee, E. (2016). Innovation ecosystems: A critical examination. *Technovation*, 54, 1–6. https://doi.org/10.1016/j.technovation.2016.02.004
Ostrom, E. (1990). *Governing the commons: The evolution of institutions for collective action.* Cambridge University Press.
Page, S. E. (2007). *The Difference: How the Power of Diversity Creates Better Groups, Firms, Schools, and Societies.* Princeton University Press.
Pancholi, S., Yigitcanlar, T., & Guaralda, M. (2018). Public space design of knowledge and innovation spaces. *Journal of Urban Design, 23*(4), 530–548.
Park, S.C., Yoon, S.J. & Lee, S.B. (2004). Effects of research and development parks on local economic development. *Asia Pacific Journal of Innovation and Entrepreneurship*, 1(1), 90-107.
Penrose, E. T. (1959). *The theory of the growth of the firm.* Basil Blackwell.
Phan, P.H., Siegel, D.S. & Wright, M. (2005). Science parks and incubators: Observations, synthesis and future research. *Journal of Business Venturing*, 20(2), 165-182.
Phenomics Australia. (2022). *National research infrastructure working together to improve the health of Australians.* https://phenomicsaustralia.org.au/national-research-infrastructure-working-together-to-improve-the-health-of-australians/
Philips. (2024). *Philips steunt Brainport Partnerfonds…* https://www.philips.nl/a-w/about/news/archive/standard/about/news/press/2024/philips-steunt-brainport-partnerfonds-voor-ontwikkeling-innovatieve-en-inclusieve-regio.html (Philips)
Pine, B. J., II, & Gilmore, J. H. (2019). *The experience economy* (Updated ed.). Harvard Business Review Press. https://hbsp.harvard.edu/product/10254-PBK-ENG hbsp.harvard.edu
Piqué, J. M., Miralles, F., & Berbegal-Mirabent, J. (2019). Areas of innovation in cities: The evolution of 22@Barcelona. *International Journal of Knowledge-Based Development, 10*(1), 3–25.
Pisano, G. P. (2010). The evolution of science-based business. *Industrial and Corporate Change, 19*(2), 465–482.
Pisano, G. P., & Shih, W. C. (2009). Restoring American competitiveness. *Harvard Business Review*, 87(7–8), 114–125. Retrieved from https://hbr.org/2009/07/restoring-american-competitiveness
Porter, M. E. (1990). *The competitive advantage of nations.* Free Press.
Porter, M. E. (1998). Clusters and the new economics of competition. *Harvard Business Review, 76*(6), 77–90.
Porter, M.E. (2005). The economic performance of regions. In B.R. Scott & G.C. Lodge (Eds.), *US competitiveness in the world economy* (pp. 549-578). Harvard Business School Press.
Powell, W. W., Koput, K. W., & Smith-Doerr, L. (2005). Interorganizational collaboration and the locus of innovation. *Administrative Science Quarterly, 41*(1), 116–145.
Powell, W. W., Packalen, K., & Whittington, K. (2012). The emergence of organizations and markets. In *The Emergence of Organizations and Markets* (pp. 434–465). Princeton University Press.
Ranga, M., & Etzkowitz, H. (2013). Triple helix systems: An analytical framework for innovation policy and practice in the knowledge society. *Industry and Higher Education*, 27(4), 237–262. https://doi.org/10.5367/ihe.2013.0165
Regional Development Australia, Hunter 2016. *Regional Innovation Smart Specialisation Strategy (RIS3) for the Hunter Region.* Download Paper
Research Triangle Park. (n.d.). *The Foundation.* https://www.rtp.org/the-foundation/ (Research Triangle Park)
Rhodes, R. A. W. (1997). *Understanding governance: Policy networks, governance, reflexivity and accountability.* Open University Press.
Rice Management Company. (n.d.). *About Ion District.* https://iondistrict.com/about/ (Ion District)
Ritala, P., & Almpanopoulou, A. (2017). In defence of 'eco' in innovation ecosystem. *Technovation, 60–61*, 39–42.
Romer, P. M. (1990). Endogenous technological change. *Journal of Political Economy, 98*(5, Part 2), S71–S102.
Ruta N. (n.d.). *About us.* https://www.rutanmedellin.org/ (Ruta N)
Saxenian, A. (2006). *The New Argonauts: Regional Advantage in a Global Economy.* Harvard University Press.
Schot, J., & Steinmueller, W. E. (2018). Three frames for innovation policy: R&D, systems of innovation and transformative change. *Research Policy, 47*(9), 1554–1567. https://doi.org/10.1016/j.respol.2018.08.011
Schumpeter, J. A. (1994). *The theory of economic development* (J. E. Elliott, Trans.). Transaction Publishers. (Original work published in 1934)
Schumpeter, J. A. (2010). *Capitalism, socialism and democracy.* Routledge. (Original work published 1942)
Sörlin, S. (2021). *Policy, Populism and Sustainability: Reclaiming Science and Innovation as a Public Good.* Palgrave Macmillan.
Spigel, B. (2017). The relational organization of entrepreneurial ecosystems. *Entrepreneurship Theory and Practice*, 41(1), 49–72. https://doi.org/10.1111/etap.12167
Station F. (2025). *Station F — the world's biggest startup campus* [Website]. https://stationf.co/ (Station F)

Station F. (2025). *Visiting Station F: The Insider Tour.* https://faq.stationf.co/hc/en-us/articles/115003119945-Visiting-STATION-F-The-Insider-Tour

Stoker, G. (1998). Governance as theory: Five propositions. *International Social Science Journal, 50*(155), 17–28.

Tech Council of Australia. (2022). *Getting to 1.2 million: Our roadmap to create a thriving Australian tech workforce* [Report]. https://techcouncil.com.au/wp-content/uploads/2022/08/2022-Getting-to-1.2-million-report.pdf

Tecnológico de Monterrey. (2025). *Innovation District Monterrey.* https://distritoinnovacionmonterrey.tec.mx/en

Tel Aviv-Yafo Municipality, Tel Aviv University, & Atidim Park. (n.d.). *CityZone — About.* https://www.city-zone.co/about (CityZone)

The Dynamic MaRS Discovery District in Downtown Toronto, BrainStation, December 19, 2017, https://brainstation.io/magazine/killer-spaces-mars-discovery-district-toronto

The Financial Times. (2025). *Station F: flying the flag for France tech and AI development.* https://www.ft.com/content/0e19802c-765d-4b2e-b00c-f974b38003fa (Financial Times)

The State of Victoria, *Australia: Self Evaluation Report: OECD Reviews of Higher Education in City and Regional Development.* IMHE, 2009. Download Paper

Therapeutic Goods Administration. (2022). *Clinical trials in Australia: Overview of the CTN and CTX schemes.* https://www.tga.gov.au/resources/resource/guidance/clinical-trials-australia

Therapeutic Innovation Australia. (2024). *Pipeline Accelerator Guidelines 2024–25* [PDF]. https://therapeuticinnovation.com.au/wp-content/uploads/2024/10/Pipeline-Accelerator-Guidelines-2024-25-Round-1-.pdf

Therapeutic Innovation Australia. (n.d.). *About TIA* [Network summary]. https://therapeuticinnovation.com.au/technical-feasibility-advice-part-of-expanded-tia-support-schemes/

Thornton, P., Ocasio, W., & Lounsbury, M. (2012). *The Institutional Logics Perspective.* Oxford University Press.

Tödtling, F., & Trippl, M. (2005). One size fits all? Towards a differentiated regional innovation policy approach. *Research Policy*, 34(8), 1203–1219. https://doi.org/10.1016/j.respol.2005.01.018

Treasury. (2024, April). *Future Made in Australia: An overview* [Info kit]. https://treasury.gov.au/sites/default/files/2024-04/p2024-498464-information-kit.pdf

Trippl, M., Zukauskaite, E., & Healy, A. (2015). Shaping smart specialisation: The role of place-specific factors in advanced, intermediate and less-developed European regions. *Regional Studies*, 49(9), 1565–1579. https://doi.org/10.1080/00343404.2013.816414

van den Berg, L. (2017). Innovation in European cities: Lessons from Eindhoven. *European Planning Studies, 25*(6), 965–984.

Victorian Auditor-General's Office. (2025). *Developing Fishermans Bend.* https://www.audit.vic.gov.au/report/developing-fishermans-bend (Victorian Auditor-General's Office)

Von Hippel, E. (2005). *Democratizing innovation.* MIT Press.

Vulcan Real Estate. (2020). *How South Lake Union became a world-class innovation hub.* https://vulcanrealestate.com/story/how-south-lake-union-became-a-world-class-innovation-hub/ (Vulcan Real Estate)

Wessner, C. (2022). Research Parks Revisited: Sustainable Innovation and Policy Evolution. National Academies Press.

Wessner, C.W. & National Research Council (US) Advisory Committee on Technology Innovation. (2012). *Best practices for industry-university collaboration.* National Academies Press.

Wilson, E., & Lupton, D. (2023). *Data Selves: More-than-Human Perspectives.* Polity Press.

World Economic Forum. (2023). *Innovation ecosystems: Fostering cross-sector collaboration to drive resilience and growth.* WEF Report.

Zuboff, S. (2019). *The Age of Surveillance Capitalism.* PublicAffairs.

Zucker, L.G, (1994). Biotechnology and the boundary spanning research organization in the evolving innovation system. In B. Bozeman, J. Crow & A. Link (Eds.), *Strategic research partnerships: Proceedings from an NSF workshop* (pp. 117-138). National Science Foundation.

ABOUT THE AUTHOR

Dr John H Howard brings a unique perspective to innovation ecosystems, drawing from over four decades of experience spanning universities, business, and government sectors. His career exemplifies the cross-sector collaboration essential to successful innovation systems.

John earned his PhD from The University of Sydney in 2004, focusing on knowledge transfer in support of innovation, specifically examining how knowledge generated in universities is adopted and applied by business, government, and community organisations. This research foundation underpins his professional expertise in connecting academic research with real-world application.

His academic credentials include serving as Pro Vice-Chancellor Innovation and Engagement at the University of Canberra from 2008 to 2011, where he built relationships and partnerships with industry and government, securing over $50 million in government capital grants. He has held research positions at The University of Sydney's Australian Centre for Innovation and the University of Western Sydney's Centre for Innovation and Industry Studies.

In the business sector, John has been a management consultant since 1984, initially establishing John H Howard and Associates before becoming a Partner at Ernst & Young in 1989 and later at Coopers & Lybrand in 1995. Following the formation of PricewaterhouseCoopers, he re-established Howard Partners in 1998 as a specialised public policy research and management consulting firm. His commercial experience includes serving on the board of the Canberra Business Council for many years.

John's government sector experience encompasses policy research and analysis for Commonwealth departments, including Treasury, Finance, and the former Department of Urban and Regional Development. He has maintained broad involvement with local government since 1984, serving as Research Director to the National Inquiry into Local Government Finance and subsequently providing advisory services to multiple reviews and councils across Australia.

His expertise in regional innovation systems is evidenced in foundation board membership of the Australian Centre of Excellence for Local Government, where he led the Innovation and Best Practice Program. His 2012 report *Innovation, Ingenuity and Initiative: The Adoption and*

Application of New Ideas in Australian Local Government remains a significant contribution to the field.

John has provided research, advisory and consultancy services to major organisations, including the Business Council of Australia, the Australian Chamber of Commerce and Industry, and the OECD. His work spans science and technology policy, industry and regional economics, and innovation management—the interconnected elements that form effective innovation ecosystems.

Currently based in Sydney, Australia, John continues his work through the Acton Institute for Policy Research and Innovation, applying decades of cross-sector experience to help organisations navigate the complexities of modern innovation systems.

ACKNOWLEDGEMENTS

In drawing the Handbook to a close I would like to acknowledge and thank the people who have been involved in putting it all together.

Firstly, I thank the people who have written or co-written *Innovation Insights* included as chapters to book: Craig Fowler (*Productivity and Innovation Need Business Own Investment in Skills*); Rajesh Gopalakrishnan Nair (*Innovation for Sustainability: Lessons From Australian and Californian Agrifood Systems*); Roy Green (*Large Scale Industrial Legacy Adaptation: Transforming the Worlds' Largest Coal Port*); Paul X McCarthy and colleagues (*The Personality Science of Start-up Success: Policy Insights for Australia's Innovation Economy*); Peter Moar (*Why Australia Needs Fraunhofer Institutes*); Victor Pantano (*Absorptive Capacity: The Missing Link in Australia's R&D Collaboration*), Todd Williams and Renu Agarwal (*Smart Specialisation: An Engagement Framework In the Hunter, New South Wales*).

In addition the following people contributed to and verified the case study profiles published in the Handbook: Tom Bentley (*Melbourne North Innovation District*), Amanda Caples (*Parkville Biomedical Precinct*), Lou Conway and David Lamb (*UNE Smart Farm*s and Incubator), Philipp Dautel (*Tonsley Innovation Precinct*), Gavin Keeley and Dalene Wray (Sunshine Coast Innovation District), Jodie Eastwood (*London Knowledge Quarter*), Roy Green (*Newcastle and Dublin Docklands*), Davor Jozic (Macquarie Park Innovation District) Ashley Langton (*Bradfield City and Aerotropolis*), Michael Priddis and Greg Miller (*MaRS Discovery District*), Josep Pique *22@Barcelona*), Matthew Proft (*Tech Central*), Craig Rowsell (*Gold Coast Health and Innovation Precinct*), and Leon Sandler (*Kendall Square)*

I also thank the many people who have assisted supported the development of the project, and especially Petr Adamek, Antonio Balaguer, Camerom Begley, Amanda Caples, Lachlan Chislett, Mark Dodgson, Callum Drummond, Dom English, Craig Fowler, Luke Garner, Rajesh Gopalakrishnan Nair, Roy Green, Glenn Hassett, Mat Howard, Tim Howard, Narelle Kennedy, Heleen Kruger, Michael Lester, Tim Lester, Swee Mak, Julie Martin, Stewart Marshall, Mark Matthews, Nick McNaughton, Carol Mills, Rohan Nelson, Jane O'Dwyer, Mirjana Prica, Matthew Proft, James Riley, Peter Roberts, Don Scott-Kemmis, Prabhu Sivabalan, Sharath Sriram, David Thodey, and Cori Stewart.

Made in the USA
Monee, IL
08 November 2026

68106864R00306